Principles of Optical Interferometry

Principles of Optical Interferometry

Edited by:

Olga Moreira

ARCLER
PRESS

www.arclerpress.com

Principles of Optical Interferometry

Olga Moreira

Arcler Press
224 Shoreacres Road
Burlington, ON L7L 2H2
Canada
www.arclerpress.com
Email: orders@arclereducation.com

ISBN: 978-1-77469-181-6 (Hardcover)

DECLARATION

ABOUT THE EDITOR

Olga Moreira is a Ph.D. in Astrophysics and B.Sc. in Physics and Applied Mathematics. She is an experienced technical writer and researcher which former fellowships include postgraduate positions at two of the most renown European institutions in the fields of Astrophysics and Space Science (the European Southern Observatory, and the European Space Agency). Presently, she is an independent scientist working on projects involving machine learning and neural networks research as well as peer-reviewing and edition of academic books.

TABLE OF CONTENTS

LIST OF CONTRIBUTORS

Jesús Muñoz Maciel
University of Guadalajara, University Center of Lagos, Lagos de Moreno, Mexico

Francisco Javier Casillas Rodríguez
University of Guadalajara, University Center of Lagos, Lagos de Moreno, Mexico

Miguel Mora González
University of Guadalajara, University Center of Lagos, Lagos de Moreno, Mexico

Francisco Gerardo Peña Lecona
University of Guadalajara, University Center of Lagos, Lagos de Moreno, Mexico

Víctor Manuel Duran Ramírez
University of Guadalajara, University Center of Lagos, Lagos de Moreno, Mexico

Ulises H. Rodriguez-Marmolejo
Departamento de Ciencias Exactas y Tecnología, Centro Universitario de los Lagos, Universidad de Guadalajara, 47460 Lagos de Moreno, JAL, Mexico
Departamento de Ingeniería Eléctrica y Electrónica, Instituto Tecnológico de Aguascalientes, 20256 Aguascalientes, AGS, Mexico

Miguel Mora-Gonzalez
Departamento de Ciencias Exactas y Tecnología, Centro Universitario de los Lagos, Universidad de Guadalajara, 47460 Lagos de Moreno, JAL, Mexico

Jesus Muñoz-Maciel
Departamento de Ciencias Exactas y Tecnología, Centro Universitario de los Lagos, Universidad de Guadalajara, 47460 Lagos de Moreno, JAL, Mexico

Tania A. Ramirez-delreal
Departamento de Ciencias Exactas y Tecnología, Centro Universitario de los Lagos, Universidad de Guadalajara, 47460 Lagos de Moreno, JAL, Mexico
Departamento de Posgrado e Investigación, Universidad Politécnica de Aguascalientes, 20342 Aguascalientes, AGS, Mexico

Julio C. Estrada
Centro de Investigacion en Matematicas (CIMAT), Computer Sciences, Guanajuato, 36023, Mexico.

Jose L. Marroquin
Centro de Investigacion en Matematicas (CIMAT), Computer Sciences, Guanajuato, 36023, Mexico.

Orlando M. Medina
Universidad Politecnica de Guanajuato (UPG), Robotics, Guanajuato, 38483, Mexico.

Dainis Dravins
Lund Observatory, Lund University, Lund SE-22100, Sweden.

Tiphaine Lagadec
Lund Observatory, Lund University, Lund SE-22100, Sweden.

Paul D. Nuñez
Collège de France, 11 Place Marcelin Berthelot, Paris FR-75005, France.
Laboratoire Lagrange, Observatoire de la Côte d'Azur, BP 4229, Nice FR-06304, France.

Dongfang Li
School of Engineering, Brown University, Providence, Rhode Island, 02912, United States.

Jing Feng
School of Engineering, Brown University, Providence, Rhode Island, 02912, United States.

Domenico Pacifici
School of Engineering, Brown University, Providence, Rhode Island, 02912, United States.

Florian Kaiser
Université Côte d'Azur, Institut de Physique de Nice (INPHYNI), Nice 06108, France
Now at 3. Physikalisches Institut, Universität Stuttgart, Stuttgart 70569, Germany
Now at Center for Integrated Quantum Science and Technology (IQST), Stuttgart 70569, Germany

Panagiotis Vergyris
Université Côte d'Azur, Institut de Physique de Nice (INPHYNI), Nice 06108, France

Djeylan Aktas
Université Côte d'Azur, Institut de Physique de Nice (INPHYNI), Nice 06108, France

Charles Babin
Université Côte d'Azur, Institut de Physique de Nice (INPHYNI), Nice 06108, France
École Normale Supérieure de Lyon, Lyon 69364, France

Laurent Labonté
Université Côte d'Azur, Institut de Physique de Nice (INPHYNI), Nice 06108, France

Sébastien Tanzilli
Université Côte d'Azur, Institut de Physique de Nice (INPHYNI), Nice 06108, France

Bartolomeo Trentadue
Dipartimento di Meccanica, Matematica e Management, Politecnico di Bari, Viale Japigia 182, 70126 Bari, Italy

Giuseppe Illuzzi
Dipartimento di Meccanica, Matematica e Management, Politecnico di Bari, Viale Japigia 182, 70126 Bari, Italy

Hendrik Spahr
Institute of Biomedical Optics, University of Lübeck, Peter-Monnik-Weg 4, 23562, Lübeck, Germany.
Medical Laser Centre Lübeck GmbH, Peter-Monnik-Weg 4, 23562, Lübeck, Germany.

Clara Pfäfe
Institute of Biomedical Optics, University of Lübeck, Peter-Monnik-Weg 4, 23562, Lübeck, Germany.
Medical Laser Centre Lübeck GmbH, Peter-Monnik-Weg 4, 23562, Lübeck, Germany.

Sazan Burhan
Institute of Biomedical Optics, University of Lübeck, Peter-Monnik-Weg 4, 23562, Lübeck, Germany.

Lisa Kutzner
Institute of Biomedical Optics, University of Lübeck, Peter-Monnik-Weg 4, 23562, Lübeck, Germany.
Medical Laser Centre Lübeck GmbH, Peter-Monnik-Weg 4, 23562, Lübeck, Germany.

Felix Hilge
Institute of Biomedical Optics, University of Lübeck, Peter-Monnik-Weg 4, 23562, Lübeck, Germany.
Medical Laser Centre Lübeck GmbH, Peter-Monnik-Weg 4, 23562, Lübeck, Germany.

Gereon Hüttmann
Institute of Biomedical Optics, University of Lübeck, Peter-Monnik-Weg 4, 23562, Lübeck, Germany.
Medical Laser Centre Lübeck GmbH, Peter-Monnik-Weg 4, 23562, Lübeck, Germany.
Airway Research Center North (ARCN), Member of the German Center for Lung Research (DZL), 22927 Großhansdorf, Germany.

Dierck Hillmann
Institute of Biomedical Optics, University of Lübeck, Peter-Monnik-Weg 4, 23562, Lübeck, Germany.
Thorlabs GmbH, MariaGoeppert-Straße 9, 23562, Lübeck, Germany.

Cheng-Chih Hsu
Department of Photonics Engineering, Yuan Ze University, Yuan-Tung Road, Chung-Li, Taiwan

Guanhao Wu
State Key Laboratory of Precision Measurement Technology and Instruments, Department of Precision Instrument, Tsinghua University, Beijing, 100084, China.

Lei Liao
State Key Laboratory of Precision Measurement Technology and Instruments, Department of Precision Instrument, Tsinghua University, Beijing, 100084, China.

Shilin Xiong
State Key Laboratory of Precision Measurement Technology and Instruments, Department of Precision Instrument, Tsinghua University, Beijing, 100084, China.

Guoyuan Li
Satellite Surveying and Mapping Application Center, NASG, Beijing, 100048, China.

Zhijian Cai
College of Physics, Optoelectronics and Energy, Soochow University, Suzhou, 215006, China.

Zebin Zhu
State Key Laboratory of Precision Measurement Technology and Instruments, Department of Precision Instrument, Tsinghua University, Beijing, 100084, China.

Hongjun Wang
Xi'an Technological University, International Center for Optical Research and Education (iCORE), No. 2 Xuefu Middle Road, Xi'an, Shaanxi 710021, China

Xueliang Zhu
Xi'an Technological University, International Center for Optical Research and Education (iCORE), No. 2 Xuefu Middle Road, Xi'an, Shaanxi 710021, China

Bingcai Liu
Xi'an Technological University, International Center for Optical Research and Education (iCORE), No. 2 Xuefu Middle Road, Xi'an, Shaanxi 710021, China

Ailing Tian
Xi'an Technological University, International Center for Optical Research and Education (iCORE), No. 2 Xuefu Middle Road, Xi'an, Shaanxi 710021, China

YusukeTsujiie
Department of Intelligent Mechanical Engineering, Faculty of Engineering, Fukuoka Institute of Technology, 3- 30-1 Wajirohigashi, Higashiku, Fukuoka, 811-0295, Japan

Yoshiyuki Kawamura
Department of Intelligent Mechanical Engineering, Faculty of Engineering, Fukuoka Institute of Technology, 3- 30-1 Wajirohigashi, Higashiku, Fukuoka, 811-0295, Japan

P. Manju
Atomlaser and Quantum Sensors Group, Department of Quantum Science, Research School of Physics, The Australian National University, Canberra 2601, Australia

K. S. Hardman
Atomlaser and Quantum Sensors Group, Department of Quantum Science, Research School of Physics, The Australian National University, Canberra 2601, Australia

P. B. Wigley
Atomlaser and Quantum Sensors Group, Department of Quantum Science, Research School of Physics, The Australian National University, Canberra 2601, Australia

J. D. Close
Atomlaser and Quantum Sensors Group, Department of Quantum Science, Research School of Physics, The Australian National University, Canberra 2601, Australia

N. P. Robins
Atomlaser and Quantum Sensors Group, Department of Quantum Science, Research School of Physics, The Australian National University, Canberra 2601, Australia

S. S. Sziget
Atomlaser and Quantum Sensors Group, Department of Quantum Science, Research School of Physics, The Australian National University, Canberra 2601, Australia

Heonoh Kim
Department of Physics, Pusan National University, Geumjeong-Gu, Busan 609-735, Korea.

Sang Min Lee
Department of Physics, Pusan National University, Geumjeong-Gu, Busan 609-735, Korea.
Present address: Korea Research Institute of Standards and Science Daejeon 305-340 Korea

Han Seb Moon
Department of Physics, Pusan National University, Geumjeong-Gu, Busan 609-735, Korea.
Present address: Korea Research Institute of Standards and Science Daejeon 305-340 Korea

ChangWoo Lee
Department of Physics Education, Kongju National University, Gongju 32588, South Korea

Jae Hoon Lee
Korea Research Institute of Standards and Science, Daejeon 34113, South Korea

Hyojun Seok
Department of Physics Education, Kongju National University, Gongju 32588, South Korea

Daodang Wang
College of Metrology and Measurement Engineering, China Jiliang University, Hangzhou, China

Rongguang Liang
College of Optical Sciences, University of Arizona, Tucson, AZ, USA

Naiting Gu
Institute of Optics and Electronics, Chinese Academy of Sciences, China

Qun Luo
National Key Laboratory of Science and Technology on Blind Signal Processing, China

Ramón José Pérez Menéndez
UNED, Lugo, Spain

Kyunghyun Lee
Department of Electrical Engineering, Sungkyunkwan University, Suwon, Republic of Korea

Hyungkwan Kwon
Department of Electrical Engineering, Sungkyunkwan University, Suwon, Republic of Korea

Kwanho You
Department of Electrical Engineering, Sungkyunkwan University, Suwon, Republic of Korea

LIST OF ABBREVIATIONS

AOM	Accousto-optical modulator
APRIL	Adjacent pulse repetition interval length
AFM	Atomic force microscope
ALD	Atomic layer deposition
ATR	Attenuate total reflection
BS	Balanced beamsplitter
BC	Birefrigent crystals
BEC	Bose–Einstein condensate
CIO	Centro de Investigaciones en Optica
CHI	circular heterodyne interferometer
CW	continuous wave
CR	corner reflector
CMT	coupled mode theory
DGHI	diffraction grating heterodyne interferometry
DSPI	digital speckle pattern interferometry
DLS	dynamic light scattering
EIM	effective index method
ESPI	electronic speckle interferometry
EOM	electro-optical modulator
EUVL	extreme ultraviolet lithography
FFT	fast Fourier transform
FBG	fiber Bragg grating
FOPRRI	fiber optic passive ring resonator interferometer
FIBE	film with the focused ion-beam etching
FEM	finite element method
FFCHI	Full-field circular heterodyne interferometry
FWHD	full width at half depth
GA	genetic algorithm

GHI	grating heterodyne interferometry
GVD	group velocity dispersion
HWP	half-wave plate
HWR	half-wave retarder
HMCR	Harmony Memory Considering Rate
HM	Harmony Memory
HMS	Harmony Memory Size
HSO	harmony search optimization
IOPRRI	integrated optical passive ring resonator interferometer
LC SLM	liquid crystal spatial light modulator
MZI	Mach–Zehnder interferometer
MPE	maximum permissible exposure
MRA	micro-retarder array
NICHI	normal incident circular heterodyne interferometer
NVAR	number of decision variables
OA	optical axis
OCE	optical coherence elastography
OPLD	optical path length difference
OTF	optical transfer function
PSO	particle swarm optimization
PPLN	periodically-poled lithium niobate
PBS	phosphate buffer solution
PZT	Piezoelectric Transducer
PAR	Pitch Adjusting Rate
PDI	point diffraction interferometer
PBS	polarization beam splitter
PWB	Printed Wiring Board
QMFS	quantum–mechanics-free subsystem
QND	quantum nondemolition
QCOP	quasi-common optical path
QPM-SPDC	quasi-phase-matched spontaneous parametric down-conversion
RSI	Radial shearing interferometer
RI	real interferogram
RIU	refractive index unit

RMSE	Root Mean Square Error
SUT	sample under test
SPM	scanning probe microscopy
SNR	signal-to-noise ratio
SOI	silicon on insulator
SBAS	small baseline subset technique
SQL	standard quantum limit
SPR	surface plasmon resonance
SAR	synthetic aperture radar
TEC	temperature control
TSSA	temporally separated and spatially anti-bunched
TSSB	temporally separated and spatially bunched
TIR	total internal reflection
TV	Total Variation
TSSA	two different spatial modes
WS	wavefront simulator
WMCHI	wavelength-modulation circular heterodyne interferometer
WLI	white-light interferometry

PREFACE

The basic principle behind interferometry is associated with the way waves, light waves in particular, propagate through space. The double-slit experiment (a.k.a. Young's experiment) demonstrates that the light can be diffracted. Monochromatic light coming from a distant point source passes through two slits and form an interference pattern. This interference pattern is due to the wave nature of light. The electrical field at each slit results from with different relative optical paths, producing a destructive and constructive interference fringes pattern.

The Young's experiment is an idealized model which is reproduced in an interferometer. In astronomy, the double-slit experiment is at the basis of long baseline optical interferometers design, where telescopes play the same role as the slits. The light received at two or more separated telescopes is brought together in a beam-combination facility for interference. Optical interferometry has evolved beyond astronomical imaging. In modern times it has expanded to other fields such as chemistry, biomedical and mechanical engineering, as well as material science.

The first part of this book consists of 3 articles (Chap. 1 - 3) focused on interference fringe analysis. Chapter 1 includes digital processing techniques applied to interferometric measurements for obtaining the phase from fringe patterns based on interferogram filtering, phase-shifting interferometry, Fourier method, and phase unwrapping. Chapter 2 is centred on a combination of frequency guide sequential demodulation and harmony search optimization algorithms as method for modelling closed fringes patterns. Chapter 3 is concerned with the reconstruction of a frequency field of wrapped phases by using adaptive Gabor filters.

The second group of articles (Chap. 4 - 6) is focused on the applications of optical interferometry principles in the fields of astronomy, medical imaging, and optical material science. Chapter 4 focuses on the applications of intensity interferometry to large telescope arrays. Chapter 5 focuses on optical interferometry based on surface plasmon polaritons as an extended application to the field of medical imaging. Chapter 6 focuses on white-light interferometry as a methodology for determining the properties of optical materials.

The third group of articles (Chap. 7 - 8) is centred on the applications of speckle Interferometry. Speckle interferometry technique are based on the extraction of spatial information from short exposure images in spite of blurring introduced by atmospheric turbulence. This technique relies on the cyclic correlation analyses of speckle patterns. Chapter 7 describes the application of a speckle interferometry technique for determining the thermo-mechanical behaviour of Printed Wiring Board (PWB) circuits. Chapter

8 applies the idea of speckle interferometry to phase-sensitive imaging in biological tissues.

The fourth group of articles (Chap. 9 -10) is centred on heterodyne interferometry. A heterodyne interferometer usually consists of a single path interferometer that utilize laser frequencies instead of one laser frequency as in a homodyne interferometer. Chapter 9 overviews the principles of heterodyne detection, the different types of heterodyne interfermeters and their applications. Chapter 10 presents a synthetic-wavelength based heterodyne interferometer of optical frequency combs for absolute distance measurement.

The fifth group of articles (Chap. 11 - 15) is centred on 3 different types of interferometers: the Fizeau, the Fabry-Perot (FP), and the Mach-Zehnder (MZ) interferometers. Chapter 11 consists of an application of Fizeau interferometry for the reduction of noise of a rotating diffuser. Chapters 12 and 13 are concerned with recent developments in FP interferometry and their applications. Chapter 12 describes a micro FP interferometer developed for optical measurements in various micro sensing systems, while chapter 13 describes an atomic analogue of the FP interferometer. Chapter 14 and 15 focus on the applications of MZ interferometer in the study of a two-photon interference and in the detection of squeezed-light-driven force, respectively.

The last group of articles (Chap. 16 -19) is centred on modern optical interferometry techniques such as Point Diffraction, Radial shearing, Fiber-optic Passive Ring Resonator, and Neural Network-Based Laser interferometry.

CHAPTER 1

DIGITAL PROCESSING TECHNIQUES FOR FRINGE ANALYSIS

Jesús Muñoz Maciel, Francisco Javier Casillas Rodríguez, Miguel Mora González, Francisco Gerardo Peña Lecona, and Víctor Manuel Duran Ramírez

University of Guadalajara, University Center of Lagos, Lagos de Moreno, Mexico

ABSTRACT

Digital image processing techniques are needed in order to recover the object information encoded in fringe patterns generated in a determined interferometric setup. Main fringe analysis techniques are reviewed in order to give the reader the most fundamental insights for the interpretation of interferograms. Phase shifted, open fringe, lateral shear and other types of interferograms make use of specific procedures to correctly retrieving

Citation: (APA): Kaiser, F., Vergyris, P., Aktas, D., Babin, C., Labonté, L., & Tanzilli, S. (2018). Maciel, J. M., Rodríguez, F. J. C., González, M. M., Lecona, F. G. P., & Ramírez, V. M. D. (2017). Digital Processing Techniques for Fringe Analysis. Optical Interferometry, 1. (25 pages)

the searched phase. Here, algorithms are described and tested in numerical simulations and real data.

Keywords: fringe analysis, phase recovery, phase unwrapping

INTRODUCTION

Interferometric techniques are widely used for measuring a wide range of physical variables including refraction index, deformations, temperature gradients, etc. Typically, an interferometer is used to generate one or several interferometric fringe patterns that contain the information of the physical variable that is being measured. Those images must be interpreted in order to recover the parameters that are encoded in the fringe patterns generated by the interferometric setup. Thus, fringe analysis methods deal with the problem of a three-dimensional reconstruction (the object information) from a two-dimensional intensity patterns (interferograms) acquired by a CCD camera. Digital interferometry became extensively used since the development of lasers and CCD devices. In those years, however, the resulting interferograms had to be interpreted visually, and only qualitative results were often achieved. Visual interpretation of an interferogram with only straight or circular fringes is not difficult, but things become more complicated for a fringe pattern that combines several regions with circular, straight and crossed fringes of varying density. Rapidly, it was recognized the need of automatic methods for fringe analysis. The first great advance arises with the development of the phase-shifting techniques. With those procedures, a set of interferograms is acquired with a phase shift among them. The phase shifts are usually introduced by a piezoelectric transducer moving the reference mirror in such way that the phase difference between two consecutive interferograms is a constant term. With phase-shifting techniques, it is possible to isolate the sine and cosine of the phase allowing the calculation of the wrapped phase distribution and consequently the continuous phase with an unwrapping algorithm. Another great success came with the method proposed by Takeda (also referred as the Fourier method) performing a band-pass filtering in the Fourier domain. The method of Takeda works only with interferograms that contain open fringes (patterns that consist in nearly straight fringes). In order to generate such interferograms, the reference beam (e.g., in a two arm interferometer) is tilted introducing a large carrier function to the phase. The Fourier transform of these interferograms is composed of three lobules, one at the center that corresponds to the background term and two lobules located symmetrically respect to the origin. One of this lobules and the one that is located at the origin are filtered out. The remaining spectrum is transformed back to the spatial domain from which the so-called wrapped phase can be calculated. A final step is to apply a phase unwrapping technique to recover the continuous phase. Interferometric measurements and fringe analysis techniques are a growing and fast-changing field of research. Through this chapter, we will review the most known procedures.

INTERFEROGRAM ACQUISITION

The wave nature of light can be studied theoretically by a homogeneous partial differential equation of second order, which satisfies the superposition principle:

$$\nabla U(x,y,z,t) = \frac{1}{c^2}\frac{\partial U(x,y,z,t)}{\partial t}. \tag{1}$$

If two waves of the same frequency are superimposed on a point in space, they excite oscillations in the same direction:

$$U_1(t) = A_1\cos(\omega t + \varphi_1), \tag{2}$$

and

$$U_2(t) = A_2\cos(\omega t + \varphi_2). \tag{3}$$

In the preceding equations and the subsequent ones in this section, we will drop the spatial dependence for displaying purposes. The amplitude of the resulting oscillation at that point is determined by the equation:

$$A^2 = {A_1}^2 + {A_2}^2 + 2A_1A_2\cos(\delta), \tag{4}$$

where

$$\delta = \varphi_1 - \varphi_2. \tag{5}$$

If the phase difference, δ, of the oscillations excited by waves remains constant with time, these waves are coherent. In the case of noncoherent waves, δ varies continuosly, taking any values with equal probability, so the average value of $cos(\delta)$ is zero. Therefore,

$$A^2 = A_1^2 + A_2^2. \tag{6}$$

So, we can conclude that the intensity observed in the superposed point by noncoherent waves equals the sum of the intensities, which create each separately. However, if the difference $\delta = \varphi_1 - \varphi_2$ is constant, the $cos(\delta)$ will also have a constant value over time, but own for each point in space, so that:

$$I = I_1 + I_2 + 2\sqrt{I_1 I_2}\cos(\delta) \quad (7)$$

where $I_1 = A_1^2$ and $I_2 = A_2^2$. This superposition at a point in space results in a different sum of the intensities of the separate components intensity [1]. This phenomenon is known as interference. The light and dark areas that observed on screen placed in the region of interference are called interference fringes, and the fringes are intensity which changes from minimum to maximum, which together form a pattern commonly called interference pattern or interferogram, as can be seen in Figure 1 .

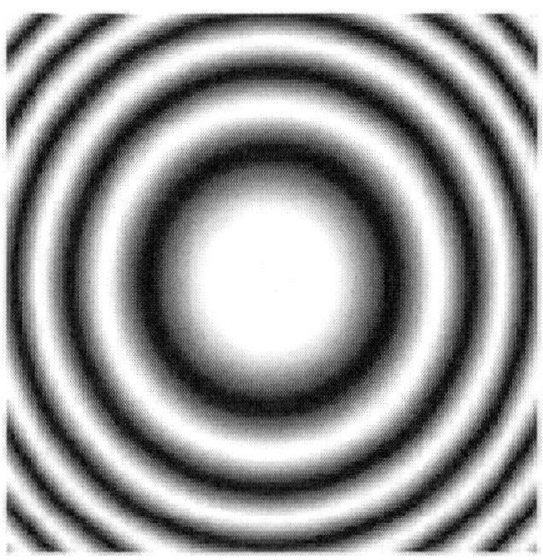

Figure 1. Interference pattern or interferogram showing circular fringes.

The interference of two or more electromagnetic waves can be usually achieved in two ways: by division of the wave front and by division of the amplitude. A mechanism used for the division of the wave front is, for example, the Young's experiment or double slit that is showed in Figure 2 .

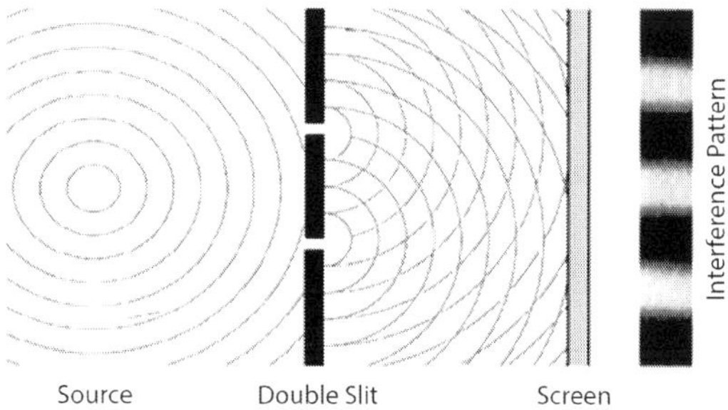

Figure 2. Young's experiment, where a double slit produces two wave fronts that interfere on a screen.

A mechanism used for dividing the amplitude of the wave is, for example, the Michelson interferometer that is shown in Figure 3 .

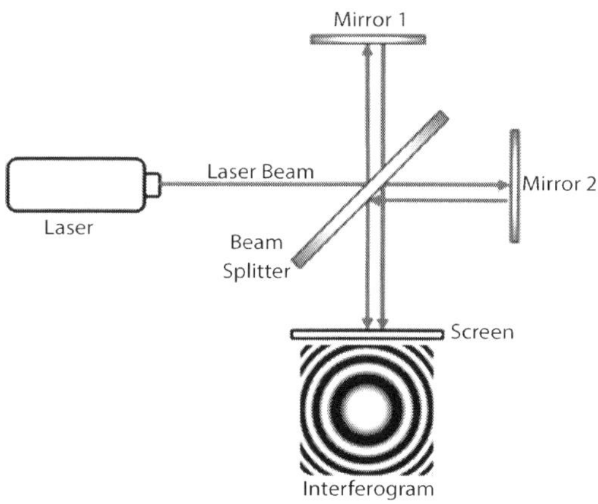

Figure 3. Michelson interferometer.

Both mechanisms produce a pattern of light and dark intensities in the plane of interference fringes. The image resulting from the interference is known as interferogram. When the interferogram is captured by a recording medium, that is, a photographic film or a CCD camera, the process commonly involves some optical system, which introduces imperfections with respect to the ideal image. Such imperfections are known as aberrations. Aberrations can be classified as chromatic and monochromatic. Chromatic aberrations are present to illuminate the object with white light or polychromatic light, that is, light with different wavelengths. These aberrations are the only ones that can be predicted by the theory of the first order, which states that an optical system consisting of lenses has different focal lengths for different wavelengths. These variations are related to the change of refractive index with respect to wavelength causing that both the position and the image size are different for each wavelength. Monochromatic aberrations occur when the objcct is illuminated with monochromatic light (i.e., light of a single wavelength), and the reflected or transmitted light is registered by a recording medium. This type of aberration causes that the captured image of a punctual object is no longer a point, but a blurred point. Monochromatic aberrations can be calculated roughly in the third-order theory, using the first two terms of the expansion in power series of the $sin\theta$ and $cos\theta$. Another alternative to calculate more accurately is to make the exact trigonometric trace rays through the system, where the deviation of the rays is calculated. These aberrations were studied in detail first by Ludwig von Seidel, hence often called Seidel aberrations, and are sphericity, coma, astigmatism, field curvature and distortion. These aberrations have a distinctive fringe distribution that appears frequently in the testing of a variety of objects with an interferometer. As the interferogram acquisition is only the capture of the interference phenomenon or interference pattern in a recording medium, it is convenient to study correcting aberrations problems. From these patterns, it can be known as the aberration coefficients using Zernike polynomials. Zernike polynomials have been successfully used in the recognition of patterns and image processing. Additionally, these have been used in astronomy to describe wavefront aberrations due to atmospheric turbulence and to describe wavefront aberrations in the human eye. This is because of the Seidel aberrations are related to the Zernike polynomials. Due to this relationship, polynomials are used to describe wavefront aberrations in order to calculate the aberration coefficients of a set of interferograms generated by an adaptive lens. These coefficients enable to describe the behavior of the aberrations present in the lens.

INTERFEROGRAM FILTERING

In order to enhance the interferogram images and reduce the noisy caused by external factors at the interference phenomenon in an interferometer (interference produced by two or more controlled wave fronts), low-pass techniques are used as preprocessing filtering step. Interferogram smoothing and denoising are the principal purposes for the application of low-pass filters in interferometry. Filtering techniques can be grouped in spatial or frequency domains, where the spatial filtering is directly applied to the interferogram image, pixel to pixel, while the frequency filtering is usually performed in the Fourier domain. Band-pass and band-stop are some filtering techniques that can be applied in the Fourier domain, and these filters are used to attenuate some frequencies of specific noise.

Spatial Filtering

The mathematical entity applied in spatial filtering is the convolution operation, also known as windowing [2], written as follows:

$$f(x,y)*g(x,y)=\sum_{m=-\frac{M+1}{2}}^{\frac{M+1}{2}}\sum_{n=-\frac{N+1}{2}}^{\frac{N+1}{2}} f(x-m,y-n)g(m,n), \tag{8}$$

where f, g, (x, y), (m, n) and $M\times N$ are the image to be filtered, the convolution mask, the original image coordinates, the coordinates where the convolution is performed and the size of the convolution mask, respectively. The kind of convolution filter is determined by the chosen convolution filter (averaging, Gaussian, quadratic, triangular, trigonometric, etc.). These convolution functions are presented in a mathematical form as follows: [3]

$$g(m,n)=\frac{1}{\sum_{m=1}^{M}\sum_{n=1}^{N}\omega_{m,n}}\begin{bmatrix}\omega_{1,1} & \cdots & \omega_{1,N}\\ \vdots & \ddots & \vdots\\ \omega_{M,1} & \cdots & \omega_{M,N}\end{bmatrix}, \tag{9}$$

where

$$\omega_{m,n}=\begin{cases} Ae^{(-B(m^2+n^2))}, & \text{Gaussian}\\ A-B(m^2+n^2), & \text{quadratic}\\ \frac{A}{2}+\frac{A}{4}[\cos(Bm)+\cos(Bn)], & \text{trigonometric}\\ A, & \text{average}\end{cases} \tag{10}$$

Here A, B and ω are the amplitude, the width factor and the weight of the spatial filter function, respectively. The study about the magnitude spectrums of Eq. (10) (low-pass masks) was reported in Ref. [2], where the Gaussian and quadratic masks delivered the best results. Low frequencies were conserved by these filters, while the high frequencies were attenuated. However, the filtering results may vary depending on the interferogram under process, as can be seen in Figure 4 , where the results of the filter on a ronchigram (a particular kind of interferogram) can be seen in Figure 4a . The four kinds of masks presented in Eq. (10) were employed to generate the filtered fringe patterns shown in Figure 4b . The parameters used by the filters were $A = 1$ and $B = 0.1$, with mask sizes of 3×3, 5×5 and 7×7 pixels.

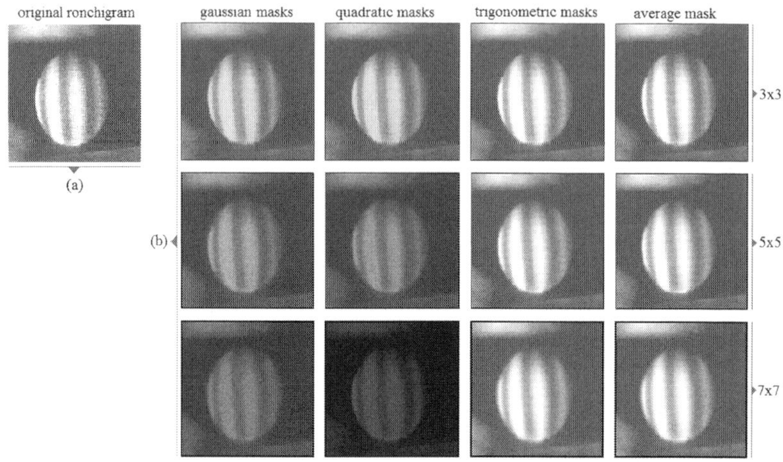

Figure 4. Spatial low-pass filters: (a) original ronchigram, (b) convolution image filtering with Gaussian, quadratic, trigonometric and average masks, as well as 3 × 3, 5 × 5 and 7 × 7 convolution mask sizes.

Frequency Filtering

Frequency filtering is usually performed in the Fourier domain. The Fourier transform represents the change from spatial to frequency domain. Eq. (11) and Eq. (12) represent a pair of discrete Fourier transforms in two dimensions [2]

$$F(u,v) = \mathcal{F}\{f(x,y)\} = \sum_{x=1}^{U}\sum_{y=1}^{V} f(x,y)e^{-2\pi i\left(\frac{ux}{U}+\frac{vy}{V}\right)}, \tag{11}$$

and

$$f(x,y) = \mathcal{F}^{-1}\{F(u,v)\} = \sum_{u=1}^{U}\sum_{v=1}^{V} F(u,v)e^{2\pi i\left(\frac{ux}{U}+\frac{vy}{V}\right)} \tag{12}$$

where (u, v), $U \times V$, $\mathcal{F}$ and $\mathcal{F}^{-1}$ are the frequency coordinates, the image size in pixels, the Fourier transform and the inverse Fourier transform operators, respectively. A significant attribute obtained from the Fourier transform is that it gives the frequencies content of the image. Due to this property, frequency filter design is a very straight forward task for image processing. Low frequencies are located into the Fourier domain around the central coordinates; as frequencies gradually increase, they spread out from the center in a radial form. This characteristic is ideal for frequency filtering (low-pass, high-pass, band-pass and band-stop) [3]. The frequency filtering development consists in the multiplication between the Fourier transform with some kind convolution function. The kind of convolution mask will determine the class of filtering to be performed. The following is a summary of the most usual filtering masks: a white centered circle for low-pass; a black centered circle for high-pass; a white centered ring for band-pass and a black centered ring for band-stop.

In Figure 5 , it is showed the masks described above along with the results of the filtering process. Low-pass filtering masks and results are presented in Figure 5b , band-pass filters and filtered images are seen in Figure 5c and finally, band-stop filters and filtering results can be appreciated in Figure 5d . The kind of filter or the size of geometrical mask depends directly in the image and in its noise content. There is no ideal filter; the kind of applied filter to process an image is dependent of the characteristics that are pretended to enhance or eliminate.

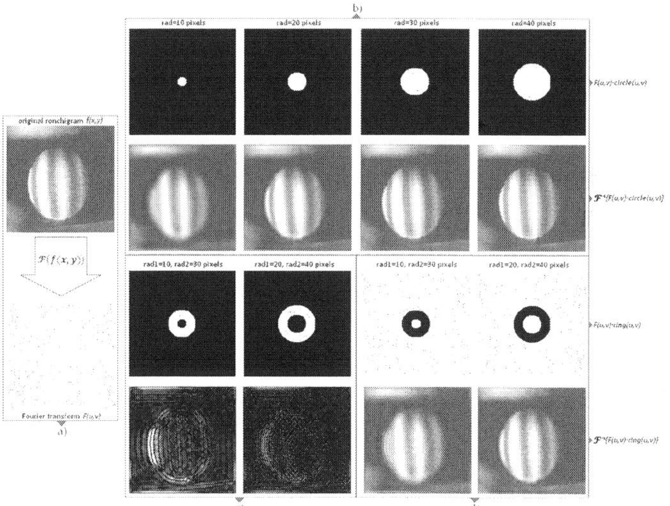

Figure 5. Frequency filtering: (a) Fourier transform of the original ronchigram, (b) low-pass filtering with different radius of binary circle mask, (c) band-pass filtering with two size of binary ring mask, and (d) band-stop filtering with two size of binary ring mask.

PHASE-SHIFTING INTERFEROMETRY

Interference fringe patterns obtained by means of interferometric techniques can be evaluated by using digital image processing techniques for the estimation of phase map distributions. The intensity of an image in an interference fringe pattern, according with Eq. (7), can be represented by Eq. (13)

$$I(x,y) = a(x,y) + b(x,y)cos[\phi(x,y)] \tag{13}$$

where $\phi(x, y)$ is the phase difference between the reference and testing wave fronts that interfere, $a(x, y) = I_1(x, y) + I_2(x, y)$ is the mean intensity and $b(x, y) = 2(I_1 I_2)^{1/2}$ is the intensity modulation registered by each pixel (x, y) of a CCD camera. When analyzing a single image for the estimation of the relative phase difference in applied metrology, the sign of the phase cannot be assessed because of the argument in the sinusoidal function of the modulation term in Eq. (13) results in an identical interferogram for $\pm\phi(x, y)$ values. In order to overcome this difficulty, multiple interferograms can be sequentially registered introducing a known small phase change amount which is linear in time; this method is well known as phase shifting [4]. In practical phase shifting, a piezoelectric device PZT is included in the interferometric system to produce the phase-shifted interferograms. In general, when a voltage signal is used to polarize a PZT actuator, this electrical signal is converted directly into linear displacement motion. Then, phase-displaced intensity images can be represented by

$$I(x,y) = a(x,y) + b(x,y)cos[\phi(x,y) + \theta]. \tag{14}$$

Since there are three unknown terms in the representation of the interference intensity equation, then the measurement of at least three interferograms at known phase shifts is needed to determine the relative phase difference. However, one of the simplest modes to determinate the phase considers the use of four interferograms equally spaced by $\theta = \pi/2$, obtaining:

$$I_1 = a + b\cos(\phi), \tag{15}$$

$$I_2 = a - bsin(\phi), \tag{16}$$

$$I_3 = a - bcos(\phi), \tag{17}$$

and

$$I_4 = a + bsin(\phi). \tag{18}$$

Rearranging the simultaneous equation system from the above formulas, in which the spatial dependence (x, y) was not included, the phase distribution from intensity images can be calculated with:

$$tan(\phi) = \frac{I_4 - I_2}{I_1 - I_3}. \quad (19)$$

The phase difference estimated from the four equations is determined in the range between $-\pi$ and π when using the arctangent function, hence producing a wrapped phase distribution. In order to analyze an interference pattern by the phase-shifting techniques, a Twyman-Green interferometer is considered for the testing of optical elements; the basic arrangement is shown in Figure 6 . A laser diode is used in the interferometer as illumination source, and then the emerging laser beam is collimated by a lens to obtain a plane wave front that propagates throughout a 50:50 beam splitter to produce two wave fronts of same amplitude. One beam is deviated to the reference mirror M1, and the second beam, to the mirror M2 under test. Next, the beams are reflected back toward the beam splitter, and part of the intensity overlaps on the observation plane, where a CCD camera sensor is placed to register an image of the resulting interferogram that is then stored in a PC for subsequent processing.

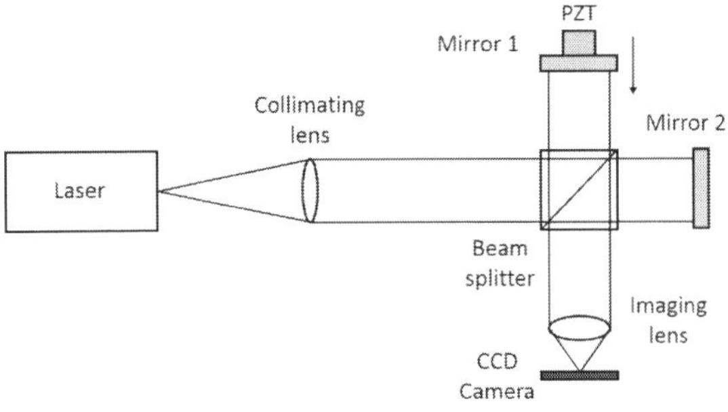

Figure 6. Twyman-Green interferometer for the testing of optical elements.

In Figure 7 , a set of four phase-shifted interferograms with phase shifts of $\pi/2$ is shown. A first interferogram with $\theta = 0$ is seen in Figure 7a , and subsequent interferograms with $\theta = \pi/2$, $\theta = \pi$ and $\theta = 3\pi/2$ are observed in Figure 7b – d , respectively. The wrapped phase is showed in Figure 7e , and the unwrapped phase related with the shape of the optical element being tested is shown in Figure 7f . A three-dimensional representation of the unwrapped phase seen in Figure 7f is presented in Figure 8 .

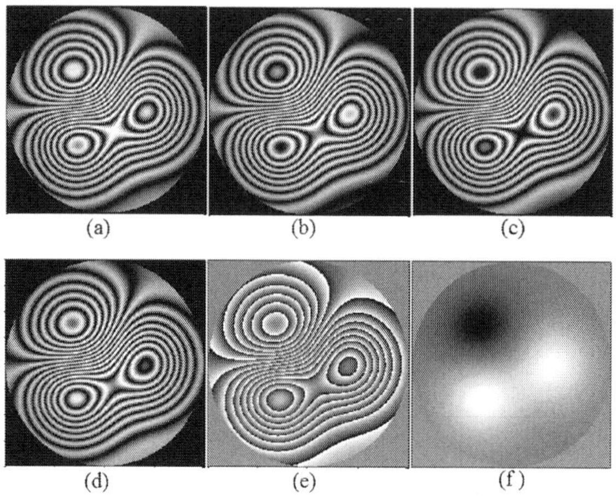

Figure 7. Four-step phase-shifted interferograms from (a) to (d), the wrapped phase (e) and the unwrapped phase (f).

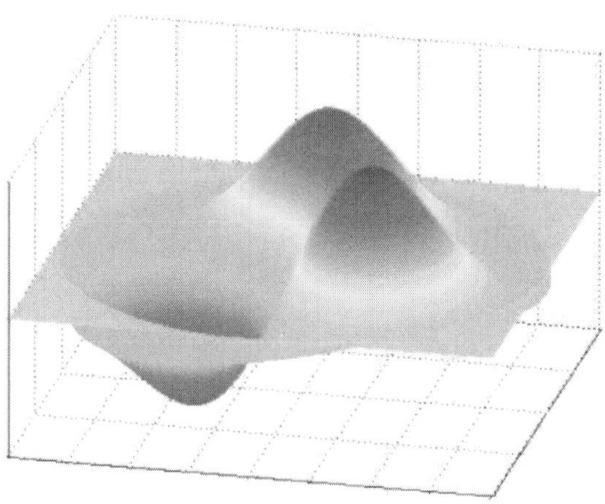

Figure 8. The unwrapped continuous phase obtained from a set of four phase-shifted fringe patterns and an unwrapping method.

In applied phase-shifting interferometry, there are concerns about the presence of errors that may affect the accurate phase extraction from phase-shifted interferograms. A typical systematical source of error introduced PZT arises when there is miscalibration of the phase-shifting actuator, causing detuning in the phase extraction process. The inclusion of phase-shifting algorithms with more than three or four interferograms can be implemented in order to reduce this systematic error [5].

THE FOURIER METHOD

The method of Takeda or the Fourier method was developed in 1982 [6]. Unlike phase-shifting methods, see Ref. [7], a single interferogram with open fringes can be analyzed. This is useful when the object under study changes dynamically or environmental disturbances (vibrations or air turbulence) do not allow the use of phase-shifting methods unless special, and often, very expensive hardware is used to acquire several images simultaneously. However, in general, the accuracy and the dynamical range of the phase that can be measured are reduced. The Fourier method makes use of the fast Fourier transform technique to separate, in the frequency domain, the background and phase terms of the interferogram. Employing complex notation, an interferogram can be written as follows:

$$I(x,y) = a(x,y) + \frac{1}{2}b(x,y)e^{i[\phi(x,y)+\varphi(x,y)]} + \frac{1}{2}b(x,y)e^{-i[\phi(x,y)+\varphi(x,y)]}, \tag{20}$$

where $a(x, y)$ and $b(x, y)$ are the background intensity and the modulation term, respectively. The phase term is denoted by $\phi(x, y)$ and finally, the symbol $\varphi(x, y)$ denotes the lineal carrier function or tilt that is introduced usually by tilting the reference mirror in a two arm interferometer. The Fourier transform of the above expression can be written as:

$$\mathcal{F}\{I(x,y)\} = \tilde{I}(u,v) = \delta(u,v) + E(u+\alpha, v+\beta) + E^{*}(u-\alpha, v-\beta), \tag{21}$$

where (u, v) is the coordinates in the frequency domain, $\delta(u, v)$ is a delta function and $E(u + \alpha, v + \beta)$ and $E^*(u-\alpha, v-\beta)$ are complex conjugate functions that correspond to the transforms of the second and third terms of Eq. (20), respectively. The introduction of the linear carrier function, $\varphi(x, y) = \alpha x + \beta y$, shifts the terms $E(u, v)$ and $E^*(u, v)$ in opposite directions in the frequency spectrum as can be seen noting that:

$$E(u,v) = \mathcal{F}\left\{\frac{1}{2}b(x,y)e^{i\phi(x,y)}\right\}, \tag{22}$$

and

$$E(u+\alpha, v+\beta) = \mathcal{F}\left\{\frac{1}{2}b(x,y)e^{i\phi(x,y)}e^{i\varphi(x,y)}\right\}, \tag{23}$$

similarly

$$E^*(u,v) = \mathcal{F}\left\{\frac{1}{2}b(x,y)e^{-i\phi(x,y)}\right\}, \tag{24}$$

and

$$E^*(u-\alpha,v-\beta) = \mathcal{F}\left\{\frac{1}{2}b(x,y)e^{-i\phi(x,y)}e^{-i\varphi(x,y)}\right\}. \tag{25}$$

The separation of the terms in the Fourier spectrum due to the introduction of a linear carrier can be observed in Figure 9 . An open-fringe interferogram and its Fourier spectrum can be seen in Figure 9a and b . It can be noted that the three terms of Eq. (21) are clearly separated. The central peak corresponds to the delta function $\delta(u, v)$ while the adjacent lobules correspond to $E(u + \alpha, v + \beta)$ and $E^*(u - \alpha, v - \beta)$. The interferogram was constructed as follows:

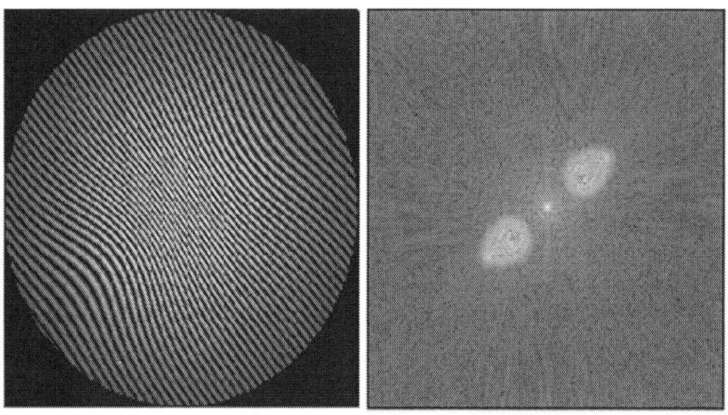

(a) (b)

Figure 9. Interferogram with open fringes (a) and its Fourier spectrum (logarithm of the module) (b).

$$a(x,y) = 120e^{-(x^2+y^2)} \tag{26}$$

$$b(x,y) = 124e^{-(x^2+y^2)}, \tag{27}$$

$$\phi(x,y) = 2\pi\left\{4e^{-4[(x-0.2)^2+(y-0.3)^2]}-5e^{-6[(x+0.2)^2+(y+0.3)^2]}\right\}, \tag{28}$$

and

$$\varphi(x,y) = 2\pi(16x+20y) \tag{29}$$

where x and y vary from 1 to −1 along the vertical and horizontal directions and the interferogram was multiplied by a circular function of radius one.

In order to recover the phase, we need to isolate one of the lateral lobules of the Fourier domain. To this end, we employ a band-pass filter. The filtered spectrum is then transformed back to the spatial domain to obtain $I_f(x, y)$ and the wrapped phase is found with the arctangent function of the ratio of the imaginary and real parts of $I_f(x, y)$ as shown in Figure 10 . Figure 10a shows the band-pass filter $H(u, v)$, the filtered spectrum can be observed in Figure 10b and finally the wrapped phase is shown in Figure 10c .

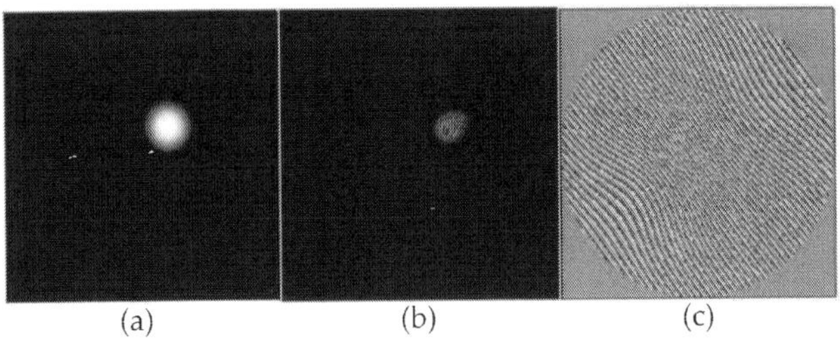

Figure 10. Band-pass filtering process of the Fourier spectrum seen in Figure 9 . Pass-band filter (a), filtered spectrum (b) and recovered wrapped phase (c).

The band-pass filter has the following form:

$$H(u,v) = e^{-3000\left[(u-0.2)^2+(v-0.2)^2\right]^2}, \tag{30}$$

where u and v vary from 1 to −1 in vertical and horizontal directions, respectively. The filtered spectrum becomes,

$$\tilde{I}_f(u,v) = \tilde{I}(u,v)H(u,v). \tag{31}$$

Transforming back to spatial domain we obtain:

$$I_f(x,y) = \mathcal{F}^{-1}\left\{\tilde{I}_f(u,v)\right\}. \tag{32}$$

The wrapped phase is found by:

$$\psi_w(x,y) = atan2\left\{imag[I_f(x,y)], real[I_f(x,y)]\right\} \tag{33}$$

where the *atan2()* function accepts two arguments corresponding to the sine and cosine and returns the results modulo 2π. This wrapped phase, however, is not the desired one because it contains the introduced tilt that is not part of the object information. The wrapped desired phase is found with:

$$\phi_w(x,y) = atan2\left\{sin[\psi_w(x,y)-\varphi(x,y)], cos[\psi_w(x,y)-\varphi(x,y)]\right\}. \tag{34}$$

The final step is to apply an unwrapping method to obtain the continuous phase related with the object under study. This last procedures are shown in Figure 11a and b where the wrapped phase φ_w (x, y) and the unwrapped reconstructed phase φ_r (x, y) can be appreciated, respectively.

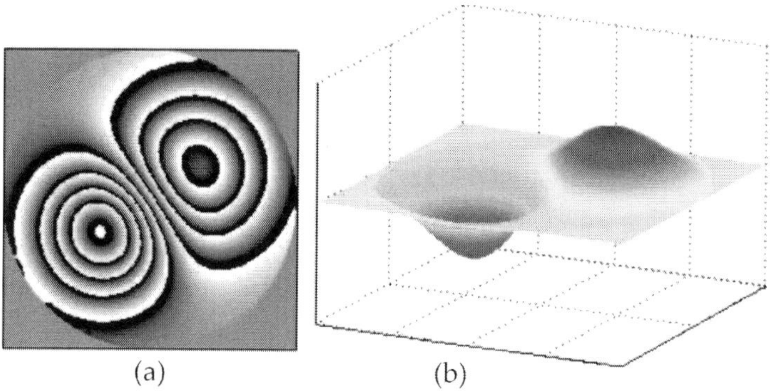

Figure 11. Phase reconstruction results. Wrapped phase (a) and unwrapped continuous phase (b).

The Fourier method is not the unique procedure for phase retrieval from one interferogram with open fringes. Besides the Fourier approach there are other procedures in the spatial domain including phase locked loop [8] and spatial carrier phase-shifting methods [9, 10].

PHASE UNWRAPPING

Phase unwrapping is a common step to finally find a continuous phase for several fringe analysis techniques such as phase shifting, Fourier, the phase synchronous and others methods that use the arctangent function of the ratio of the sine and cosine of the phase to obtain a wrapped phase. In its simplest form, phase unwrapping consists in adding or subtracting 2π terms to the pixel being unwrapped if a difference greater that π is found with a previous pixel already unwrapped [11]. The phase unwrapping problem in one dimension can be observed in Figure 12 . The wrapped phase found with the arctangent function is seen in Figure 12a and the unwrapped continuous phase is showed in Figure 12b .

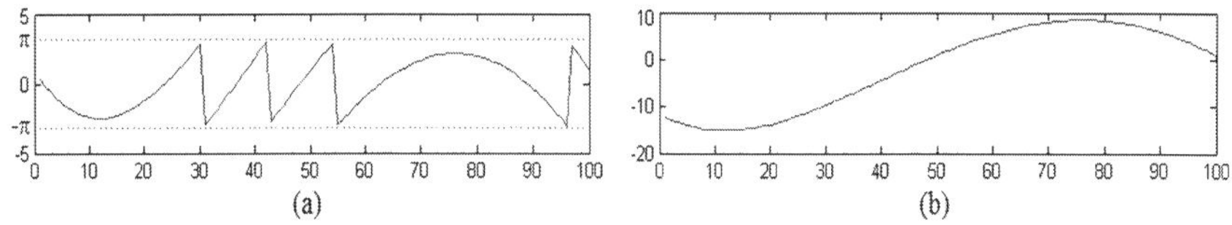

Figure 12. Wrapped (a) and unwrapped phase (b).

The described procedure works well only for wrapped phases with no inconsistencies and low noise levels, however, delivers wrong results when dealing with noisy wrapped phases or those obtained from interferograms with broken or unconnected fringes. A more consistent approach is achieved with the least squares phase unwrapping method [12]. The least square technique integrates the discretized laplacian of the phase. To this end, the laplacian of the phase is calculated as follows:

$$L_{i,j} = \phi^x_{i+1,j} - \phi^x_{i,j} + \phi^y_{i,j+1} - \phi^y_{i,j}, \tag{35}$$

where

$$\phi^x_{i,j} = atan2\left[\sin\left(\phi^w_{i,j} - \phi^w_{i-1,j}\right), cos\left(\phi^w_{i,j} - \phi^w_{i-1,j}\right)\right] p_{i,j} p_{i-1,j}, \tag{36}$$

and

$$\phi^y_{i,j} = atan2\left[\sin\left(\phi^w_{i,j} - \phi^w_{i,j-1}\right), cos\left(\phi^w_{i,j} - \phi^w_{i,j-1}\right)\right] p_{i,j} p_{i,j-1}. \tag{37}$$

In the last equations, we have used pixel subscript notation in order to limit the extension of the equations. The pupil function $p_{i,j}$ is defined as equal to one where we have valid data and zero otherwise. One may note that if $p_{i,j} = p_{i+1,j} = p_{i-1,j} = p_{i,j+1} = p_{i,j-1} = 1$, then

$$L_{i,j} = -4\phi_{i,j} + \phi_{i+1,j} + \phi_{i-1,j} + \phi_{i,j+1} + \phi_{i,j-1}. \tag{38}$$

Solving for the phase in the above equation, we obtain that the unwrapping problem under the least square approach consist in the resolution of a linear system of equations, as:

$$\phi_{i,j} = \frac{\phi_{i+1,j} + \phi_{i-1,j} + \phi_{i,j+1} + \phi_{i,j-1} - L_{i,j}}{4}. \tag{39}$$

An iterative technique that solves the above system of linear equations is the overrelaxation method in which the following equation is iterated until convergence:

$$\phi_{i,j}^{k+1} = \phi_{i,j}^{k} - \frac{(d\phi_{i,j}^{k} - \phi_{i-1,j}^{k+1} - \phi_{i+1,j}^{k} - \phi_{i,j-1}^{k+1} - \phi_{i,j+1}^{k} + L_{i,j})w}{d}, \tag{40}$$

where

$$d = p_{i-1,j} + p_{i+1,j} + p_{i,j-1} + p_{i,j+1}. \tag{41}$$

In the last equations, k is the iteration number and w is a parameter of the overrelaxation method that must be set between 1 and 2. Results of the least square technique are showed in Figure 13 . A two-dimensional wrapped phase is seen Figure 13a , the phases differences are shown in Figure 13b and c , respectively. The laplacian of the phase is showed in Figure 13d , the reconstructed phase in a two-dimensional view is observed in Figure 13e and finally, for comparison purposes the reconstructed rewrapped phase is also showed in Figure 13f . One may observe that the original wrapped phase and the reconstructed wrapped phase are slightly different, this is because the least square method recovers the phase with an arbitrary constant term, however this term is usually neglected since doesn't carry any information of the object being measured.

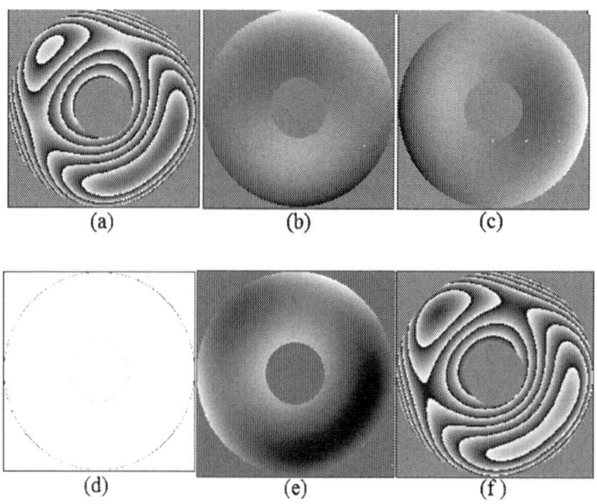

Figure 13. Least square results to unwrap the phase. Wrapped phase (a), phase differences in the x (b) and y (c) directions, laplacian of the phase (d), unwrapped phase (e) and rewrapped phase (f).

If desired, the constant term in the retrieved phase may be corrected easily in the following form:

$$\phi_c = \phi_r - atan2[\sin(\phi_r - \phi_w), \cos(\phi_r - \phi_w)]. \tag{42}$$

The wrapped phase seen in Figure 13a was constructed as follows:

$$\phi_w = atan2[\sin(\phi), \cos(\phi)], \tag{43}$$

where

$$\phi_{i,j} = 2\pi \left\{ \begin{array}{c} 16\left[(x_i)^2 + (y_j)^2\right]^2 - 18\left[(x_i)^2 + (y_j)^2\right] + \\ 5y_j(x_i)^2 - 4x_i(y_j)^2 + 3y_jx_i + 2x_i - 2y_j + 0.56 \end{array} \right\}. \tag{44}$$

In the above equations x_i and y_j are range variables that vary from −1 to 1, the wrapped phase was multiplied by an annular pupil function with an exterior radius of 198 pixels, while the interior radius was 60 pixels for an image size of 400 × 400. The convergence of the reconstructed phase seen in Figure 2e was reached after 700 iterations; the overrelaxation parameter w was set to 1.99, which is usual for large images. The reconstructed phase with the constant term corrected and the phase error, $\varepsilon = \phi = \phi_c$, can be appreciated in Figure 14a and b . The maximum error was about 0.00005 radians.

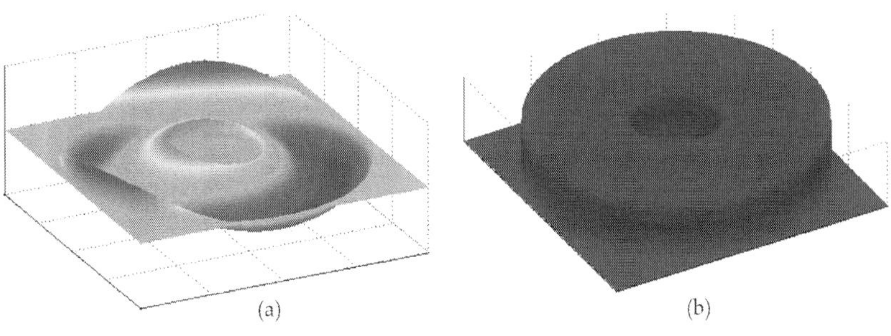

Figure 14. Unwrapped phase and phase error. Three-dimensional view of the reconstructed and corrected phase (a) and phase error (b).

Finally, results on a noisy wrapped phase are presented. Random noise with uniform distribution in the range of $-\pi/4$ to $\pi/4$ radians were added to the phase to obtain the wrapped phase seen in Figure 15a , the unwrapped phase is shown in Figure 15b , and the rewrapped reconstructed phase is observed in Figure 15c . As can be noticed the least square method is a very reliable technique that works with any pupil configuration and stands noisy measurements.

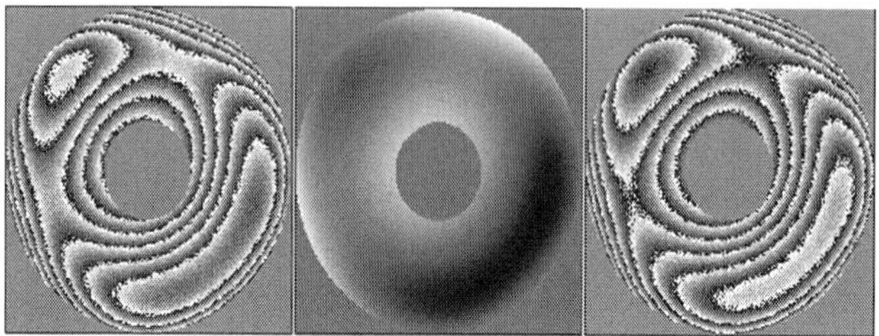

Figure 15. Results on noisy measurements. Wrapped phase with noise (a), retrieved phase (b) and wrapped retrieved phase (c).

PHASE RECOVERY FROM LATERAL SHEARING INTERFEROGRAMS

Lateral shearing interferometry is a very important field in experimental optical measurements, in which, the test beam interferes with a laterally displaced version of itself instead of a reference beam. The resulting fringe patterns are thus related with the object wavefront derivative in a given direction. This is very useful when the object information of interest is related with the derivative as in strain analysis or when the dynamical range of the object wave front is too high that cannot be measured with direct interferometry. Let us consider a laterally shear interferogram with a beam displacement in the x direction a quantity Δx, we obtain:

$$I^x(x,y) = B(x,y) + C(x,y)cos[\psi^x(x,y)], \tag{45}$$

where

$$\psi^x(x,y) = \phi(x,y) - \phi(x-\Delta x, y). \tag{46}$$

In Eq. (45) $B(x, y)$ is the background intensity and $C(x, y)$ is the modulation term. The objective is to retrieve the undisplaced phase $\varphi(x, y)$. To this end, we need, at least, another laterally shear interferogram with a beam displacement in the y direction another quantity Δy, obtaining:

$$I^y(x,y) = B(x,y) + C(x,y)cos[\psi^y(x,y)], \tag{47}$$

where

$$\psi^{y}(x,y) = \phi(x,y) - \phi(x,y-\Delta y). \tag{48}$$

Let us considered that we have retrieved the phase differences ψ^x (x, y) and ψ^y (x, y) by means of a phase-shifting technique as is depicted in Figure 16 and Figure 17 for the x and y directions, respectively. A set of four phase-shifted interferograms can be seen in Figure 16a . The wrapped phase differences ψ^{wx} (x, y) and the unwrapped phase ψ^x (x, y) are observed in Figure 16e and f , respectively. The modulation term and a sheared pupil function p^x (x, y) are shown in Figure 16g and h , respectively. The sheared pupil p^x (x, y) is found after normalization from zero to one and thresholding the normalized modulation term. A second set of phase-shifted interferograms are seen Figure 17a and d . The wrapped phase differences ψ^{wy} (x, y) and the unwrapped phase ψ^y (x, y) are observed in Figure 17e and f , respectively. The modulation term and a sheared pupil function p^y (x, y) are shown in Figure 4g and h , respectively. In a similar way to the first set of phase-shifted interferograms, the pupil p^y (x, y) is found after normalization from zero to one and thresholding the normalized modulation term. The sheared pupils p^x (x, y) and p^y (x, y) are useful to find the undisplaced pupil $p(x, y)$, where the original wave front $\varphi(x, y)$ have valid data, since p^x (x, y) $= p(x, y)p(x - \Delta x, y)$ and p^y $(x, y) = p(x, y)p((x, y) - \Delta y)$

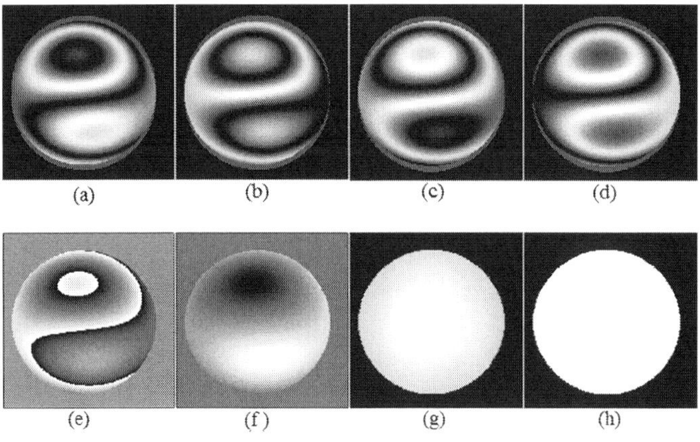

Figure 16. Recovery of the phase differences in the x direction. Set of four sheared interferograms acquired under a phase-shifting technique (a) to (d), wrapped phase differences (e), unwrapped phase differences (f), modulation (g) and recovered sheared pupil (h).

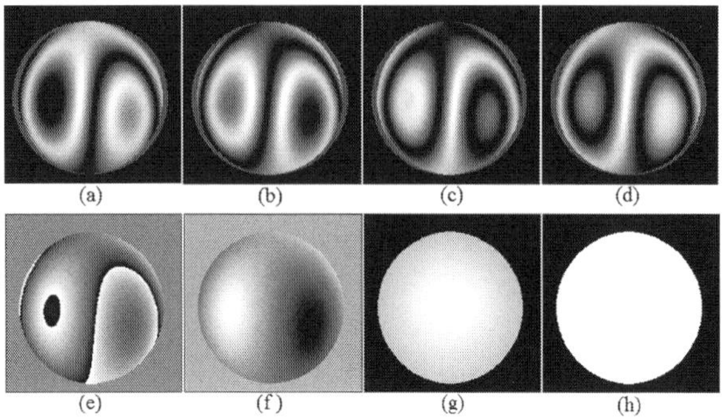

Figure 17. Recovery of the phase differences in the y direction. Set of four sheared interferograms acquired under a phase-shifting technique (a) to (d), wrapped phase differences (e), unwrapped phase differences (f), modulation (g) and recovered sheared pupil (h).

Once the phase differences $\theta^x(x, y)$ and $\theta^y(x, y)$ are known, we can use the next procedure to find the searched phase $\varphi(x, y)$ observing that:

$$\psi^x_{i,j}p_{i-a,j} - \psi^x_{i+a,j}p_{i+a,j} = \phi_{i,j}(p_{i-a,j} - p_{i+a,j}) - \phi_{i-a,j}p_{i-a,j} - \phi_{i+a,j}p_{i+a,j}, \tag{49}$$

and

$$\psi^y_{i,j}p_{i,j-b} - \psi^y_{i,j+b}p_{i,j+b} = \phi_{i,j}(p_{i,j-b} - p_{i,j+b}) - \phi_{i,j-b}p_{i,j-b} - \phi_{i,j+b}p_{i,j+b}. \tag{50}$$

In the above equations we have changed the (x, y) dependence by pixel subscript notation. As described before, $p_{i,j}$ is the undisplaced pupil and the displacement quantities Δx and Δy are rounded to the nearest integer in pixel dimensions obtaining a and b, respectively. Adding Eq. (49) and Eq. (50) and solving for the phase, we obtain:

$$\phi_{i,j} = \frac{F+G}{H}, \tag{51}$$

where

$$H = p_{i+a,j} + p_{i-a,j} + p_{i,j+b} + p_{i,j-b}, \tag{52}$$

$$\psi^x_{i,j}p_{i-a,j} - \psi^x_{i+a,j}p_{i+a,j} + \psi^y_{i,j}p_{i,j-b} - \psi^y_{i,j+b}p_{i,j+b}, \tag{53}$$

and

$$G = \phi_{i-a,j}p_{i-a,j} + \phi_{i+a,j}p_{i+a,j} + \phi_{i,j-b}p_{i,j-b} + \phi_{i,j+b}p_{i,j+b}. \tag{54}$$

Eq. (51) represents a linear system of equations; however, it is ill posed because there are more unknowns than equations due to the effects of the sheared pupils. Nevertheless, a regularization term may be aggregated to overcome this problem [13–15]. The regularization term is in the form of discrete Laplacians of the phase among adjacent pixels. The following equation that incorporates the regularization term is iterated until convergence:

$$\phi_{i,j}^{k+1} = \phi_{i,j}^{k} - \frac{H\phi_{i,j}^{k} - (F+G) + \alpha(L_{i+1,j}^{x} - 2L_{i,j}^{x} + L_{i-1,j}^{x} + L_{i,j+1}^{y} - 2L_{i,j}^{y} + L_{i,j-1}^{y})}{H}, \tag{55}$$

where

$$L_{i,j}^{x} = (\phi_{i+1,j} - 2\phi_{i,j} + \phi_{i-1,j})p_{i+1,j}p_{i-1,j}, \tag{56}$$

and

$$L_{i,j}^{y} = (\phi_{i,j+1} - 2\phi_{i,j} + \phi_{i,j-1})p_{i,j+1}p_{i,j-1}. \tag{57}$$

Here, α is a parameter that controls the effects of the regularization term. The phase reconstruction is seen in Figure 18 . A two-dimensional view of the retrieved phase is observed in Figure 18a , the same phase but in a three-dimensional perspective is shown in Figure 18b and the phase error (the actual phase minus the reconstructed one) can be appreciated in Figure 18c . This reconstruction was achieved after 800 iterations with a parameter α = 0.1 obtaining a maximum error of about 0.0004 radians.

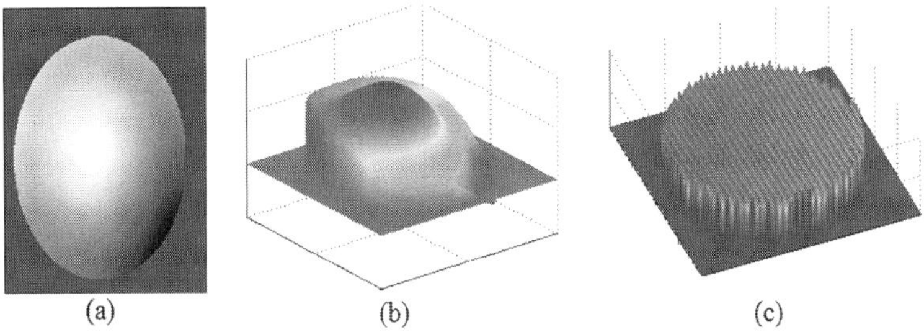

Figure 18. Phase reconstruction from lateral sheared interferograms. Reconstructed phase (a), three-dimensional view of the reconstructed phase (b) and the phase error (c).

The actual phase was constructed as follows:

$$\phi_{i,j} = 2\pi \left\{ \begin{array}{c} 20\left[(x_i)^2 + (y_i)^2\right]^2 - 24\left[(x_i)^2 + (y_i)^2\right] - 6(x_i)^2 y_j \\ +8x_i(y_j)^2 + 5x_i y_j + 4x_i - 5y_j + 0.45 \end{array} \right\}, \tag{58}$$

where x_i and y_j are range variables that vary from −1 to 1 in both vertical and horizontal directions. The pupil function is a circular one with a radius of 170 pixels. The shear distances were set to $\Delta x = a = 12$ pixels and $\Delta y = b = 12$ pixels.

CONCLUSIONS

Digital processing techniques applied to interferometric measurements allow to obtain the phase from fringe patterns. The fringe analysis methods described here can be used to recover the phase that is associated with the physical variable under study. Under controlled conditions, phase-shifting techniques are the most used methods to retrieve the wrapped phase. If experimental conditions suffer from vibrations, air turbulences or the object changes dynamically, among other factors, then a Fourier method may be preferable to analyze an open-fringe interferogram. Those procedures deliver a wrapped phase. Then, an unwrapping algorithm is needed to reconstruct a continuous phase related with the object being studied. The aim of this chapter is to present, to the reader, the fundamentals of principal fringe analysis techniques. Numerical simulations are provided, in such way that the reader can reproduce them by its own. The extension of the chapter is insufficient to introduce many important techniques. However, the methods presented here were described as clearly and briefly as we could. We hope that the reader finds this information useful in the interpretation of interferograms obtained in the study of some object or phenomena by using an interferometric setup.

REFERENCES

1. Born M., Wolf E. Principles of optics. 7th ed. New York, USA: Cambridge University Press; 2002. 952 p.
2. Mora-González M., Casillas-Rodríguez F.J., Muñoz-Maciel J., Martínez-Romo J.C., LunaRosas F.J., de Luna-Ortega C.A., Gómez-Rosas G., Peña-Lecona F.G. Reducción de ruido digital en señales ECG utilizando filtraje por convolución. Investigación y Ciencia. 2008;16(040):26–32.
3. Mora-Gonzalez M., Muñoz-Maciel J., Casillas F.J., Peña-Lecona, Chiu-Zarate R., PerezLadron de Guevara H. Image procesing for optical metrology. In: Ionescu C.M., editor. MATLAB–a ubiquitous tool for the practical engineer. 1st ed. Rijeka, Croatia: InTech; 2011. pp. 523–546. doi:10.5772/820
4. Gåsvik K.J. Optical metrology. 3rd ed. West Sussex, UK: John Wiley and Sons, Ltd.; 2002. 372 p.
5. Malacara D., editor. Optical shop testing. 3rd ed. NJ, USA: John Wiley and Sons, Inc.; 2007. 888 p.
6. Takeda M., Ina H., Kobayashi S. Fourier-transform method of fringe-pattern analysis for computer-based topography and interferometry. Journal of the Optical Society of America. 1982;72:156–160.
7. Bruning J.H., Herriott D.R., Gallagher J.E., Rosenfeld D.P., White A.D., Brangaccio D.J. Digital wavefront measuring interferometer for testing optical surfaces and lenses. Applied Optics. 1974;13:2693–2703.
8. Servín M., Rodríguez-Vera R. Two-dimensional phase locked loop demodulation of interferogram. Journal of Modern Optics. 1993;40:2087–2094.
9. Shough D.M., Kwon O.Y., Leary D.F. High speed interferometric measurement of aerodynamic phenomena. In: Ulrich P.B., Wilson L.E., editors. Propagation of high-energy laser beams through the earth's atmosphere. Proc. SPIE, LA, USA; 1990. pp. 394–403. doi:10.1117/12.18359
10. Kujawinska M., Wójciak J. Spatial-carrier phase shifting technique of fringe pattern analysis. In: Jueptner W.P., editor. Industrial applications of holographic and speckle measuring techniques. Proc. SPIE, The Hague, Netherlands; 1991. doi:10.1117/12.47089

11. Robinson D.W. Phase unwrapping methods. 1st ed. Bristo, UK: CRC Press; 1993. 302 p.
12. Ghiglia D.C., Pritt M.D. Two-dimensional phase unwrapping: theory algorithms and software. 1st ed. New York, USA: John Wiley and Sons, Inc.; 1998. 512 p.
13. Legarda R., Rivera M., Rodríguez R., Trujillo G. Robust wave-front estimation from multiple directional derivatives. Optics Letters. 2000;25:1089–1091. doi:10.1364/ OL.25.001089
14. Servín M., Malacara D., Marroquín J.L. Wave-front recovery from two orthogonal sheared interferograms. Applied Optics. 1996;35:4343–4348. doi:10.1364/AO.35.004343
15. Villa J., García G., Gómez G. Wavefront recovery in shearing interferometry with variable magnitude and direction shear. Optics Communications. 2001;195:85–91. doi:10.1016/ S0030-4018(01)01304-9

CHAPTER 2

FSD-HSO OPTIMIZATION ALGORITHM FOR CLOSED FRINGES INTERFEROGRAM DEMODULATION

Ulises H. Rodriguez-Marmolejo,[1,2] Miguel Mora-Gonzalez,[1] Jesus Muñoz-Maciel,[1] and Tania A. Ramirez-delreal[1,3]

[1]Departamento de Ciencias Exactas y Tecnología, Centro Universitario de los Lagos, Universidad de Guadalajara, 47460 Lagos de Moreno, JAL, Mexico
[2]Departamento de Ingeniería Eléctrica y Electrónica, Instituto Tecnológico de Aguascalientes, 20256 Aguascalientes, AGS, Mexico
[3]Departamento de Posgrado e Investigación, Universidad Politécnica de Aguascalientes, 20342 Aguascalientes, AGS, Mexico

ABSTRACT

Due to the physical nature of the interference phenomenon, extracting the phase of an interferogram is a known sinusoidal modulation problem. In order to solve this problem, a new hybrid mathematical optimization model for phase extraction is established. The combination of frequency guide

Citation: (APA): Rodriguez-Marmolejo, U. H., Mora-Gonzalez, M., Muñoz-Maciel, J., & Ramirez-Delreal, T. A. (2016). Fsd-hso optimization algorithm for closed fringes interferogram demodulation. Mathematical Problems in Engineering, 2016. (12 pages)

sequential demodulation and harmony search optimization algorithms is used for demodulating closed fringes patterns in order to find the phase of interferogram applications. The proposed algorithm is tested in four sets of different synthetic interferograms, finding a range of average relative error in phase reconstructions of 0.14–0.39 rad. For reference, experimental results are compared with the genetic algorithm optimization technique, obtaining a reduction in the error up to 0.1448 rad. Finally, the proposed algorithm is compared with a very known demodulation algorithm, using a real interferogram, obtaining a relative error of 1.561 rad. Results are shown in patterns with complex fringes distribution.

INTRODUCTION

Interferometry includes a series of techniques used in the measurement of aberrations, deformations, flatness, and perturbations. It may also be applied to measure variables as temperature gradients, strain analysis, depth measurement, and so forth. Widely known techniques to demodulate interferograms are phase shifting algorithms for several images and the Fourier method for a single linear carrier fringe pattern [1]. However, these techniques are difficult to implement when the object under study changes fast and continuously, and the dynamic range of the phase does not allow the use of a large linear carrier without infringing the sampling theorem. In those situations a single interferogram with closed and possibly complex fringe distribution must be analyzed to recover the phase information related to the physical phenomena being measured. This is known as a difficult task since there are many solutions that are compatible with the measured data but lacks physical meaning. The accuracy of measurements carried out from a single fringe pattern that includes closed fringes is thus intensely dependent on the phase distribution of the recorded interferogram being estimated. Recently, many phase recovering methods have been developed as combination of genetic algorithms and parametric methods [2–4], soft computing techniques applied to Zernike polynomials [5], combination of genetic algorithms and frequency guided sequential demodulation [6], particle swarm optimization [7], unwrapping of phase maps with sign changes [8], two-dimensional regularized phase-tracking technique [9], and so forth. In general, there are not particular processes that succeed in obtaining the phase for any given interferogram, but all of them are limited to specific features of the fringe pattern.

The phase demodulation trouble has been formulated as an optimization challenge, where soft computing procedures may be used to find the phase solution that best matches the nonlinear equation represented by fringe patterns. Few years ago, genetic algorithms have been tested [2]; the authors developed a parametric method for fringe pattern demodulation using a genetic algorithm (GA). A parametric estimation of the coefficients of a 15th degree Zernike polynomial is used in order to approximate the phase; a population of chromosomes is programmed within the coefficients to calculate the phase. A cost function is then employed considering the number of the observed fringes and the fringes that result from the recovered phase match, the phase softness, and the prior knowledge of the object. Normally, the final solution of the GA is based on a cost function, which is stated as the comparison between the better individual in the population and the target (real fringes); a population evolution process is allowed until a cost function average threshold is achieved. The authors reported a root mean square (rms) error of 0.12 radians. This method was applied to noisy fringe patterns and to a single closed fringe image. Additional improvements and variations of this work were subsequently presented by the same research team [3, 4].

Another soft computing technique used for phase reconstruction is particle swarm optimization (PSO). This algorithm was introduced by Kennedy and Eberhart in 1995 [10], as an evolving optimization technique. In 2012 Jiménez et al. [7] used PSO for phase recovery; they compared a GA and a PSO for phase recovery on several fringe patterns, obtaining errors of 0.4281 and 0.313 rad., respectively, showing an improvement in accuracy of PSO over GA; processing time improvements were announced, but no results were shown.

As mentioned before, the demodulation of a single interferogram often involves a combination of methods (GA + Zernike, PSO + Zernike, Neuronal networks + others, etc.). In 2009 Wang and Kemao reported a new hybrid method; they used frequency guided sequential demodulation (FSD) as interferogram demodulator, combined with Levenberg Marquardt (LM) optimization [11], method implemented by their quickness and efficiency in fringes demodulation.

In this work a FSD with harmony search optimization (HSO) is investigated in order to test the performance in a single interferogram with closed fringes. The main motivation is the advantages of the HSO technique over other soft computing techniques already reported. The HSO technique was inspired in the observation of musical composition to search a perfect harmony and was introduced by Geem et al. in 2001 [12] and has found its way in several applications as diverse as engineering, math, industrial process, biology, and so forth [13–19]. An excellent recent review and categorization of the applications of HSO was conducted by Manjarres et al. in 2013 [20]. Some advantages of this method are that it uses simple algebraic equations and real values, while the derivative information is unnecessary unlike GA and other optimization techniques.

In the following section, the physical theory of the interferograms is presented as well as the concepts of the HSO and FSD algorithms. In the next section, the image-processing techniques and the experimental setup used to implement the soft computing proposed method are described. Finally, in the last two sections the results and the conclusions are presented, respectively.

THEORY

Metrology has techniques such as fringe projection profilometry and optical interferometry to measure physical quantities in many areas of engineering and science, but the importance of these methods lies in the fact that they are noninvasive procedures [21]. Recently, advances in computational techniques have the potential to extend the measuring capabilities of optical metrology applications. In the present section the optical metrology basis, the harmony search optimization model, and the frequency guided sequential demodulation process are shown.

Optical Interferometry

Interferometry studies the engagement of two or more light waves, where one of them has suffered a modification by one characteristic of an object being tested [22]. Demodulation of the phase is the most important task in interferometry measurements; the phase is related to a physical quantity to be measured.

The optical arrangement, shown in Figure 1, is a Twyman-Green interferometer setup (a Michaelson interferometer modification). The interference is produced by the difference of optical path between the two arms of the interferometer. The interferogram is reordered by a photo detector array (e.g., a charged coupled display camera) and then digitized for show on a monitor or stored for further processing with computational algorithms like unwrapped phase, digital filtering, demodulation phase, and so forth [23]. The optical components Lc, Lf, and Bs are a positive collimating lens, a positive focusing lens, and a beam splinter, respectively. The fringe pattern intensity is modeled by

$$I(x,y) = a(x,y) + b(x,y)\cos(\varphi(x,y)), \quad (1)$$

where $I(x, y)$, $a(x, y)$, $b(x, y)$, $\varphi(x, y)$, and (x, y) are the interference fringe pattern intensity, the background illumination, the modulation amplitude, the phase term, and the spatial coordinates of the surface under test, respectively.

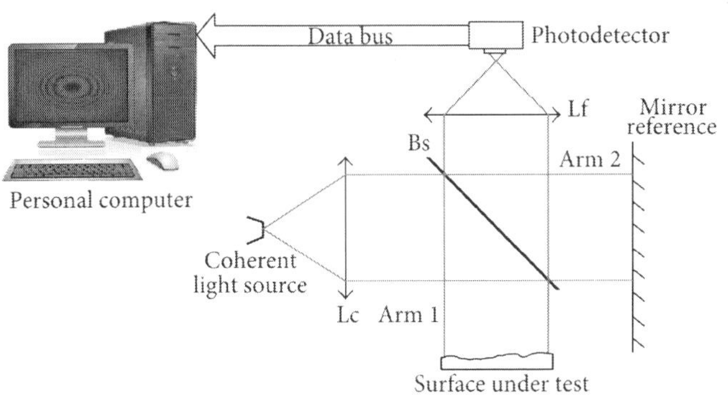

Figure 1. Interferogram acquisition system with a Twyman-Green interferometer.

Harmony Search Optimization

The harmony search is a metaheuristic optimization algorithm; it was created by Lee and Geem in 2005 [24]; it was inspired in music harmony and is a powerfully soft computing tool to solve many optimization problems. The HSO algorithm is structured in five steps [25], as shown in Figure 2. Those steps are the following: (1) initialize the problem and algorithm parameters; (2) initialize the Harmony Memory (HM) array; (3) improvise a new harmony from HM; (4) update the HM; and (5) check the stopping criterion. These steps are described as follows.

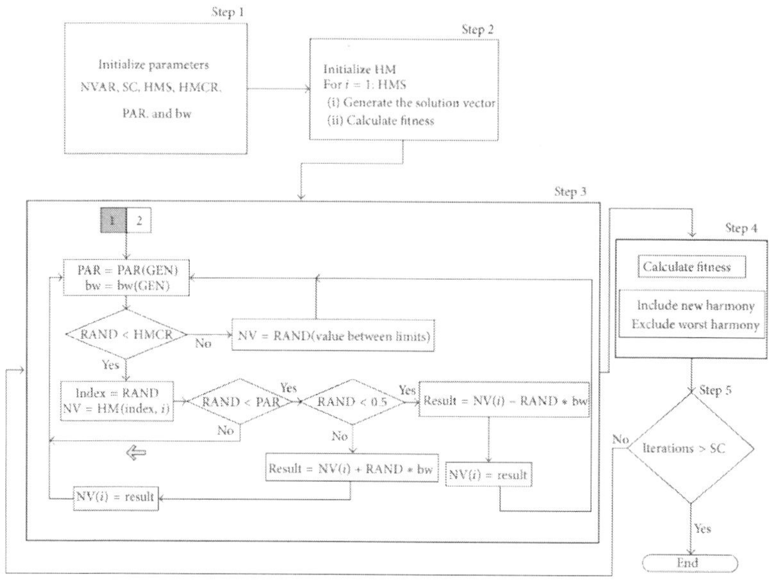

Figure 2. Flow diagram to indicate the structured HSO algorithm.

Step 1. The optimization problem is specified and initialized its parameter values as follows:

$$\text{Mininize} \quad f(x)$$
$$\text{subject to} \quad x_i \in X_i = 1, 2, \ldots, \text{NVAR}, \tag{2}$$

where $f(x)$, x, and X_i are the objective function, the set of each decision variable x_i, and the set of the possible range of values for each decision variable, respectively. Also, the search range of X_i is $x_i^L \leq X_i \leq x_i^U$, with x_i^L and x_i^U as the lower and upper bounds for each decision variable, respectively. The parameters to initialize the algorithm are the number of decision variables (NVAR); the number of improvisations or stopping criterion (SC); the number of solution vectors in the Harmony Memory or the Harmony Memory Size (HMS); Harmony Memory Considering Rate (HMCR); Pitch Adjusting Rate (PAR) for each generation; and the arbitrary distance bandwidth for each generation (bw). In the following steps the role of some of these parameters is explained.

Step 2. The Harmony Memory (HM) array is complete with a random solution vectors as the HMS, as can be seen in the following:

$$\text{HM} = \begin{bmatrix} x_1^1 & x_2^1 & \cdots & x_N^1 \\ x_1^2 & x_2^2 & \cdots & x_N^2 \\ \vdots & \vdots & \ddots & \vdots \\ x_1^{\text{HMS}} & x_2^{\text{HMS}} & \cdots & x_N^{\text{HMS}} \end{bmatrix}, \tag{3}$$

where HM acts as a dynamical memory address, where the sets of decision variables (all the solution vectors) are saved. This HM is similar to the genetic array in the genetic algorithm (GA) [26].

Step 3 (improvise a new harmony with parameters adjustment). For this improvisation, a new harmony vector $x' = (x'_1, x'_2, \ldots, x'_N)$ is generated taking into account three basic considerations [27]: memory consideration, pitch adjustment, and random selection.

In the memory consideration, the value of the first decision variable x'_1 for the new vector is chosen from any of the values in the specified HM. The values of other decision variables $x'_2, \ldots, x'_N$ are chosen in the same manner. The HMCR has a range of values from 0 and 1, while 1-HMCR is the possibility of creating a random HM member [27].

In the second consideration, a pitch adjusting is performed only after a value is chosen from the HM. The PAR considers the possibility of change of some elements of HM.

In the last consideration, a new vector x'_i is generated with a pitch adjustment threshold of 0.5, and the definition of this vector is described by

$$x'_i = x_i \pm \text{RAND}\,() * \text{bw}, \tag{4}$$

where bw has an uniform distribution between −1 and 1. The function RAND( ) is fed with a random number between 0 and 1 to adjust the PAR. The sign of the addition in (4) depends of the value obtained with RAND function; if the value is less than 0.5 the sign is positive, and negative in other case.

Other parameters used in this step are NV and GEN, a new vector to improvise HM and for the depuration of the actual generation, respectively.

Step 4 (update HM). If the new harmony vector is better than the worst harmony vector in HM, then this latter is replaced by the new harmony vector. The HM is then sorted by the objective function value.

Step 5 (check the stopping criterion). If the SC is satisfied, computation is terminated. Otherwise, Steps 3 and 4 are repeated.

The HMCR and PAR parameters introduced in Step 3 help the algorithm to find locally and completed improved solutions, respectively. PAR and bw in HS algorithm are important parameters in fine-tuning of the optimized solution vectors and can be potentially useful in adjusting convergence rate of the algorithm to an optimal solution. The HSO pseudocode is shown in Algorithm 1.

Algorithm 1. The improvisation procedure of HSO.

```
Generate the HM (initial harmonics as real number arrays)
Define the PAR
Define the HMCR
while (Max number of iterations > SC)
   Create new harmonics by accepting better harmonics
   Adjust PAR to take new solutions (harmonics)
      if (rand > HMCR)
         Choose an existing harmonic randomly
      else-if (rand > PAR)
         Adjust the pitch randomly within a bandwidth (bw)
      else
         Create new harmonics via randomization
      end if
   Accept the new solutions (harmonics) if better
end while
```

The traditional HS algorithm uses fixed value for both PAR and bw. In the HS method PAR and bw values adjusted in initialization step (Step 1) and cannot be changed during new generations. Large bw values with small PAR values can produce poor performance of the algorithm and increased iterations are needed to find an optimum solution. Although in early generations bw must take a bigger value to enforce the algorithm to increase the diversity of solution vectors, small bw values in final generations increase the fine-tuning of solution vectors. Also, small bw values with large PAR values usually cause the improvement of best solutions in final generations in which the algorithm converges to the optimal solution vector [28].

Frequency Guided Sequential Demodulation

The frequency guided sequential demodulation (FSD) method was created by Kemao and Soon in 2007 [29], and it is used for recovery from the phase of closed fringe patterns. The algorithm of this method is explained in the following six sentences.

The first sentence is fringe denoising. The noise of (1) will be removed (low-pass filtering).

The second sentence is normalization. The parameters of resulting closed interferogram are normalized [30]; this is the background (x, y) and the amplitude $b(x, y)$; in effect, (1) is modified in order to find the interference fringe pattern normalized intensity

$$I_n(x, y) = \cos(\varphi(x, y)). \tag{5}$$

The third sentence is extracting the phase. The direct phase is extracted by the following equation:

$$\widehat{\varphi}_{wa}(x, y) = \cos^{-1}(I_n(x, y)) \in [0, \pi], \tag{6}$$

which is used to calculate the iteration frequencies or intermediate frequencies ξ_{xa} and ξ_{ya}; those frequencies are temporal, whilst they refine optimization in order to find the final frequencies ω_x and ω_y; this is, $(\omega_x, \omega_y) = \widehat{\omega}(x, y)$; it is feasible to use exhaustive search algorithm that guarantees a global minimum [31].

The fourth sentence is about the extraction of local frequency. The true phase is to be considered locally linear, as

$$\widetilde{\varphi}_a(x, y; u, v, \widetilde{\mathbf{p}}) = \widehat{\varphi}_{wa}(x, y) + \xi_{xa} \cdot (x - u) + \xi_{ya} \cdot (y - v), \tag{7}$$

where (u, V), (x, y), (ξ_{xa}, ξ_{ya}) and the operator "·" are the coordinates of the pixel under study, neighbors pixel coordinates, the internal frequencies while the optimization process is being performed, and the dot product, respectively. In other words $\widetilde{\mathbf{p}} = (\xi_{xa}, \xi_{ya})^T$ is the intermediate parameter vector. Then, a virtual fringe pattern can be generated as

$$\tilde{I}(x, y; u, v, \tilde{\mathbf{p}}) = \cos\left[\tilde{\varphi}_a(x, y; u, v, \tilde{\mathbf{p}})\right]. \tag{8}$$

The fitness or energy cost function to be used by the HSO algorithm is equivalent to the squared difference between the virtual and the real fringe pattern; that is,

$$E(u, v; \tilde{\mathbf{p}}) = \sum_{y=-\infty}^{\infty} \sum_{x=-\infty}^{\infty} \left[g(x-u, y-v)\right.$$

$$\left. \cdot \left[\tilde{I}(x, y; u, v, \tilde{\mathbf{p}}) - I_n(x, y)\right]^2\right], \tag{9}$$

where $g(x, y)$ is a subimage window function briefly studied. The local frequency is estimated by minimizing the fitness function:

$$\hat{\mathbf{p}}(u, v) = \arg \min_{\tilde{\mathbf{p}}} E(u, v; \tilde{\mathbf{p}}). \tag{10}$$

The fifth sentence is the extraction of estimated frequencies. Finally, the two coefficients found for the pixel under study or the estimated frequencies by HSO are

$$\hat{\mathbf{p}}(u, v) = \left[\hat{\omega}_{xa}(u, v), \hat{\omega}_{ya}(u, v)\right]^T. \tag{11}$$

The sixth sentence is *eliminated the ambiguity error*. In order to determine the frequency guided sign, all subimages were processed to correct the problem of ambiguity sign. Palpably $\hat{\mathbf{p}}(u, v)$ and $\hat{\varphi}_{wa}(x, y)$ have sign ambiguity and the sign function (x, y) can be determined by forcing the local frequencies to be continuous; this is,

$$s(x_i, y_i) = \begin{cases} 1, & \text{if } \hat{\boldsymbol{\omega}}_a(x_i, y_i) \cdot \hat{\boldsymbol{\omega}}(x_{i-1}, y_{i-1}) \geq 0 \\ -1, & \text{if } \hat{\boldsymbol{\omega}}_a(x_i, y_i) \cdot \hat{\boldsymbol{\omega}}(x_{i-1}, y_{i-1}) < 0, \end{cases} \tag{12}$$

and the ambiguity is indicated by the subindex "a." The sign determination is then continued with a pixel that adjoins pixel (x_i, y_i) until all of the pixels have been processed. Once the sign field (x, y) is determined, both $\hat{\omega}(x, y)$ and $\hat{\varphi}(x, y)$ can be determined [31] by

$$[\widehat{\omega}(x,y), \widehat{\varphi}_w(x,y)] = s(x,y)\,[\widehat{\omega}_a(x,y), \widehat{\varphi}_{wa}(x,y)]. \tag{13}$$

The Levenberg-Marquardt optimization is another algorithm that has been applied for this kind of recovery phase; this algorithm uses the advantages of Gauss-Newton and gradient-descent methods based on adaptive rules [11].

FRINGE PATTERN DEMODULATION WITH FSD-HSO ALGORITHM

As explained above, (5) expresses the normalized interferogram and (6) represents the direct phase extracted. The process to demodulate a closed fringe pattern applying the FSD-HSO algorithm consists in splitting the complete closed fringe pattern into subimages of 5 × 5 pixels (see Figure 3). Additionally, a pixel is demodulated with the nearest eight neighbors, and the HSO algorithm is implemented after selecting the pixel in order to recover the phase for that element.

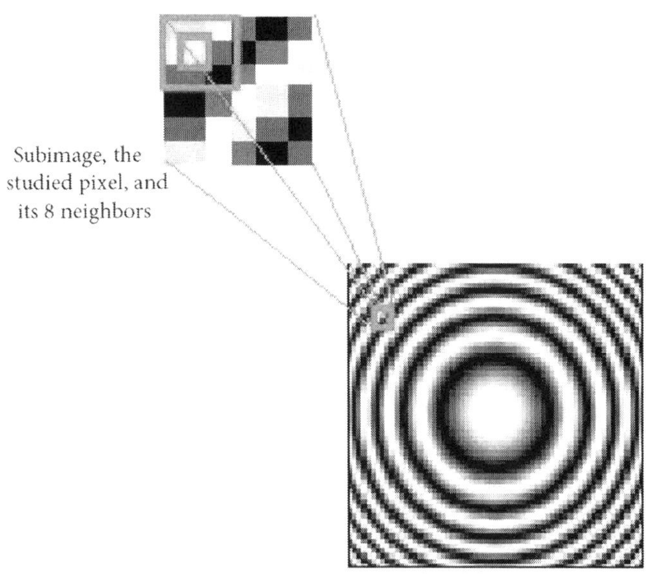

Figure 3. Interferogram divided in 5 × 5-pixel subimages.

Further, the technique FSD is solved applying the HSO algorithm and taking into consideration (7), where the true phase is locally linear. Equation (11) expresses the two calculated frequencies by the HSO algorithm. Equation (8) describes the interferogram obtained with the estimated phase. Equation (9) depicts the fitness function for the HSO; it is the square of the subtraction of the real and the calculated fringe patterns; hence the proper frequencies cause an energy cost function minimization. The above process is applied in all pixels for the subimage. When the process is finished for all the subimages, (13) corrects the ambiguity error. Figure 4 shows the diagram of the demodulation process for each pixel applying HSO and FSD.

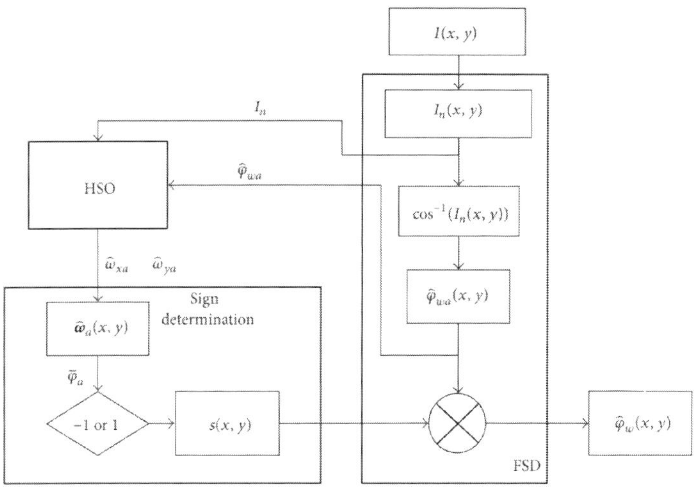

Figure 4. Demodulation process diagram applied to each pixel, using the proposed FSD-HSO algorithm.

The pseudocode algorithms necessary to implement the proposed FSD-HSO procedure are shown in Algorithms 1, 2, 3, 4. First algorithm is the HSO method; second algorithm is the sign determination, in order to find the sign ambiguities of the estimate frequencies; third algorithm implements the procedure to clean the interferograms; fourth algorithm is the local frequencies calculation.

Algorithm 2. Sign determination.

```
for x = 1,...,n
  for y = 1,...,m
    Sign determination by coercing the continuity in the local frequencies:
      Calculate s(x, y) through neighboring pixels (x_i, y_i)
      Calculate [ω̂(x, y), φ̂_w(x, y)] with the correction of sign
  end of y
end of x
```

Algorithm 3. Fringe denoising.

```
for x = 1,...,n
  for y = 1,...,m
      Fringe denoising (remove the noise without distorting the fringes)
      Background removal and amplitude normalization I_n(x, y) between [-1, 1]
      Phase extraction directly by cos^-1(I_n(x, y))
  end of y
end of x
```

Algorithm 4. Frequency extraction.

```
for x = 1,...,n
  for y = 1,...,m
    for u = 1,...,5 (small window)
      for v = 1,...,5 (small window)
          Calculate the energy cost or fitness function, E(u, v; p̃)
      end of v
    end of u
  end of y
end of x
Minimizing the energy function, p̂(u, v)
Estimate the local frequencies by [ω̂_xa(u, v), ω̂_ya(u, v)]^T
```

RESULTS OF CLOSED FRINGE PATTERN DEMODULATION WITH FSD-HSO ALGORITHM

A variety of physical quantities can be measured through the phase; it can be using interferometric techniques. The method to calculate the phase for a closed fringe pattern is explained below. Synthetic interferograms (closed fringe) are demodulated using the FSD-HSO algorithm. The physical variable associated with the obtained phase is smooth and continuous.

The HSO algorithm begins its search with HM and in this case is created by n rows of individuals (HMS = n) with two elements (NVAR), each with random values that correspond to the beginning of estimated frequencies by (14); this is

$$\mathrm{HM} = \begin{bmatrix} \omega_x^1 & \omega_y^1 \\ \omega_x^2 & \omega_y^2 \\ \omega_x^3 & \omega_y^3 \\ \vdots & \vdots \\ \omega_x^{\mathrm{HMS}} & \omega_y^{\mathrm{HMS}} \end{bmatrix}. \quad (14)$$

The values of each parameter in HSO (NVAR, SC, HMS, PAR, HMCR, and bw) are according to the interferogram to be demodulated. In this work, four distinct simulated closed fringe patterns (synthetic interferograms, SIs) and a real interferogram (RI) are exposed. The first example, synthetic-interferogram-one (SI1), corresponds to Newton rings interferogram, that is, classical defocus interference generated in a Twyman-Green or a Fabry-Perot interferometer; the synthetic-interferogram-two (SI2) is a pair of symmetrical positive lobules; the synthetic-interferogram-three (SI3) is a pair of asymmetrical positive lobules; finally, the synthetic-interferogram-four (SI4) is a pair of inverse lobules; RI is a noisy interferogram with defocus aberration. These five examples are a compendium of different difficulties of closed fringe demodulation. The normalization of the SIs and RI were calculated and shown in Figure 5(a). The normalized interferogram $I(x, y)$ was partitioned into subimages N_{xy} of 5 × 5 pixels and also is considering the linear phase to be local. The direct phase is created by inverse cosine of normalized image; see Figure 5(b). Each subimage was processed and estimated the frequencies ω_x and ω_y, along x and y directions, respectively; see Figures 5(c) and 5(d). Equation (10) represents the cost function; it is decreased close to the selected pixel (x_1, y_1) and its eight neighbors. A small value in cost function implies good performance in the demodulation. The direct phase $\widehat{\varphi}_{wa}(x, y)$ was processed by the FSD-HSO algorithm described in flow diagram of Figure 4; as seen in Figure 5(e), the sign ambiguity creates discontinuities in this step of phase reconstruction process. In Figure 4 a block of sign determination was included because of the phase ambiguity problem presented in the function cosine; this is $\cos(\varphi_e(x, y)) = \cos(-\varphi_e(x, y))$; details of this method can be found in [31]. Once the subimage is completed, the HSO starts with other subimages; when the last subimage is demodulated, the procedure finishes. At last, the estimated phases $\widehat{\varphi}_w(x, y)$ are shown in Figure 5(f).

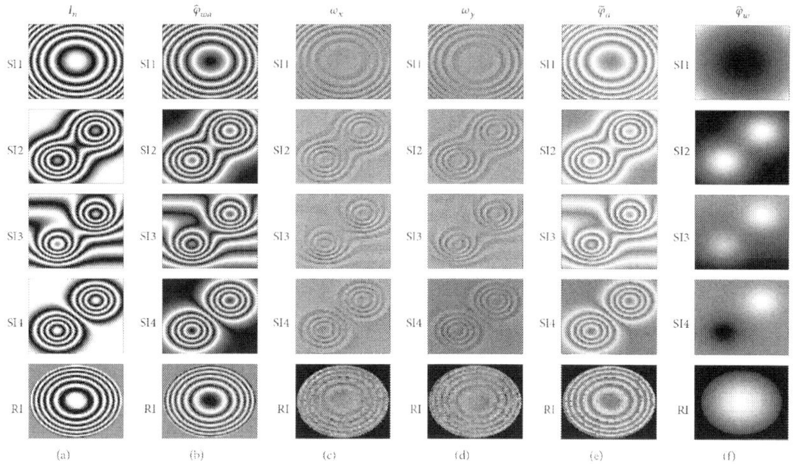

Figure 5. FSD-HSO algorithm applied to demodulate synthetic and real interferograms: (a) normalized interferograms; (b) direct phases of interferograms; (c-d) calculated frequencies along (x, y) axis, respectively; (e) demodulated phases with ambiguity errors; and (f) correct demodulated phases without ambiguity error.

In order to observe minutiae details of the synthetically estimated phases, a series of 3D graphs are presented in Figure 6. The estimated phase of SI1 is observed as a single negative paraboloid, as shown in Figure 6(a); in counterpart, the other three SIs estimated phases have two peaks. The estimated phase of SI2 is a pair of positive symmetrical peaks; see Figure 6(b). However, the respective SI3 estimated phase, is observed as two positive asymmetrical peaks; see Figure 6(c). At last, two inverse peaks are represented in Figure 6(d) as SI4 estimated phase. This last interferogram was the most difficult to demodulate, because the search process of the HSO algorithm was slow due to the increase in iterations for the PAR and bw adjustment. It is seen that the combination of the HSO and FSD methods can handle interferograms with more complex fringe distribution than circular fringes.

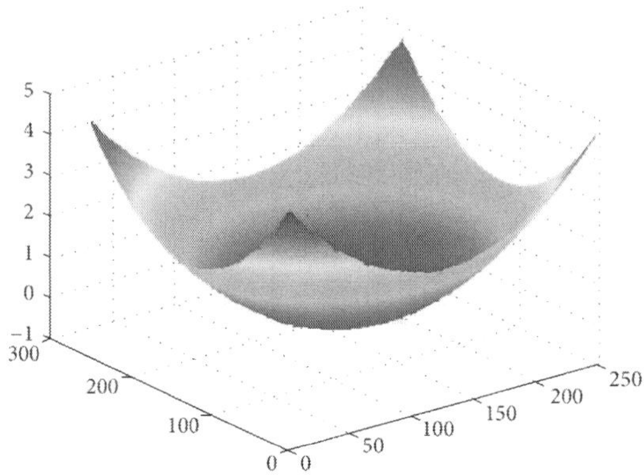
5
4
3
2
1
0
−1
300
200
100
0
0
50
100
150
200
250

(a)

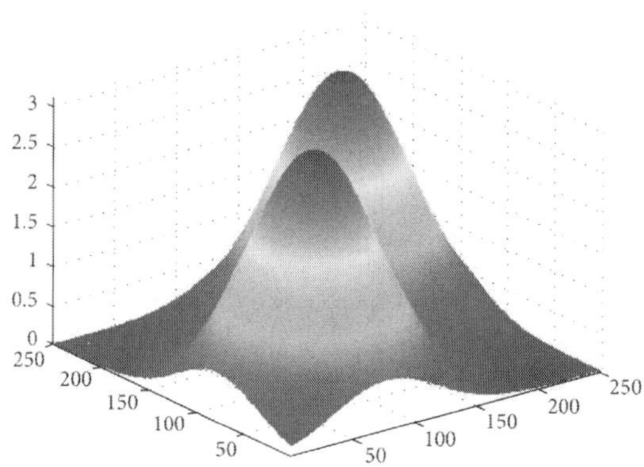
3
2.5
2
1.5
1
0.5
0
250
200
150
100
50
50
100
150
200
250

(b)

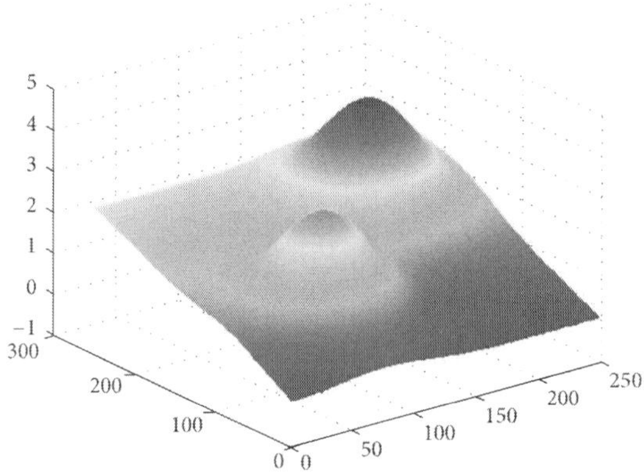
5
4
3
2
1
0
−1
300
200
100
0
0
50
100
150
200
250

(c)

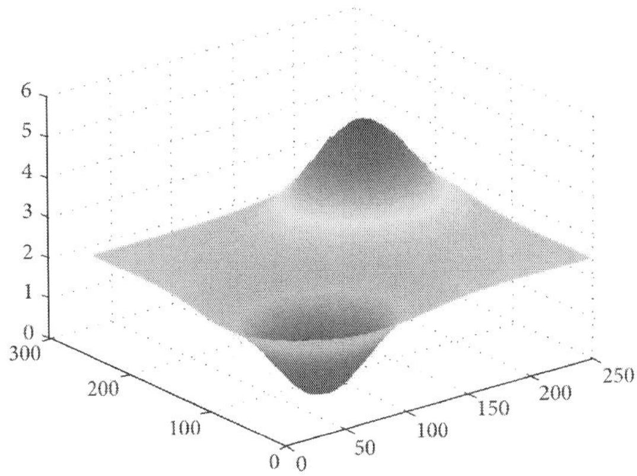

Figure 6. 3D estimated phases of SIs demodulations: (a) $\widehat{\varphi}_w$ of SI1; (b) $\widehat{\varphi}_w$ of SI2; (c) $\widehat{\varphi}_w$ of SI3; and (d) $\widehat{\varphi}_w$ of SI4.

The average relative error was calculated by the following equation [3]:

$$\% \text{ error} = \frac{100}{MN}\sum_x\sum_y\left|\frac{\varphi_o(x,y)-\widehat{\varphi}_w(x,y)}{\max(\widehat{\varphi}_w(x,y))-\min(\widehat{\varphi}_w(x,y))}\right|, \qquad (15)$$

where φ_o, MN are the original interferogram phase and the interferogram size, respectively.

In addition to the synthetic analysis, the RI estimated phase demodulated with FSD-HSO was compared with a robust unwrapped phase method, implemented by [8], for this occasion that method is taken to calculate the original interferogram phase. In order to compare the two results visually, their 3D phases are shown in Figures 7(a) and 7(b); these two graphs are very similar; to complement the analysis, the variance between the original phase and the calculated phase $|\varphi_o(x,y)-\widehat{\varphi}_w(x,y)|$ was obtained, for this RI whose difference was very small (around ~0.5 radians), as shown in Figure 7(c). For SI analysis, the original phase is the phase with which was programmed each synthetic interferogram.

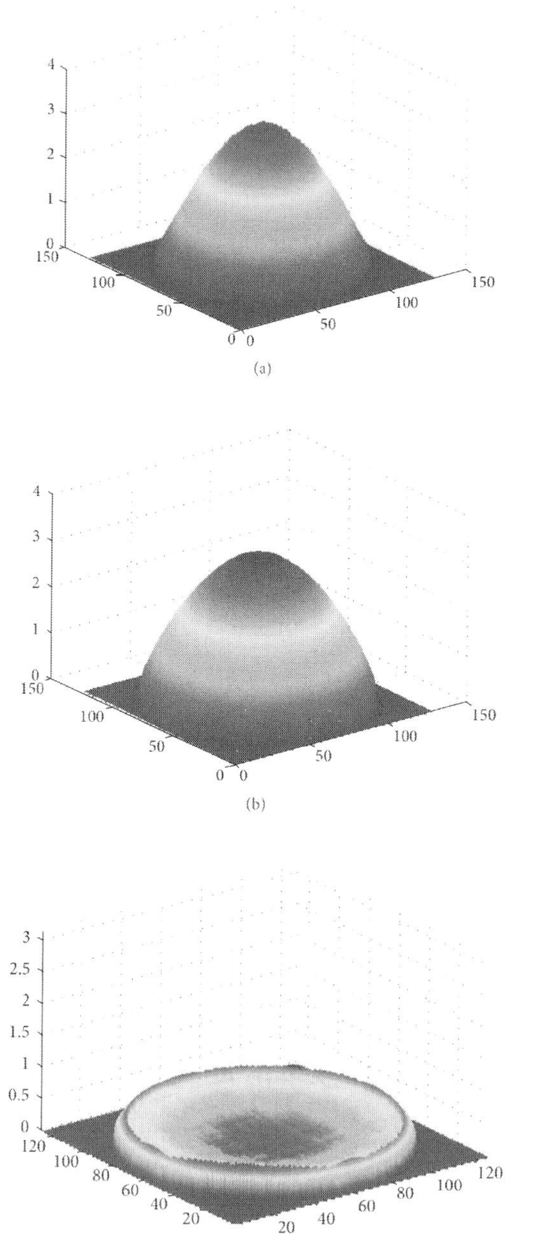

Figure 7. 3D analysis of RI demodulation phase: (a) $\widehat{\varphi}_w$ of RI with FSD-HSO demodulation; (b) φ_0 of RI with unwrapped phase method; and (c) phases difference $|\varphi_0 - \widehat{\varphi}_w|$.

Additionally, the results of HSO algorithm are compared with genetic algorithm (GA) method, both using FSD as demodulation technique. Results of this comparison can be seen in Table 1, where column one contains the five different interferograms used in the present work. The second column of Table 1 contains the peak to valley (p-v) deviations, where column represents the maximal and minimal values obtained in the respective estimated phases. The third column of the same table presents the rms deviations of the estimated phases. In p-v and rms calculus significant differences cannot be observed between GA and HSO methods, but these data are presented because of their physical significance. The average relative errors were presented in fourth column; it can be observed that HSO algorithm has smaller errors than GA algorithm; the difficulties presented in SI4 and RI (pronounced sign peaks asymmetry and/or very noisy image) increment a little the error in both methods; the biggest error was 1.561 rad.; therefore, the robustness of FSD-HSO is confirmed. The last column presents the time processing, where the FSD-HSO demonstrated as a fast method with symmetrical sign peaks interferograms demodulation (until 1.5 minutes), but the time processing increases as different sign peak lobules appear in the interferogram. Results of Table 1 are supported by the plot of Figure 8, in which the number of iterations for the proposed method is about one half to achieve convergence.

Table 1. Comparison between GA and HSO optimization techniques. The demodulation method used in both cases was FSD.

Interferogram	**p-v deviation (radians)**		**rms deviation (radians)**		**% error (radians)**		**Time processing (seconds)**	
	AG	**HSO**	**AG**	**HSO**	**AG**	**HSO**	**AG**	**HSO**
SI1	4.8740	4.8639	1.9526	1.9009	1.1729	0.1700	398.63	234.50
SI2	3.0328	3.0031	1.3571	1.3043	2.1474	0.2123	444.83	247.13
SI3	4.3100	4.3090	2.1649	2.1110	1.3505	0.1448	407.88	239.93
SI4	4.9980	4.9980	2.7042	2.6344	1.2058	0.3984	5412.98	3026.59
RI	3.485	3.1491	1.346	1.3471	5.2806	1.5610	192.59	87.54

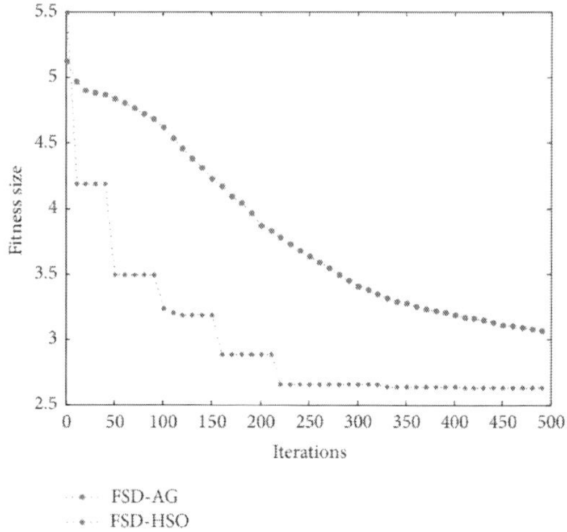

Figure 8. Convergence comparison for SI1 interferogram pixels.

The selected parameters used in this work are shown in Table 2. The entire SIs and RI images had MN size of 250 × 250 pixels and 128 × 128 pixels, respectively. In order to obtain consistent results, all the calculus is made with the same computational system: Intel Celeron B815 1.6 GHz, with 2 GB of RAM, 32-bit operative system, Windows 7, and Matlab R2011a 7.120.635. The FSD-GA at its best operation, used to compare the proposed algorithm, has the following parameters: individuals 1500–2500; bits by gen 12–15; gens 2; generations 1000–1500; limits of each coefficient [−0.3, 0.3]– [−0.5, 0.5].

Table 2. Parameters used in FSD-HSO optimization algorithm.

Interferogram	NVAR	SC	HMS	HMCR	PAR	bw	Limits
SI1	2	800	1000	0.9	0.75	0.05	[−0.3, 0.3]
SI2	2	1000	1500	0.9	0.7	0.08	[−0.2, 0.2]
SI3	2	1000	1800	0.85	0.75	0.07	[−0.5, 0.5]
SI4	2	1500	2200	0.9	0.9	0.1	[−0.5, 0.5]
RI	2	1500	2000	0.85	0.7	0.09	[−0.5, 0.5]

Experimentally it is observed that the proposed algorithm's tuning parameters depend on the interferogram characteristics. The best resolution in the image demodulation was found experimentally where PAR must approach its upper limit (one) and bw tends to its lower limit (zero).

CONCLUSIONS

The use of FSD algorithm to demodulate the phase is a simple technique of soft computing, because just two variables for this work, ω_x and ω_y, need to be calculated compared to other polynomial approaches that calculate more variables or coefficients.

Instead, the HSO is a relatively new optimization technique that requires a few input parameters. The fitness function used by the HSO is small and the data search is relational, so convergence is fast and has a low percentage of error.

Therefore, the combined algorithm FSD-HSO is an uncomplex algorithm because of its algebraic nature, and, used in the demodulation of closed fringe patterns, its convergence is fast. The error calculated in synthetic interferograms simulated was 0.14–0.39 which is a very good approximation. The method is limited only to demodulation of closed fringe patterns.

The proposed algorithm is relatively fast compared to the FSD-AG algorithm; it converges with less iterations. It only presents difficulties in interferograms with different sign peak lobules, but this disadvantage is present in both methods.

Other methods to demodulation also shed low error rates, such as the AG-Zernike algorithm, but, using these two methods for example, the algorithm becomes very complex, since it is necessary to perform a lot of computational processes such as data mutations, data changes from binary to real, and a lot of variables to be estimated (a minimum of 7 variables).

ACKNOWLEDGMENTS

Mr. Rodriguez-Marmolejo would like to acknowledge *Centro Universitario de los Lagos* and *Instituto Tecnológico de Aguascalientes* authorities for the support in his Ph.D. studies.

REFERENCES

1. D. Malacara, M. Servin, and Z. Malacara, *Interferogram Analysis for Optical Testing*, CRC Press, Boca Raton, Fla, USA, 2nd edition, 2005.
2. F. J. Cuevas, J. H. Sossa-Azuela, and M. Servin, "A parametric method applied to phase recovery from a fringe pattern based on a genetic algorithm," *Optics Communications*, vol. 203, no. 3–6, pp. 213–223, 2002.
3. F. J. Cuevas, F. Mendoza, M. Servin, and J. H. Sossa-Azuela, "Window fringe pattern demodulation by multi-functional fitting using a genetic algorithm," *Optics Communications*, vol. 261, no. 2, pp. 231–239, 2006.
4. L. E. Toledo and F. J. Cuevas, "Optical metrology by fringe processing on independent windows using a genetic algorithm," *Experimental Mechanics*, vol. 48, no. 4, pp. 559–569, 2008.
5. L. E. Mancilla Espinosa, J. M. Carpio Valadez, and F. J. Cuevas, "Demodulation of interferograms of closed fringes by zernike polynomials using a technique of soft computing," *Engineering Letters*, vol. 15, no. 1, pp. 99–104, 2007.
6. U. H. Rodriguez-Marmolejo, T. A. Ramirez-delreal, J. Muñoz-Maciel, and M. Mora-Gonzalez, "Combination of genetic algorithms and FSD applied to fringe pattern demodulation," in *Proceedings of the Image Reconstruction from Incomplete Data VIII*, vol. 9600 of *Proceedings of SPIE*, p. 7, San Diego, Calif, USA, August 2015.
7. J. Jiménez, H. Sossa, F. Cuevas, and L. Gómez, "Demodulation of interferograms based on particle swarm optimization," *Polibits*, vol. 45, pp. 83–91, 2012.
8. J. Muñoz-Maciel, F. J. Casillas-Rodríguez, M. Mora-González, F. G. Peña-Lecona, V. M. Duran-Ramírez, and G. Gómez-Rosas, "Phase recovery from a single interferogram with closed fringes by phase unwrapping," *Applied Optics*, vol. 50, no. 1, pp. 22–27, 2011.
9. M. Servin, J. L. Marroquin, and F. J. Cuevas, "Demodulation of a single interferogram by use of a two-dimensional regularized phase-tracking technique," *Applied Optics*, vol. 36, no. 19, pp. 4540–4548, 1997.
10. J. Kennedy and R. Eberhart, "Particle swarm optimization," in *IEEE International Conference on Neural Networks*, vol. 4, pp. 1942–1948, Perth, Australia, November-December 1995.

11. H. Wang and Q. Kemao, "Frequency guided methods for demodulation of a single fringe pattern," *Optics Express*, vol. 17, no. 17, pp. 15118–15127, 2009.
12. Z. W. Geem, J. H. Kim, and G. V. Loganathan, "A new heuristic optimization algorithm: harmony search," *Simulation*, vol. 76, no. 2, pp. 60–68, 2001.
13. G. Ingram and T. Zhang, "Overview of applications and developments in the harmony search algorithm," in *Music-Inspired Harmony Search Algorithm*, Z. W. Geem, Ed., vol. 191 of *Studies in Computational Intelligence*, pp. 15–37, Springer, New York, NY, USA, 2009.
14. F. Deng, S. Tuo, L. Yong, and T. Zhou, "Construction example for algebra system using harmony search algorithm," *Mathematical Problems in Engineering*, vol. 2015, Article ID 836925, 15 pages, 2015.
15. Q.-K. Pan, P. N. Suganthan, J. J. Liang, and M. F. Tasgetiren, "A local-best harmony search algorithm with dynamic subpopulations," *Engineering Optimization*, vol. 42, no. 2, pp. 101–117, 2010.
16. M. Sulaiman, A. Salhi, B. I. Selamoglu, and O. B. Kirikchi, "A plant propagation algorithm for constrained engineering optimisation problems," *Mathematical Problems in Engineering*, vol. 2014, Article ID 627416, 10 pages, 2014.
17. G. Li and Q. Wang, "A cooperative harmony search algorithm for function optimization," *Mathematical Problems in Engineering*, vol. 2014, Article ID 587820, 12 pages, 2014.
18. M. Jaberipour and E. Khorram, "A new harmony search algorithm for solving mixed-discrete engineering optimization problems," *Engineering Optimization*, vol. 43, no. 5, pp. 507–523, 2011.
19. X. Z. Gao, V. Govindasamy, H. Xu, X. Wang, and K. Zenger, "Harmony search method: theory and applications," *Computational Intelligence and Neuroscience*, vol. 2015, Article ID 258491, 10 pages, 2015.
20. D. Manjarres, I. Landa-Torres, S. Gil-Lopez et al., "A survey on applications of the harmony search algorithm," *Engineering Applications of Artificial Intelligence*, vol. 26, no. 8, pp. 1818–1831, 2013.
21. K. J. Gassvik, *Optical Metrology*, John Wiley & Sons, New York, NY, USA, 1987.

22. E. Hecht, *Optics*, Addison-Wesley, San Francisco, Calif, USA, 4th edition, 2002.

23. D. Malacara, "Twyman-Green interferometer," in *Optical Shop Testing*, D. Malacara, Ed., chapter 2, pp. 46–96, John Wiley & Sons, Hoboken, NJ, USA, 3rd edition, 2007.

24. K. S. Lee and Z. W. Geem, "A new meta-heuristic algorithm for continuous engineering optimization: harmony search theory and practice," *Computer Methods in Applied Mechanics and Engineering*, vol. 194, no. 36–38, pp. 3902–3933, 2005.

25. X. Z. Gao, X. Wang, and S. J. Ovaska, "Harmony search methods for multi-modal and constrained optimization," in *Music-Inspired Harmony Search Algorithm: Theory and Applications*, vol. 191 of *Studies in Computational Intelligence*, pp. 39–51, Springer, Berlin, Germany, 2009.

26. Y. Xin-She, *Engineering Optimization. An Introduction with Metaheuristic Applications*, John Wiley & Sons, Hoboken, NJ, USA, 2010.

27. L. dos Santos Coelho and D. L. A. Bernert, "A harmony search approach using exponential probability distribution applied to fuzzy logic control optimization," in *Recent Advances in Harmony Search Algorithm*, Z. W. Geem, Ed., vol. 270 of *Studies in Computational Intelligence*, pp. 77–88, Springer, 2010.

28. S. Tuo, L. Yong, and T. Zhou, "An improved harmony search based on teaching-learning strategy for unconstrained optimization problems," *Mathematical Problems in Engineering*, vol. 2013, Article ID 413565, 29 pages, 2013.

29. Q. Kemao and S. H. Soon, "Sequential demodulation of a single fringe pattern guided by local frequencies," *Optics Letters*, vol. 32, no. 2, pp. 127–129, 2007.

30. J. A. Quiroga, J. Antonio Gómez-Pedrero, and Á. García-Botella, "Algorithm for fringe pattern normalization," *Optics Communications*, vol. 197, no. 1–3, pp. 43–51, 2001.

31. Q. Kemao, *Windowed Fringe Pattern Analysis*, SPIE Press, Bellingham, Wash, USA, 2013.

CHAPTER 3

RECONSTRUCTION OF LOCAL FREQUENCIES FOR RECOVERING THE UNWRAPPED PHASE IN OPTICAL INTERFEROMETRY

Julio C. Estrada[1], Jose L. Marroquin[1], and Orlando M. Medina[2]

[1]Centro de Investigacion en Matematicas (CIMAT), Computer Sciences, Guanajuato, 36023, Mexico.
[2]Universidad Politecnica de Guanajuato (UPG), Robotics, Guanajuato, 38483, Mexico.

ABSTRACT

In optics, when interferograms or digital holograms are recorded and their phase is recovered, it is common to obtain a wrapped phase with some errors, noise and artifacts such as singularities due to the non linearities of the demodulation process. This paper shows how to reconstruct the frequency field of the wrapped phase by using adaptive Gabor filters. Gabor filters are Gaussian quadrature filters tuned in at a certain frequency. We adapt these Gabor filters by tuning them locally and estimating the frequency using wrapped finite differences of the estimated phase. Doing this process

Citation: (APA): Estrada, J. C., Marroquin, J. L., & Medina, O. M. (2017). Reconstruction of local frequencies for recovering the unwrapped phase in optical interferometry. Scientific reports, 7(1), 1-10. (10 pages)

iteratively, the frequency estimation is refined and smoothed. The unwrapped phase is easily recovered by integrating the recovered frequency field using, for example, a simple line raster integration. We don't have problems with phase inconsistencies or residues while integrating the phase, because these are removed. The obtained unwrapped phase is clean, consistent and practically error-free. We show estimation errors with simulated data and the performance of the proposed method using real-world recorded wavefronts.

INTRODUCTION

Today, in optical interferometry one can find dozens of papers showing how to recover the wrapped phase from a single interferogram with or without frequency carrier, or by using two or more phase-shifted interferograms, see refs 1,2,3 for a review. When the interferograms are free of noise, have high contrast and homogeneous illumination, these demodulation methods recover an ideal wrapped phase[4,5,6]. However under real laboratory conditions, for example when using speckle interferometry or shearography, we are far from these ideal conditions[7,8,9,10,11,12]. Instead, the wavefront recovered is a wrapped phase with imperfections like noise, miscalibration errors and artifacts that introduce phase inconsistencies or residues (singularities). These inconsistencies make the phase unwrapping process difficult, because the unwrapping is an integration of wrapped phase differences and this integration becomes path-dependent[13,14,15,16]. For the wrapped phase to be consistent, its wrapped differences must obey the following second order cross difference relation:

$$\Delta_y\left[\Delta_x\left[\phi_w\right]\right] = \Delta_x\left[\Delta_y\left[\phi_w\right]\right] \tag{1}$$

where Δ_x and Δ_y are the wrapped difference operator along x and y respectively. The wrapped difference operator, at a certain site x, is defined as the modulus 2π of the phase difference between the phase at x and its neighbor $x-1$, assuming backward differences. When the wrapped differences obey Eq. (1), the phase unwrapping process is path-independent. The question is how can we estimate the wrapped differences in such a way that inconsistencies or residues are removed. What we are going to propose in this paper is based on the reconstruction of the noise-free frequency field, which is taken as the wrapped differences of the phase along x and y. Thus, given the wrapped phase with noise and miscalibration errors, we want to obtain the best possible estimation of the frequency field without noise and inconsistencies, in such a way that its integration becomes path-independent, so that one is able to unwrap the phase with simple line scanning integration.

There are two published methods that are related to the approach presented here. One is the work of ref. 16 in which the residual error between the observed wrapped phase and an estimated unwrapped and re-wrapped one, is iteratively approximated using a smoothed frequency field (where the smoothing is performed using a Total Variation (TV) scheme) which is then integrated to obtain the phase. This permits one to obtain a good approximation to the unwrapped phase, although the unwrapped phase is recovered with noise and inconsistencies. The other is the work of ref. 17 in which an estimation for the frequency field of an observed interferogram is obtained via a fixed bank of Gabor quadrature filters which filter the noise, and permit the estimation of the local frequency as the tuning frequency of the filter with maximal response. These two methods will be analyzed in more detail in the Discussion section.

In the method presented here, we also focus in the processing of the frequency field. The main difference in our work is that we propose the use of a linear filter whose tuning frequency is continuously adapted to the local frequency field, which allows for a better noise elimination, so that stable results may be obtained at very low signal to noise ratios. In particular, our proposal is to use a linear system whose impulse response is a Gabor filter tuned at the local frequency that is obtained from the wrapped phase. In the following sections we will show how the frequency field is estimated and used to tune the Gabor filters at each pixel and how the frequency reconstruction and smoothing is accomplished by using an iterative process involving the application of the adaptive Gabor filters and a smooth estimation of the frequency field using wrapped differences. We remark that just a few iterations of this process are needed to obtain a clean and consistent frequency field. Finally, the unwrapped phase is easily obtained after integration.

METHOD

Let us start with the model of a typical wrapped phase obtained after demodulation:

$$\phi_W(x, y) = ([\phi(x, y) + c \sin[2\phi(x, y)] + \eta(x, y)] \bmod 2\pi) - \pi, \tag{2}$$

where $\phi_W(x, y)$ is the obtained wrapped phase, $\phi_W(x, y)$ is the ground-truth phase or modulated phase, $c\sin[2\phi(x, y)]$ is the miscalibration or tune-in error introduced by the demodulation algorithm used, and $\eta(x, y)$ the induced noise. The noise is not necessarily normally distributed, for example in the case of speckle noise produced by coherent sources. The constant value $c>0$ is directly proportional to the tune-in error of the algorithms used to demodulate the phase[18]. As the phase is affected by noise and wrapped modulus 2π, if $\Delta[\phi(x, y)+\eta(x, y)]>|\pi|$ its wrapped differences do not obey the cross differences Equation (1), due to aliasing, and it translates to singularities and phase inconsistencies that difficult the phase unwrapping[13]. One must get rid of the noise and miscalibration error in order to obtain a perfect frequency reconstruction and phase unwrapping. The idea we use here is to apply an adaptive linear system to the following complex signal constructed from the wrapped phase

$$g(x, y) = \exp[i\phi_W(x, y)], \tag{3}$$

where $i = \sqrt{-1}$. The frequency field $\omega(x, y)=[u(x, y), v(x, y)]$ for this complex signal can be obtained from the wrapped differences of $\phi_W(x, y)$ as:

$$\begin{aligned} u(x, y) &= \Delta_x[\phi_W(x, y)] \\ v(x, y) &= \Delta_y[\phi_W(x, y)], \end{aligned} \tag{4}$$

where the wrapped phase difference Δ_x and Δ_y can be given as

$$\Delta_x[\phi(x, y)] = ([\phi(x, y) - \phi(x-1, y) + \pi] \mod 2\pi) - \pi \tag{5}$$

and

$$\Delta_y[\phi(x, y)] = ([\phi(x, y) - \phi(x, y-1) + \pi] \mod 2\pi) - \pi. \tag{6}$$

The complex signal $g(x, y)$ inherits the imperfections of the wrapped phase $\phi_W(x, y)$ such as the noise and tune-in errors. To remove them effectively, it is not sufficient to use a linear shift-invariant system. We need an adaptive system that controls its tuning frequency for each site (x, y) and removes the phase noise letting pass only the local frequency of that site. To this end, we propose to use Gabor filters, which are defined as

$$h_{\sigma,\omega_0}(x, y) = \exp\left(-\frac{x^2+y^2}{2\sigma^2}\right)\exp[-i\omega_0 \cdot (x, y)], \tag{7}$$

where (x, y) is the spatial domain, σ controls the bandwidth and $\omega_0=(u_0, v_0)$ is the tuning frequency. The corresponding frequency response of the filter is:

$$H(u, v)_{\sigma,\omega_0} = \exp\left(-\frac{(u - u_0)^2 + (v - v_0)^2}{2\sigma_\omega^2}\right), \tag{8}$$

where $\sigma_\omega = 1/\sigma$. Applying this filter in the frequency domain is equivalent to the convolution of the complex signal (3) with kernel (7), so that its output is expressed as

$$f(x, y) = \sum_{w,z} g(x - w, y - z) h_{\sigma,\omega_0}(w, z). \tag{9}$$

The standard application of this Gabor filtering, as depicted by this last equation, is to choose a tune-in frequency ω_0 and apply the filter to all the image, i.e., the filter is shift-invariant. However, if we consider the first order local approximation around a site (x_0, y_0) of the signal phase, it is a plane with local frequencies (slopes) u_0 and v_0. Around this site (x_0, y_0), its Fourier transform is an impulse located at frequency $\omega_0=(u_0, v_0)$, as illustrated in one-dimension in Fig. 1. What we want is to adapt to this site (x_0, y_0) the Gabor filter according to its local frequency, in such a way that we are letting pass only the frequency information around $\omega_0=(u_0, v_0)$, as shown in Fig. 1. Then, for each pixel (x_0, y_0) we have an $\omega_0=(u_0, v_0)$ and the output is

$$f(x_0, y_0) = \sum_{w,z} g(x_0 - w, y_0 - z) h_{\sigma,\omega_0}(w, z); \tag{10}$$

where the tune-in frequency $\omega_0=(u_0, v_0)$ of the Gabor filter is not constant, but depends on the site (x_0, y_0), i.e., the filter is not shift-invariant. In this way, we are filtering out noise in an adaptive way.

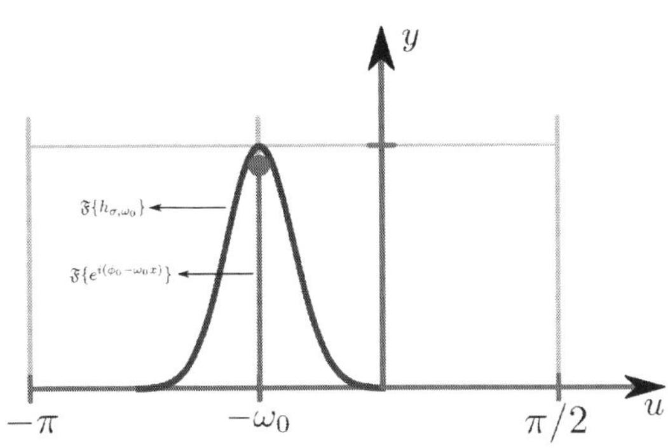

Figure 1. The red line represents an impulse at frequency ω_0, which is the Fourier transform of a complex signal whose phase is a line with slope ω_0. The blue line is the one-dimensional Fourier transform of the Gabor filter given by

Eq. (7). The Gabor filter is tuned (or adapted) at the frequency ω_0, passing only the information around this frequency ω_0.

Reconstruction Process

Given the wrapped phase ϕ_W, which we assume is modeled as in (2), we take the first approximation of the frequency field as in Eq. (4), which give us the local frequency for each pixel (x, y) of the wrapped phase image. Then, we apply the adaptive Gabor filter using Eq. (10), so that for each site (x_0, y_0) the complex signal $g(x_0, y_0)$ is convolved with its own Gabor filter tuned in at $\omega_0=[u(x_0, y_0), v(x_0, y_0)]$. Then, we take the argument of the complex output $f(x, y)$ to obtain a filtered phase as

$$\hat{\phi} = \arg[f(x, y)]. \tag{11}$$

An improved frequency field estimation is obtained as

$$\begin{aligned} u(x, y) &= \Delta_x[\hat{\phi}(x, y)] \\ v(x, y) &= \Delta_y[\hat{\phi}(x, y)]. \end{aligned} \tag{12}$$

As we are processing the wrapped phase corrupted by noise and miscalibration errors, phase differences increase the noise power and introduce singularities with frequency jumps of 2π. Therefore, it is necessary to filter the estimated frequency field so that these noise-induced inconsistencies are removed. This may be achieved by low-pass filtering the current local estimates given by Eq. (4), using a Gaussian kernel. In particular we compute the filtered frequency estimates as:

$$\begin{aligned} \hat{u}(x, y) &= \sum_{\forall(m,n)} u(m, n)\exp\left[-\frac{(n-x)^2+(m-y)^2}{2s^2}\right] \\ \hat{v}(x, y) &= \sum_{\forall(m,n)} v(m, n)\exp\left[-\frac{(n-x)^2+(m-y)^2}{2s^2}\right]. \end{aligned} \tag{13}$$

To set the s parameter, one may consider that if F_M is the maximum frequency magnitude that one wants to preserve in the spatial variation of the frequency fields, then the frequency response of this smoother should be such that at F_M one does not get too much attenuation. In particular, we set this attenuation so that the s_ω parameter of the Gaussian frequency response of the smoother is $s_\omega=F_M/3$, which in the space domain gives $s=3/F_M$. We have found experimentally that for a sample of images with a resolution of 200 × 200 pixels, the best value for F_M is about $\pi/4.2$. We also found that using the same value for the σ parameter for the adaptive Gabor filter gives good results; we note that the precise value for these parameters is not important for the performance of the method, as long as it is within the appropriate range. If the resolution of the images vary, however, one must make a correction, since the frequency content of the phase image scales linearly with the resolution (taken here as the width N of the image). Therefore, one gets the following formula for the parameter values, which is the one we use in all the experiments presented here:

$$s = \sigma = \frac{3 \times 4.2 \times N}{200\pi}. \qquad (14)$$

The frequencies $\hat{u}(x, y)$ and $\hat{v}(x, y)$ represent a clean and refined estimation of the frequency field. Thus, we can use them again to improve the output $f(x, y)$, obtaining a cleaner phase $\hat{\phi}(x, y)$ and estimate again a smooth frequency field $\hat{u}(x, y)$ and $\hat{v}(x, y)$ from this using (13). After few iterations of this process, the obtained smooth frequency field $\hat{u}(x, y)$ and $\hat{v}(x, y)$ is integrated to obtain the unwrapped phase using a simple line scanning as follows:

- First set $\phi(0, 0) = 0$. Then integrate line 0 using the estimated frequency $\hat{u}(x, y)$:

$$\phi(x, y) = \phi(x - 1, y) + \hat{u}(x, y) \qquad (15)$$

starting with $x = 1$.

- Then, integrate the remaining lines (starting with the line corresponding to $y = 1$) using the phase estimate at the previous line and the estimated frequency $\hat{v}(x, y)$:

$$\phi(x, y) = \phi(x, y - 1) + \hat{v}(x, y) \qquad (16)$$

The block diagram for the complete algorithm is shown in Fig. 2.

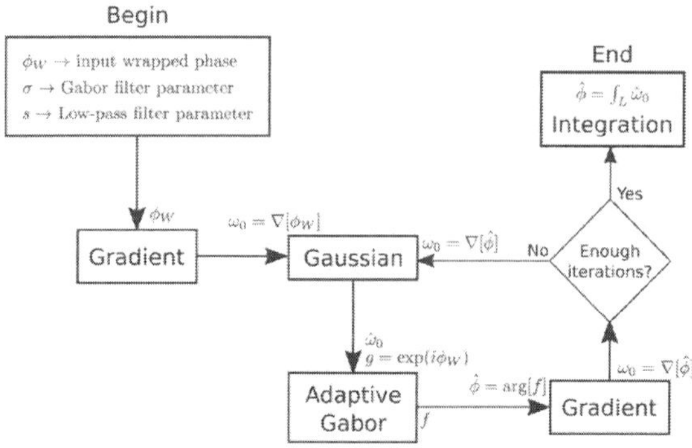

Figure 2. Block diagram of the method presented here. The gradient is taken by means of finite differences as in (4). Convergence takes about 6 iterations (see text).

Data Availability

In the results presented here, we use both simulated and real experimental data for the tests. For the simulated data the model for generating them is given in the description of each test. The real experimental data were obtained by O.M. Medina at the *Centro de Investigaciones en Optica* (CIO), Mexico, and may be available by requesting them directly to the authors.

RESULTS

For the simulated tests we present here, we take the Root Mean Square Error (RMSE) to measure the error as:

$$\text{RMSE} = \sqrt{\frac{1}{M \times N}\sum_{x,y}\left([\hat{u}(x,y) - u(x,y)]^2 + [\hat{v}(x,y) - v(x,y)]^2\right)}, \tag{17}$$

where $\hat{u}(x, y)$, $u(x, y)$, $\hat{v}(x, y)$, and $v(x, y)$, are the frequency estimation and corresponding ground truth along x and y, respectively. M and N denote the number of rows and columns of the lattice (x, y) for the wrapped phase images.

First we show how the method presented here removes the miscalibration errors. For example, in phase shifting interferometry, error sources of miscalibration are very common when the step of the phase shift is not given correctly or when the phase shifting algorithms are not tuned to the frequency step of the interferograms. We simulated a wrapped phase image using Eq. (2) without noise, taking $\varphi(x, y) = 1.5 \times 10^{-3}$ $((x - 128)^2 + (y -$

128)²) in a lattice of $M \times N = 256 \times 256$ pixels and miscalibration error of $c = 0.8$. In theory, the frequency fields are two ramps, one in x and another in y with slope of 3×10^{-3} each one. Usin the Eq. (14), we set the filter parameters to $\sigma = s = 5.5$ for the Gabor and Gaussian filter, respectively, and performed 5 iterations. In Fig. 3(A) we show the simulated wrapped phase and Fig. 3(B,C) shows the frequency field taken as the wrapped differences of the simulated phase using Eq. (4). In Fig. 3(D) we show the obtained phase after frequency reconstruction and smoothing. The phase is recovered unwrapped but we show it wrapped for comparison purposes. The reconstructed frequency field is shown in Fig. 3(E,F). Compare the frequency field of B and C with the reconstructed frequency field of E and F in the figure. In B and C one can see the effect of the miscalibration error which is removed in E and F. The graph on the right, shows the central row of both unwrapped phases, the one with miscalibration errors and the one recovered after integrating the reconstructed frequency field. In the table shown below the graph, we can see that the miscalibration error is reduced by at least two orders of magnitude.

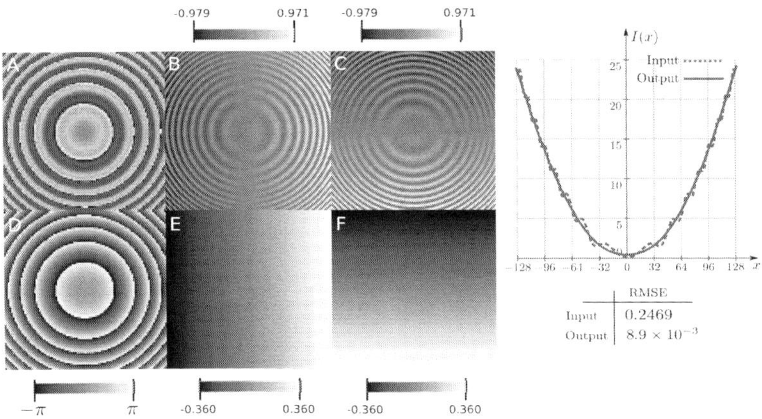

	RMSE
Input	0.2469
Output	8.9×10^{-3}

Figure 3. Wrapped phase with miscalibration errors. In (**A**) we see the wrapped phase, (**B** and **C**) show the x and y frequency fields. In (**D**). We show the recovered phase after frequency reconstruction and integration. The phase is shown wrapped for comparison proposes. In (**E** and **F**) we show the reconstructed frequencies. The graph at the right shows a plot of the central row of the unwrapped input and the unwrapped output of images (**A** and **D**) respectively. Color bars show the dynamic range of the images.

The second test shows how we can reconstruct the frequency field and obtain a clean unwrapped phase under different levels of noise, which we measure using the Signal to Noise Ratio (SNR). In this case, due to the noise, we split the iterations in two: first we perform 3 iterations with $\sigma = 1.0$ for the Gabor filter and $s = 4.3$ for smoothing the frequencies, then, we perform 3 more iterations setting $\sigma = s = 4.3$, according to Eq. (14), for doing a total of 6 iterations. We do it in this way because experimentally we found that for noisy phase maps one can obtain a better estimation for the local frequencies by relaxing the Gabor filtering with an small σ in order to get the detail of the local frequencies, then, set $\sigma = s$ of the Gabor filter to filter out noise.

For this test, we simulated a wrapped phase as

$$\phi_W(x, y) = \operatorname{atan2}(Is(x, y), Ic(x, y)), \tag{18}$$

where $Is(x, y) = \sin[\varphi(x, y)] + \eta(x, y)$ and $Ic(x, y) = \cos[\varphi(x, y)] + \eta(x, y)$. The scalar field $\eta(x, y)$ is Gaussian noise with mean zero and variance σ_n, the argument $\phi(x, y)$ is the ground true phase for which we take the same model used in ref. 16, which is:

$$\phi(x, y) = \frac{2\pi}{\lambda}\phi_0 \exp[-(x^2 + y^2)/(2a^2)] \tag{19}$$

where $x, y \in [-1, 1]^2$, and we take $\phi_0 = 20$ and $a = 0.3$, in a lattice of $M \times N = 201 \times 201$ pixels. The SNR of the signals Ic and Is is measured as $1/\sigma_n$ since the variance of the cosine and sine functions is one. As already said, the parameter σ_n corresponds to the variance of the noise, which is normally distributed. In Fig. 4 we show the result of this test for levels of SNR = 1.3, 1.0, 0.5 and 0.2. In row *Input* we show the noisy fringe pattern I_c that modulates the phase of Eq. (19), in row *Filtered* we show the wrapped phase obtained at end of the process iterations, in row *Unwrapped* we show the unwrapped phase we obtain after intergrating the reconstructed frequency field and row *Error map* shows the error obtained for each pixel as the difference between the obtained unwrapped phase and the ground true phase. Note that the error is localized mostly near the peaks of the phase, due to the reduction in dynamic range caused by the smoothing process; this reduction is in all cases less 10% of the total dynamic range of the unwrapped phase.

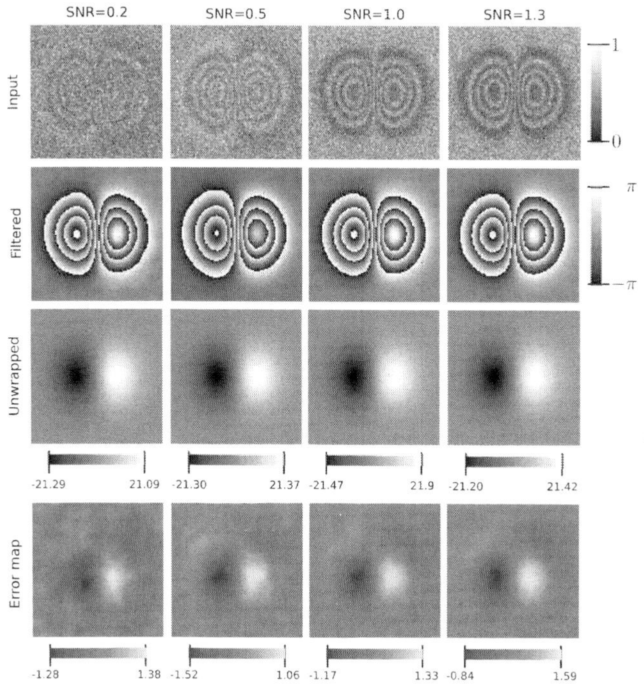

Figure 4. Test with four different levels of noise. The image columns show the test for SNR=0.2, 0.5, 1.0 and 1.3 as depicted. The rows show the noisy fringe pattern, the obtained filtered phase, unwrapped phase and the error map. The color bars indicate the dynamic range of the images. The error map is the error (in radians) we obtain for each pixel by taking the difference of the obtained unwrapped phase and the ground true phase.

To see how accurate the reconstruction is, we used the same simulated phase of Fig. 4 and tested our method with 29 different levels of SNR from SNR=0.2 to SNR=3.0. For each level of noise, we ran 100 different realizations, reconstructed the phase and took the RMSE. The mean of the RMSE and standard deviation for each level of noise is shown in the error graph of Fig. 5. There we can see that the error is small and decreases proportionally with the level of noise. The variance of the RMSE let us see the stability of our method. For the case of SNR=0.2, the variance is greater but consider that SNR=0.2 is a very high leve of noise. These tests were made in a computer with 4xIntel® Core i5-6300U CPU @ 2.4 GHz and 16 Gb of RAM. The mean processing time for each image was of 2.882 sec. The results shown here were obtained using double precision arithmetic.

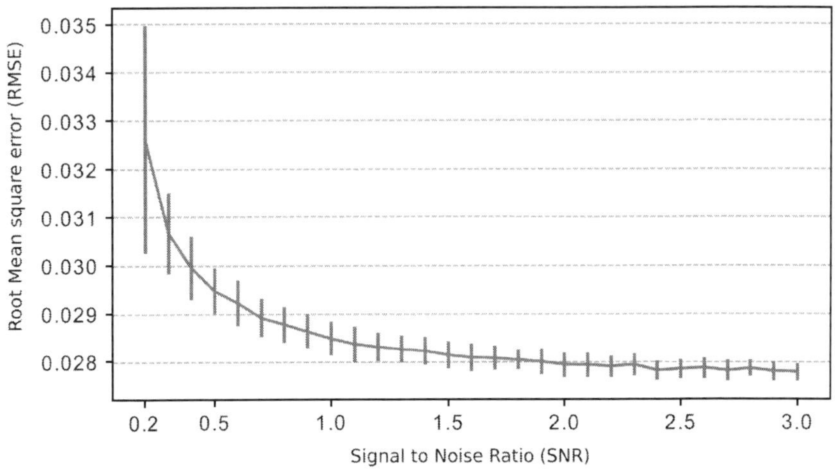

Figure 5. Graph of the Root Mean Square Error (RMSE) and standard deviation (vertical bars) for SNR levels from 0.3 to 3. The mean and standard deviation of the RMSE was taken from 100 different realizations for each level of SNR.

Special Case: Phase Shifting Interferometry

In the examples we have presented, we assume that the wrapped phase map is the only data available, and that it follows the model of Eq. (2) and showed that our method obtains stable results for SNR >0.2, as shown in Fig. 5. However, in most cases we may have access to the interferograms from which the phase was obtained. We now show that in this case the noise tolerance of our method may increase. As an example, consider the case of Phase Shifting Interferometry (PSI). In this case, we have a sequence of interferograms which may be modeled as:

$$I_n(x, y) = 1 + \cos[\phi(x, y) + \alpha n] + noise(x, y), \quad n = 0, 1, 2, \ldots, N-1. \tag{20}$$

In particular, we consider the same example used in Fig. 6 of ref. 16, so that a comparison may be established between the performance of both methodologies. Therefore, as in ref. 16 we use $N=4$, $\alpha=\pi/2$, additive white Gaussian noise, and the same definition for the phase (see Eq. (19)), to obtain a four-step phase shifting sequence of interferograms. In this case, we can obtain a complex signal by combining them in the following way (derived from the standard 4-step phase recovery algorithm[8]):

$$g(x, y) = I_0(x, y) - I_2(x, y) + i[I_1(x, y) - I_3(x, y)]. \tag{21}$$

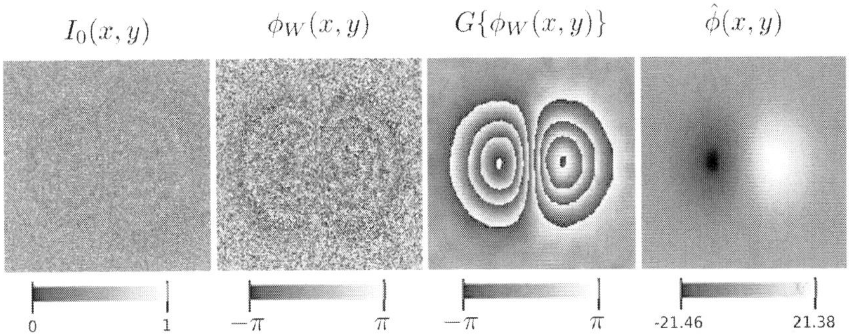

Figure 6. Result obtained by using the four interferograms of a four-step PSI sequence. The first image is the first fringe pattern of the 4-steps sequence; the second image is the wrapped phase obtained from the 4-step algorithm; the third image is the filtered wrapped phase we obtain, denoted here as $G\{\phi_W(x, y)\}$, and the fourth image is the corresponding unwrapped phase. The signal to noise ratio of interferograms is SNR=0.08. The color bars indicate the dynamic range of the images.

Using this complex signal in our reconstruction method, instead of the complex signal of Eq. (3), the method is more robust and can deal with interferograms with SNR as low as 0.08, while in the case of the experiment reported in ref. 16, their method becomes unstable for SNR=0.3. In Fig. 6, we show the first interferogram of the sequence described above with SNR=0.08, which is lower than the SNR used in Fig. 4. Using the complex signal (21) and following our method, with the same iterations and parameters as described for the test of Fig. 4, we obtained a cleaner phase (comparing with the input). This is appreciated in the last two images of Fig. 6.

Comparison with TV Filtering Approach

To perform a fair comparison with the work presented in ref. 16, we programmed that method as described in that work so that we could reproduce their results, and then compared the RMSE for as defined in Eq. (17) with our approach. We ran this test for 29 different levels of noise from $SNR=0.2$, to 3.0 with steps of 0.1 with the same test image shown in Fig. 4. For each noised simulated wrapped phase, we applied the method described in ref. 16. and the one described here. In Fig. 7 we show the RMSE in logarithmic scale of both methods for each level of noise. We used the logarithmic scale for clarity. As one can see, with the approach presented one obtains a RMSE at least two orders of magnitude smaller than the one obtained with the approach presented in ref. 16 (see Discussion section).

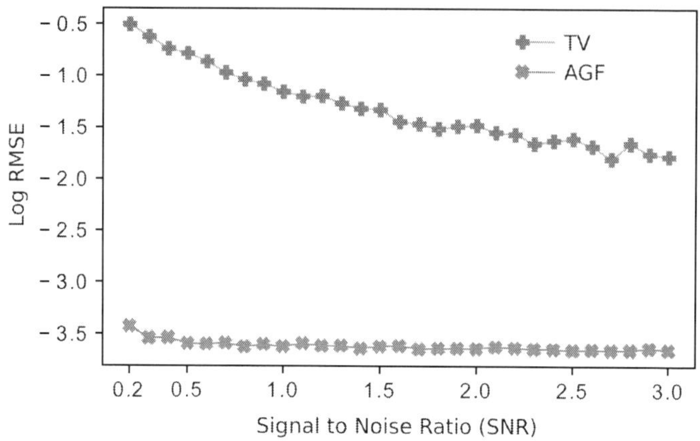

Figure 7. The graph shows the error obtained using the Adaptive Gabor Filters (AGF) approach, shown here, and the TV filtering approach shown in ref. 16. For clarity, the error is in logarithmic scale.

Tests with Real Data

In the last test, we used three different wrapped phases obtained from real optical experiments using speckle interferometry.

In Fig. 8 we show the results. The inputs for these tests are affected by speckle noise and miscalibration errors. The miscalibration error is evident in the test shown in the middle row if you compare how it is seen in Fig. 3. In the column *Reconstruction* we can see the recovered wrapped phase obtained by integrating the reconstructed frequency field. In the column *Unwrapped* we show the unwrapped phase obtained after integrating the frequency field. As one can see, the final unwrapped phase is obtained clean without detuning errors. The sizes of each wrapped phase image given as input are 246×527, 480×610 and 482×641, respectively. As with the test of Fig. 4, we made 6 iterations in total of our method. The first 3 iterations are with parameters $\sigma = 1.0$ and $s = 11.3$ and the next tree iterations are with parameters $\sigma = s = 11.3$. As before, the parameters s and σ are set according to Eq. (14), using N=527, which is the width of the first wrapped phase of Fig.

8. Note that as we said, it is not necessary to use the precise values given by Eq. 14; the intermediate value we use gives good results in all these cases.

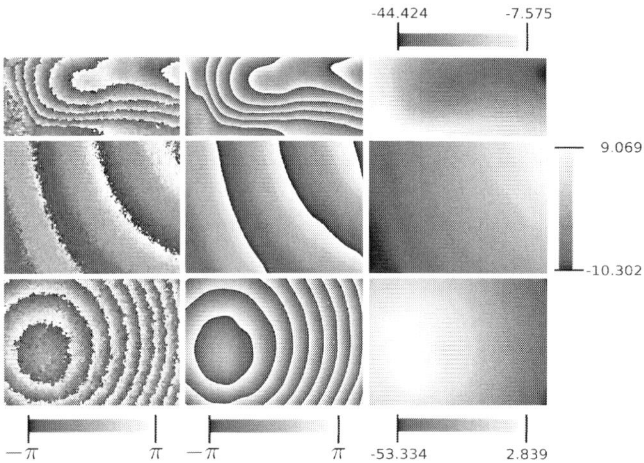

Figure 8. Test using real laboratory experiments. The first column shows the input, the second column shows the filtered wrapped phase and last column shows the corresponding unwrapped phase. All experiments are from speckle interferometry.

DISCUSSION

We have presented a simple and computationally efficient method for the accurate reconstruction of the frequency field from the noisy wrapped phase which is obtained from optical experiments. From this denoised frequency field the clean unwrapped phase may be easily obtained by path-independent integration.

The basis of the method is the use of an adaptive linear quadrature filter which permits very good noise elimination. The method makes the implicit assumption that the true unwrapped phase is smooth. If this holds, one can obtain better denoising performance than that of even non-linear wide passband methods. This is important because, as we have seen here, the problem of phase unwrapping may be complicated due to the presence of noise which induces inconsistencies in the corresponding frequency field. In[16], the approach is by approximating in an iterative way the residual error between the observed wrapped phase and an estimated one, which is obtained by integrating a TV-smoothed frequency field. To summarize the method we developed a block diagram for each iteration which is presented in Fig. 9. One problem with this method is that when one adds to the current unwrapped estimator $\hat{\phi}_W$ the final estimated wrapped residual error R_w between the observed wrapped phase and the re-wrapped $\hat{\phi}_W$, one effectively obtains an unwrapped phase that is appropriately unwrapped, but with the original (wrapped) noise added back to it. As a result, the performance of the method decreases sharply as the SNR diminishes. In contrast, in our approach all the noise is progressively eliminated from the estimated unwrapped phase by means of the adaptive Gabor filter, which explains the big difference in RMS error between the two approaches, which is observed in Fig. 7. It should be noted however, that for high SNR, the method in[16] will give very good results because adding the final residual R_w permits the full preservation of the phase dynamic range, while in our approach it may be slightly compressed due to the smoothing process. For the same reasons, the method in ref. 16 may be more adequate for the unwrapping of discontinuous phase maps, since in our approach the discontinuities may be smoothed out to some degree.

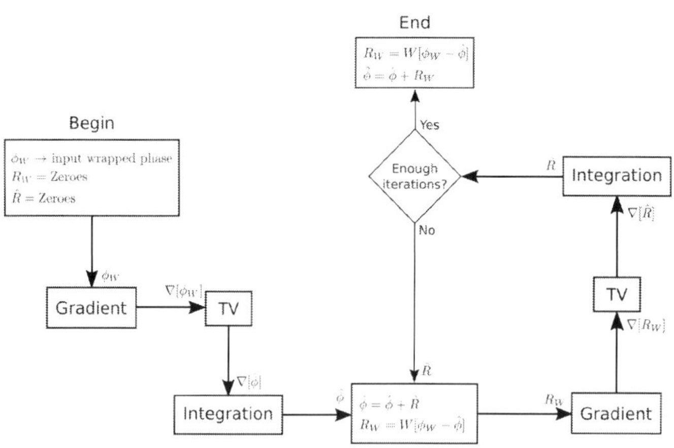

Figure 9. Block diagram of the method presented in ref. 16. As with our method, the gradient is taken by means of finite differences. As recommended in ref. 16. we use 500 iterations in the experiment reported here.

Also related to the approach presented here is the work of ref. 17. There, a local frequency map is estimated by passing an observed interferogram through a fixed bank of Gabor filters and then estimating the local frequency as the tuning frequency of the filter with maximal response. Some problems with this approach are its sensitivity to noise and the fact that since the estimated frequency is restricted to the discrete set of tuning frequencies of the filter bank, one always gets a quantization error which is inversely proportional to the number of filters in the bank, and hence, to the computational complexity. In contrast, in our approach we start with the observed wrapped phase map and estimate the local frequency as the phase gradient; this estimate is progressively refined through the iterations of the method, and used to tune local Gabor filters that eliminate noise, until one finally gets a high-precision, smoothly-varying frequency map, which is then used to find a clean, inconsistency-free unwrapped phase map.

Our method may be very useful in optical applications, such as interferometry, shearography and digital holography, as well as in other areas, such as fringe projection for 3-D reconstruction of smooth objects. In many of these instances, the reconstruction problem is difficult due to the presence of speckle noise and miscalibration errors; as we have shown, our method presents an excellent performance in these cases. However, if there are many true phase discontinuities present, as is the case, for example, in Synthetic Aperture Radar (SAR) images of urban landscapes, other methods, perhaps based on the TV functional, or dedicated algorithms[19, 20] may be more adequate.

REFERENCES

1. Takeda, M. & Mutoh, K. Fourier transform profilometry for the automatic measurement of 3-D object shapes. *Appl. Opt.* **22**, 3977–3982, http://ao.osa.org/abstract.cfm?URI=ao-22-24-3977 (1983).
2. Malacara, D., Servín, M. & Malacara, Z. *Interferogram Analysis for Optical Testing* (CRC Press, 2005).
3. Estrada, J. C., Servin, M. & Marroquín, J. L. Local adaptable quadrature filters to demodulate single fringe patterns with closed fringes. *Optics Express* **15**, 2288, https://www.osapublishing.org/oe/abstract.cfm?uri=oe-15-5-2288 (2007).
4. Servin, M., Estrada, J. C., Quiroga, J. A., Mosino, J. F. & Cywiak, M. Noise in phase shifting interferometry. *Opt. Express* **17**, 8789–8794, http://www.opticsexpress.org/abstract.cfm?URI=oe-17-11-8789 (2009).
5. Servin, M., Quiroga, J. A. & Estrada, J. C. Phase-shifting interferometry corrupted by white and non-white additive noise. *Opt. Express* **19**, 9529–9534, http://www.opticsexpress.org/abstract.cfm?URI=oe-19-10-9529 (2011).
6. Ayubi, G. A., Perciante, C. D., Martino, J. M. D., Flores, J. L. & Ferrari, J. A. Generalized phase-shifting algorithms: error analysis and minimization of noise propagation. *Appl. Opt.* **55**, 1461–1469, http://ao.osa.org/abstract.cfm?URI=ao-55-6-1461 (2016).
7. Servin, M., Quiroga, J. A., Padilla, M. *Fringe Pattern Analisys for Optical Metrology: Theory, Algorithms and Applications* (Whiley-VCH, 2014).
8. Servin, M., Estrada, J. C. & Quiroga, J. A. The general theory of phase shifting algorithms. *Opt. Express* **17**, 21867–21881, http://www.opticsexpress.org/abstract.cfm?URI=oe-17-24-21867 (2009).
9. Mingxing, L., Kai, Y. & Liyun, Z. Wavefront phase reconstruction approach of imaging interferometry based on shearography and the integrated unwrapping algorithm. *Optical and Quantum Electronics* **47**, 545–559, doi:10.1007/s11082-014-9931-1 (2015).
10. Wu, S., He, X. & Yang, L. Enlarging the angle of view in Michelson-interferometer-based shearography by embedding a 4 f system. *Appl. Opt.* **50**, 3789–3794, http://ao.osa.org/abstract.cfm?URI=ao-50-21-3789 (2011).

11. Mihaylova, E., Whelan, M. & Toal, V. Simple phase-shifting lateral shearing interferometer. *Opt. Lett.* **29**, 1264–1266 http://ol.osa.org/abstract.cfm?URI=ol-29-11-1264(2004).
12. Bhaduri, B., Mohan, N. K., Kothiyal, M. P. & Sirohi, R. S. Use of spatial phase shifting technique in digital speckle pattern interferometry (DSPI) and digital shearography (DS). *Opt. Express* **14**, 11598–11607, http://www.opticsexpress.org/abstract.cfm?URI=oe-14-24-11598 (2006).
13. Itoh, K. Analysis of the phase unwrapping algorithm. *Applied optics* **21**, 2470, http://www.ncbi.nlm.nih.gov/pubmed/20396058 (1982).
14. Estrada, J. C., Servin, M. & Quiroga, J. A. Noise robust linear dynamic system for phase unwrapping and smoothing. *Opt. Express* **19**, 5126–5133 http://www.opticsexpress.org/abstract.cfm?URI=oe-19-6-5126 (2011).
15. Estrada, J. C., Servin, M. & Vargas, J. 2D simultaneous phase unwrapping and filtering: A review and comparison. *Optics and Lasers in Engineering* **50**, 1026–1029, http://www.sciencedirect.com/science/article/pii/S0143816612000097 http://linkinghub.elsevier.com/retrieve/pii/S0143816612000097 (2012).
16. Huang, H. Y. H. *et al.* Path-independent phase unwrapping using phase gradient and total-variation (TV) denoising. *Optics Express* **20**, 14075–14089, http://www.opticsexpress.org/abstract.cfm?URI=oe-20-13-14075 (2012).
17. Perea-Vega, D. E. *Local fringe frequency estimation in synthetic aperture radar interferograms using a multiband pre-filtering approach.* Ph.D. thesis, https://open.library.ubc.ca/cIRcle/collections/831/items/1.0065357 (2000).
18. Servin, M., Estrada, J. C. & Quiroga, J. A. Phasorial analysis of detuning error in temporal phase shifting algorithms. *Optics express* **17**, 5618–5623 (2009).
19. Fornaro, G., Franceschetti, G., Lanari, R., Rossi, D. & Tesauro, M. Interferometric SAR phase unwrapping using the finite element method. *IEE Proceedings-Radar, Sonar and Navigation* **144**, 266–274 (1997).
20. Chen, C. W. & Zebker, H. A. Phase unwrapping for large SAR interferograms: Statistical segmentation and generalized network models. *IEEE Transactions on Geoscience and Remote Sensing* **40**, 1709–1719 (2002).

CHAPTER 4

OPTICAL APERTURE SYNTHESIS WITH ELECTRONICALLY CONNECTED TELESCOPES

Dainis Dravins[1], Tiphaine Lagadec[1], and Paul D. Nuñez[2,3]

[1]Lund Observatory, Lund University, Lund SE-22100, Sweden.
[2]Collège de France, 11 Place Marcelin Berthelot, Paris FR-75005, France.
[3]Laboratoire Lagrange, Observatoire de la Côte d'Azur, BP 4229, Nice FR-06304, France.

ABSTRACT

Highest resolution imaging in astronomy is achieved by interferometry, connecting telescopes over increasingly longer distances and at successively shorter wavelengths. Here, we present the first diffraction-limited images in visual light, produced by an array of independent optical telescopes, connected electronically only, with no optical links between them. With an array of small telescopes, second-order optical coherence of the sources is measured through intensity interferometry over 180 baselines between pairs

Citation: (APA): Dravins, D., Lagadec, T., & Nuñez, P. D. (2015). Optical aperture synthesis with electronically connected telescopes. Nature communications, 6(1), 1-5. (5 pages)

of telescopes, and two-dimensional images reconstructed. The technique aims at diffraction-limited optical aperture synthesis over kilometre-long baselines to reach resolutions showing details on stellar surfaces and perhaps even the silhouettes of transiting exoplanets. Intensity interferometry circumvents problems of atmospheric turbulence that constrain ordinary interferometry. Since the electronic signal can be copied, many baselines can be built up between dispersed telescopes, and over long distances. Using arrays of air Cherenkov telescopes, this should enable the optical equivalent of interferometric arrays currently operating at radio wavelengths.

INTRODUCTION

In optical astronomy, the highest angular resolution is presently offered by phase/amplitude interferometers combining light from telescopes separated by baselines up to a few hundred metres. Tantalizing results show how stellar disks start to become resolved, revealing stars as a diversity of individual objects, although until now feasible only for a small number of the largest ones. Concepts have been proposed to extend such facilities to scales of a kilometre or more, as required for surface imaging of bright stars with typical sizes of a few milliarcseconds. However, their realization remains challenging, both due to required optical and atmospheric stabilities within a fraction of an optical wavelength, and the need to span many interferometric baselines (given that optical light cannot be copied with retained phase, but must be divided and diluted by beamsplitters to achieve interference among multiple telescope pairs). While atmospheric issues could be avoided by telescope arrays in space, such are impeded by their complexity and their cost. However, atmospheric issues can be circumvented by measuring higher-order coherence of light through intensity interferometry.

In the following, we present a laboratory demonstration of a multi-telescope array to verify the end-to-end sequence of operation from observing star-like sources to reconstruction of their images.

RESULTS

Origin and Principles of Intensity Interferometry

The technique of intensity interferometry[1], once pioneered by Hanbury Brown and Twiss[5] is effectively insensitive to atmospheric turbulence and permits very long baselines at short optical wavelengths also. What is observed is the second-order coherence of light (that of intensity I, not of amplitude or phase), by computing temporal correlations of arrival times between photons recorded in different telescopes observing the same target. When the telescopes are close together, the intensity fluctuations measured in both telescopes are correlated in time, but when further apart the degree of correlation diminishes.

Although this behaviour can be understood in terms of classical optical waves undergoing random phase shifts, it is fundamentally a two-photon process measuring the degree of photon 'bunching' in time, and its full explanation requires a quantum description. The concept is normally seen as the first experiment in quantum optics[2,3,4], with its first realization already in the 1950s, when Hanbury Brown and Twiss5 succeeded in measuring the diameter of the bright and hot star Sirius. This laid the foundation for experiments of photon correlations including also states of light that do not have classical counterparts (such as photon antibunching), and has found numerous applications outside optics, particularly in particle physics since the same bunching and correlation properties apply to other bosons, that is, particles with integer quantum spin, whose statistics in an equilibrium maximum-entropy state follow the Bose–Einstein distribution. The method can also be seen as mainly exploiting the corpuscular rather than the wave nature of light, even if those properties cannot be strictly separated.

The quantity measured is $<I_1(t)\cdot I_2(t)>=<I_1(t)><I_2(t)>\ (1+|\gamma_{12}|^2)$, where $<>$ denotes temporal averaging and γ_{12} is the mutual coherence function of light between locations 1 and 2, the quantity commonly measured in phase/amplitude interferometers. Compared with randomly fluctuating intensities, the correlation between intensities I_1 and I_2 is 'enhanced' by the coherence parameter, and an intensity interferometer thus measures $|\gamma_{12}|^2$ with a certain electronic time resolution. Realistic values of a few nanoseconds correspond to light-travel distances around 1 m, and permitted error budgets relate to such a number rather than to the optical wavelength, making the method practically immune to atmospheric turbulence and optical imperfections. The above coherence relation holds for ordinary thermal ('chaotic') light where the light wave undergoes 'random' phase jumps on timescales of its coherence time but not necessarily for light with different quantum statistics[2,3,4].

Intensity Interferometry in the Modern Era

The method has not been pursued in astronomy since Hanbury Brown's[1] early two-telescope intensity interferometer in Narrabri, Australia, but new possibilities are opening up with the erection of large arrays of air Cherenkov telescopes. Although primarily devoted to measuring optical Cherenkov light in air induced by cosmic gamma rays, their optical performance, telescope sizes and distributions on the ground would be suitable for such applications[6].

Realizing that potential, various theoretical simulations of intensity interferometry observations with extended telescope arrays have been made[7,8,9,10,11], in particular examining the overall sensitivity and limiting stellar magnitudes. Different studies conclude that, with current electronic performance and current Cherenkov telescope designs, one night of observation in one single wavelength band permits measurements of hotter stars down to visual magnitude about m_v=8, giving access to thousands of sources. Such values agree with extrapolations from the early pioneering observations, also when including observational issues such as background light from the night sky[10]. Fainter sources could be reached once simultaneous measurements in multiple wavelength bands can be realized, or if full two-dimensional (2D) data are not required.

Interferometer Array in the Laboratory

In this work, we report the first end-to-end practical experiments of such types of measurements. A set-up was prepared in a large optics laboratory with artificial sources ('stars'), observed by an array of small telescopes of 25 mm aperture, each equipped with a photon-counting solid-state detector (silicon avalanche photodiode operated in Geiger-mode). The continuous streams of photon-counts were fed into electronic firmware units in real time (with 5 ns time resolution) computing temporal cross correlations between the intensity fluctuations measured across baselines between many different pairs of telescopes. The degree of mutual correlation for any given baseline provides a measure of the second-order spatial coherence of the source at the corresponding spatial frequency. Numerous telescope pairs of different baseline lengths and orientations populate the interferometric Fourier-transform plane, and the measured data provide a 2D map of the second-order spatial coherence of the source, from which its image can be reconstructed.

Experiment Set-Up

Artificial 'stars' (single and double, round and elliptic) were prepared as tiny physical holes drilled in metal (using miniature drills otherwise intended for mechanical watches). Typical aperture sizes of 0.1 mm subtend an angle of ~1 arcsec as observed from the 23 meter source-to-telescope distance in the laboratory.

The technique of intensity interferometry, in principle, requires chaotic light for its operation[2,3,4] in order for 'random' intensity fluctuations to occur and, in practice, also sources of high surface brightness (such as stars hotter than the Sun). This brightness requirement of a high effective source temperature comes from the need for telescopes to be small enough to resolve the structures in the Fourier domain, while still receiving sufficiently high photon fluxes. Of course, the same limitations apply to laboratory sources as to celestial ones. There are numerous hot stars in the sky but correspondingly hot laboratory sources are less easy to find. Ordinary sources producing chaotic light have too low effective temperatures (for example, high-pressure arc lamps of ~5,000 K) to permit conveniently short integration times while laser light cannot be directly used since it is not chaotic (ideally, undergoing no intensity fluctuations whatsoever).

Following various tests, the required brilliant illumination was produced by scattering λ=532 nm light from a 300 mW laser against microscopic particles in thermal (Brownian) motion. Those were monodispersive polystyrene spheres of 0.2 μm diameter, suspended in room-temperature distilled water, undergoing random motion due to collisions with the water molecules surrounding them. On scattering, the laser light is broadened through Doppler shifts, producing a spectral line with a Lorentzian wavelength shape, and with a chaotic (Gaussian) distribution of the electric field amplitudes[12,13]. Such light is thus equivalent in its photon statistics and intensity fluctuations to the thermal (white) light expected from any normal star, the difference being that the spectral passband is now very much narrower. The width of the scattered line was estimated from its measured temporal coherence to be ~10–100 kHz, with a brightness temperature of the order 10^5 K. The relatively long coherence times enabled convenient measurements on microsecond scales, with typical photon-count rates between 0.1 and 1 MHz, with integration times of a few minutes for each telescopic baseline. Such count rates assure negligible photon-noise contributions although there may still remain issues of systematics due to, for example, slightly inhomogeneous illumination of the artificial stars.

Here we note that the signal-to-noise ratio in intensity interferometry in principle is independent of optical bandpass, be it broadband white light or just a narrow spectral line[1]. For any broader bandpass, realistic electronic time resolutions are slower than optical coherence times, and the noisier measurement of a smaller photon flux from a narrower passband (with longer coherence time) is compensated by lesser averaging over fewer coherence times. Thus, also the present quasi-monochromatic light source can represent measurements over broader optical bandpasses.

Simulating a Large Telescope Array

Angular resolution in optical systems is often quantified by the Airy disk diffraction radius $\theta \approx 1.22\ \lambda/D$ rad. At λ=532 nm, the baseline D required to resolve 1 arcsec is ~10 cm, dictating a rather compact set-up of small telescopes. Our five units offered baselines between 3 and 20.5 cm, the former constraining the largest field to nominally 4.5 arcsec and the latter limiting the resolution to 0.65 arcsec. The zero baseline (needed for calibration) was realized with two additional separate telescopes behind one beamsplitter, both telescopes observing the same signal.

To mimic a large telescope array, the effective number of telescopes and baselines was increased by measuring across different interferometer angles. The array was mounted on a horizontal optical table and thus covered horizontal baselines only. To cover also oblique and vertical baselines in the 2D Fourier-transform plane, the position angle of the artificial 'star' was successively rotated relative to the plane of the telescopes. Various details of the overall laboratory arrangements and of measuring techniques are described elsewhere[14]. With N telescopes, $N(N-1)/2$ different baselines can be constructed. Five telescopes give 10 baselines at any given angle, and a series of 18 measurements in steps of 10° over a total of 180° produces 180 baselines. The second-order coherence in a point (u,v) of the Fourier-transform plane equals that in $(-u,-v)$, so each baseline provides two points; data for angles up to 360° are copies of the first ones. Figure 1 shows the coherence pattern for one thus measured asymmetric binary 'star', with one larger and one smaller component (flux ratio 4:1). No smoothing was applied to the measured coherence values, neither for this plot nor for the later image reconstructions. The overplotted surface merely visualizes the global measured pattern, whose individual points are too numerous to be readily displayed.

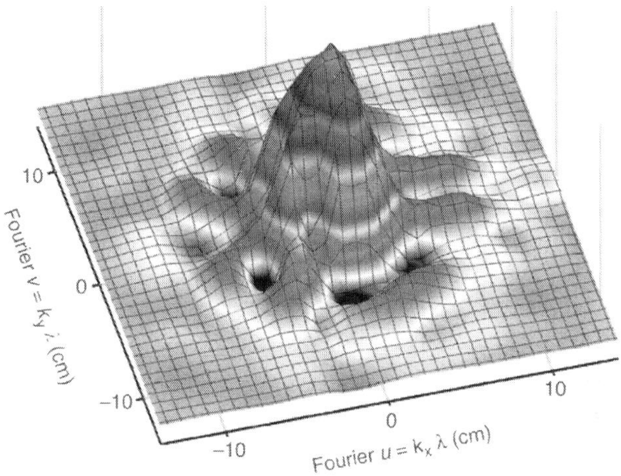

Figure 1: Second-order optical coherence surface: measurements for an artificial asymmetric binary star built up from intensity-correlation measurements over 180 baselines between pairs of small optical telescopes in the laboratory. The coordinates refer to the plane of the Fourier transform of the source image and correspond to different telescope baseline lengths and orientations.The image reconstructed from these data is in Fig. 2, right.

Image Reconstruction Principles

Image reconstruction from second-order coherence implies some challenges not present in first-order phase/amplitude techniques. While intensity interferometry has the advantage of not being sensitive to phase errors in the optical light paths, it also does not measure such phases but rather obtains the absolute magnitudes of the respective Fourier-transform components of the source image. Such data can by themselves fit model parameters such as stellar diameters or binary separations but images cannot be directly computed through merely an inverse Fourier transform.

Theoretical simulations have verified that, provided the Fourier-transform plane is well-filled with data, 2D image restoration becomes feasible[8,9,11]. The pattern of second-order coherence is equivalent to the intensity pattern produced by diffraction of light in a correspondingly sized aperture, and it is already intuitively clear that a thorough mapping of that pattern must put stringent constraints on the source geometry.

Image Reconstruction Procedures and Results

Strategies in analysing intensity interferometry data involve estimating Fourier phases from Fourier magnitudes[8,15,16]. Since the Fourier transform of a finite object is an analytic function, the Cauchy–Riemann equations can be used to find derivatives of the phase from derivatives of the magnitude, producing images that are unique except for translation and reflection. Here one could start with a 1D phase estimate along a single slice through the (u,v) plane origin. 2D coverage requires combining multiple 1D reconstructions, while ensuring mutual consistency between adjacent ones. However, these algorithms appear to require a very dense coverage of the (u,v) plane and, despite our numerous baselines, did not appear efficient with current data.

Instead, an inverse-problem approach was taken, imposing very general constraints to interpolate between Fourier frequencies ('regularization'). The Multi-aperture Image Reconstruction Algorithm, MiRA (previously tested on simulated data[9]), produced more stable results. This iterative procedure maximizes the agreement between measurements and the squared modulus of the Fourier transform of the reconstructed image. Regularization techniques can be applied to favour solutions with certain properties, for example, if something is initially known about the source. To reconstruct the elliptical star in Fig. 2, a 'compact' solution (quadratic regularization) was favoured, and a 'smooth' solution for the binary (quadratic smoothness regularization)[17].

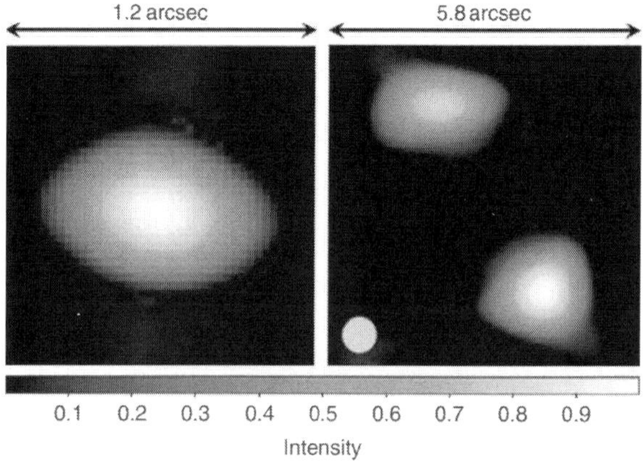

Figure 2: Optical images reconstructed from electronic intensity interferometry; measurements with 100 and 180 baselines, respectively, of a single elliptical artificial 'star' and an asymmetric binary one (flux ratio 4:1). Sufficiently

dense coverage of the interferometric Fourier-transform plane enables a full reconstruction of source images, despite the lack of Fourier phase information. In principle, the images are unique except for their possible mirrored reflections. The spatial scales are similar on both axes; colours show reconstructed linear intensities, the circle indicates the 0.65 arcsec resolution set by the longest baseline.

Figure 2 shows examples of such image reconstructions at λ=532 nm from measurements with 100 and 180 optical baselines. As far as we are aware, these are the first diffraction-limited images in visual light that have been reconstructed from an array of dispersed optical telescopes, connected by electronic software only, without any optical links between them.

Similar to any phase/amplitude interferometry or aperture synthesis, the fidelity of the reconstructed images depends on measurement noise, the density of coverage across the Fourier (u,v) plane and the character of the 'cleaning' algorithms applied. In the present measurements, the random part of the noise is essentially negligible but some systematics are likely present, for example, due to a probably somewhat inhomogeneous illumination of the artificial stars that makes their brightness shapes not identical to their geometrical apertures.

Reconstructions were tested on different sources. For the single elliptic star (Fig. 2, left), its major and minor axes could be determined to a precision of 2%, a number estimated from several image reconstructions with different initial random images. Even slightly better values were obtained for a symmetric binary. However, the present asymmetric one was chosen to be a more difficult target, because of the contrast between its components, and since the Fourier-plane coverage is marginal for both the largest and smallest scales, contributing to some image artefacts seen in Fig. 2. Still, the reconstruction precision is consistent with what is found in previous detailed theoretical simulations[8], where the reconstructed radii gradually become less precise for greater brightness ratios between both binary members.

The computing effort is modest (and not comparable to the extensive numerical efforts involved in radio aperture synthesis that involve also spectral analysis). To obtain an image of 10,000 pixels from measurements across 100 baselines, starting from a random image, requires just some 10 s on a 3 GHz processor. The exact time can be shorter if one already has some initial guess (perhaps obtained from some 'raw' reconstruction from a Cauchy–Riemann analysis), or be longer for additional baselines or more numerous pixels.

The image quality was further explored as a function of the sampling or smoothing in the (u,v) plane, affecting the pixel sizes in the reconstructed image. The best performance was found when the number of samples within the Fourier plane roughly matches the number of baseline measurements. Possibly, there could be other dependences for noise-limited data, where some kind of smoothing may be required. In general, the optimization of reconstruction for different numbers of sample points in the presence of various noise levels and incomplete Fourier-plane coverages is a somewhat complex problem that will require further studies (analogous to what in the past has been made for aperture synthesis in the radio).

Long-Baseline Optical Interferometry

The longer-term ambition is to connect widely separated telescopes in a large optical aperture synthesis array. Similar to current radio interferometry, the electronic signal from any telescope can be freely copied or even stored for possible later analysis. Realistic signal bandwidths for measured intensity fluctuations are in the order of 100 MHz, very comparable to those encountered in radio arrays. However, the requirements on the signal correlators are much more modest than for radio phase/amplitude interferometers. Optical intensity interferometry can in practice only detect spatial coherence while the wavelengths are selected by colour filters or through detector responses, not—as the case in radio—by temporal coherence measurements. Consequently, the correlator electronics can be simple table-top devices rather than massive computing facilities.

The signal-to-noise properties[1,6,7] of intensity interferometry require the telescopes to be large to reach adequately low photon noise, but their optical precision need not be higher than that of ordinary air Cherenkov telescopes. However, the limiting signal-to-noise ratios may still depend on their exact optical design. For example, how high electronic time resolutions that can be exploited may be constrained by the time spread of light paths inside the telescope. Details of such and other parameters are discussed elsewhere[6,7,10].

Obvious candidates for a kilometre-sized optical imager are the telescopes of the planned Cherenkov Telescope Array[18], foreseen to have in the order of 100 telescopes, distributed over a few square kilometres. Assuming it becomes available for also interferometric observations, the spatial resolution at the shortest optical wavelengths may approach ~30 μas. Such resolutions have hitherto been reached only in the radio, and it is awkward to predict what features on, or around, stellar surfaces will appear in the optical[6]. However, to appreciate the meaning of such resolutions, we note that a hypothetical Jupiter-size exoplanet in transit across some nearby bright star such as Sirius would subtend an easily resolvable angle of some 350 μas (ref. 14). While spatially resolving the disk of an exoplanet in its reflected light may remain unrealistic for the time being, the imaging of its dark silhouette on a stellar disk—while certainly very challenging[19]—could perhaps be not quite impossible.

ACKNOWLEDGEMENTS

This work was supported by the Swedish Research Council and The Royal Physiographic Society in Lund. Early experiments in intensity interferometry and photon correlation techniques at the Lund Observatory involved also Toktam Calvén Aghajani, Daniel Faria, Johan Ingjald, Hannes Jensen, Lennart Lindegren, Eva Mezey, Ricky Nilsson and Helena Uthas.

AUTHOR CONTRIBUTIONS

D.D. conceived this project; D.D. and T.L. together set up the laboratory experiment at the Lund Observatory, performed all measurements and their reductions; P.D.N. carried out the image reconstructions. D.D. wrote the draft manuscript but all authors contributed to editing its final version.

REFERENCES

1. Hanbury Brown, R. *The Intensity Interferometer. Its Applications to Astronomy* Taylor & Francis: London, (1974).
2. Labeyrie, A., Lipson, S. G. & Nisenson, P. *An Introduction to Optical Stellar Interferometry* Cambridge Univ. Press: Cambridge, (2006).
3. Saha, S. K. *Aperture Synthesis. Methods and Applications to Optical Astronomy* Springer (2011).
4. Shih, Y. *An Introduction to Quantum Optics. Photon and Biphoton Physics* CRC Press: Boca Raton, (2011).
5. Hanbury Brown, R. & Twiss, R. Q. A test of a new type of stellar interferometer on Sirius. *Nature* **178**, 1046–1048 (1956).
6. Dravins, D., LeBohec, S., Jensen, H. & Nuñez, P. D. for the CTA consortium, Optical intensity interferometry with the Cherenkov Telescope Array. *Astropart. Phys.* **43**, 331–347 (2013).
7. Dravins, D., LeBohec, S., Jensen, H. & Nuñez, P. D. Stellar intensity interferometry: Prospects for sub-milliarcsecond optical imaging. *New Astron. Rev.* **56**, 143–167 (2012).
8. Nuñez, P. D., Holmes, R., Kieda, D. & LeBohec, S. High angular resolution imaging with stellar intensity interferometry using air Cherenkov telescope arrays. *Mon. Not. R. Astron. Soc.* **419**, 172–183 (2012).
9. Nuñez, P. D., Holmes, R., Kieda, D., Rou, J. & LeBohec, S. Imaging submilliarcsecond stellar features with intensity interferometry using air Cherenkov telescope arrays. *Mon. Not. R. Astron. Soc.* **424**, 1006–1011 (2012).
10. Rou, J., Nuñez, P. D., Kieda, D. & LeBohec, S. Monte Carlo simulation of stellar intensity interferometry. *Mon. Not. R. Astron. Soc.* **430**, 3187–3195 (2013).
11. Dolne, J. J., Gerwe, D. R. & Crabtree, P. N. Cramer-Rao lower bound and object reconstruction performance evaluation for intensity interferometry. Proc. SPIE 9146, Optical and Infrared Interferometry IV, 914636 (2014).
12. Berne, B. J. & Pecora, R. *Dynamic Light Scattering. With Applications to Chemistry, Biology and Physics*. (Dover, 2000).
13. Crosignani, B., Di Porto, P. & Bertolotti, M. *Statistical Properties of Scattered Light* Academic Press (1975).

14. Dravins, D. & Lagadec, T. Stellar intensity interferometry over kilometer baselines: Laboratory simulation of observations with the Cherenkov Telescope Array. Proc. SPIE 9146, Optical and Infrared Interferometry IV, 91460Z (2014).
15. Holmes, R. B. & Belen'kii, M. S. Investigation of the Cauchy-Riemann equations for one-dimensional image recovery in intensity interferometry. *J. Opt. Soc. Am. A* **21**, 697–706 (2004).
16. Holmes, R., Calef, B., Gerwe, D. & Crabtree, P. Cramer-Rao bounds for intensity interferometry measurements. *Appl. Opt.* **52**, 5235–5246 (2013).
17. Thiébaut, E. Image reconstruction with optical interferometers. *New Astron. Rev.* **53**, 312–328 (2009).
18. Actis, M. et al. Design concepts for the Cherenkov Telescope Array CTA: An advanced facility for ground-based high-energy gamma-ray astronomy. *Exp. Astron.* **32**, 193–316 (2011).
19. Strekalov, D. V., Erkmen, B. I. & Yu, N. Intensity interferometry for observation of dark objects. *Phys. Rev. A* **88**, 053837 (2013).

CHAPTER 5

NANOSCALE OPTICAL INTERFEROMETRY WITH INCOHERENT LIGHT

Dongfang Li, Jing Feng, and Domenico Pacifici

School of Engineering, Brown University, Providence, Rhode Island, 02912, United States.

ABSTRACT

Optical interferometry has empowered an impressive variety of biosensing and medical imaging techniques. A widely held assumption is that devices based on optical interferometry require coherent light to generate a precise optical signature in response to an analyte. Here we disprove that assumption. By directly embedding light emitters into subwavelength cavities of plasmonic interferometers, we demonstrate coherent generation of surface plasmons even when light with extremely low degrees of spatial and temporal coherence is employed. This surprising finding enables novel

Citation: (APA): Li, D., Feng, J., & Pacifici, D. (2016). Nanoscale optical interferometry with incoherent light. Scientific reports, 6(1), 1-8. (8 pages)

sensor designs with cheaper and smaller light sources and consequently increases accessibility to a variety of analytes, such as biomarkers in physiological fluids, or even airborne nanoparticles. Furthermore, these nanosensors can now be arranged along open detection surfaces and in dense arrays, accelerating the rate of parallel target screening used in drug discovery, among other high volume and high sensitivity applications.

INTRODUCTION

Optical interferometry based on surface plasmon polaritons (SPPs) has enabled a broad range of technical advantages[1,2,3,4,5,6], including enhanced light transmission[7,8], all-optical modulation[9], light beaming[10,11,12,13,14] and medical imaging[15]. More recently, control of SPPs interference at the nanoscale has led to biochemical sensors with unparalleled sensitivity and selectivity[16,17,18,19]. However, sensing architectures based on plasmonic interferometry typically require coherent external light sources that have to be carefully aligned and periodically calibrated, which poses challenges to reliable integrated devices. Here we develop an alternative design for a biochemical sensor that does not require a coherent external light source, precise alignment, or re-calibration. We have done so by embedding solid-state light emitters directly into the nanoaperture of plasmonic interferometers, where they serve as the localized light source. It is definitively shown that the optical response of this novel sensor is independent of coherence, spectral bandwidth and incidence angle of the excitation source. This latter property is particularly advantageous when measuring chemical analytes in micro-droplets, where lensing effects can significantly vary the angle of incidence of the light beam, thus introducing artifacts in the sensor response.

Plasmonic interferometers consisting of a central hole flanked by three grooves (H-3G) with subwavelength width were fabricated on a 300 nm-thick silver film by focused ion beam milling and then coated with a thin (~40 nm) Cr^{3+}:MgO light emitting layer[20,21] (see Fig. 1a). In contrast to other implementations of this structure (also known as bullseye[10,12,13,14]), here each excited emitter (i.e. Cr^{3+} ion in MgO) within the nanohole of each interferometer has a high probability to decay by generating SPP waves (with amplitude E_{SPP}, dashed horizontal lines in Fig. 1a)[22,23]. These SPPs propagate away from the nanohole and are then partially reflected back[24,25] by the subwavelength grooves toward the nanohole, where they interfere with the directly emitted light (with amplitude E_0, red solid arrow in Fig.

1a), thus modifying the fluorescence spectra in the far field. The fluorescence spectra transmitted through the nanohole can then be modulated by varying the interferometer arm length R_G or the refractive index of the dielectric material on the surface, thus providing for a novel sensing scheme.

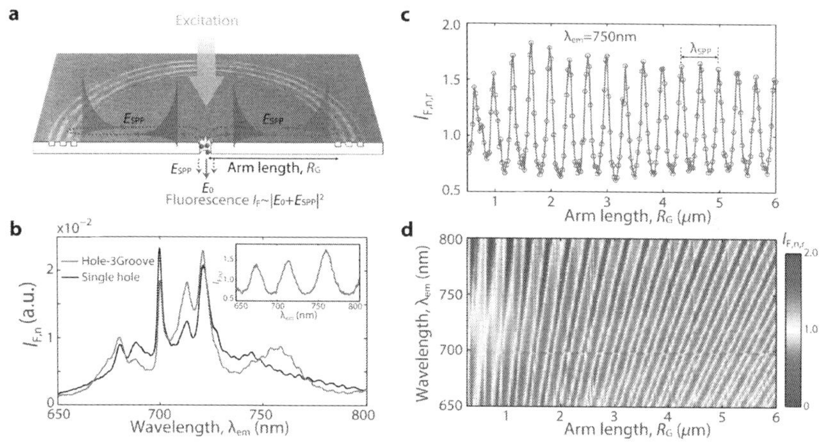

Figure 1. Active plasmonic interferometry with embedded light emitters. (**a**) Scanning electron micrograph (SEM) and schematic cross section of a circular hole-3groove (H-3G) plasmonic interferometer whose nanoaperture has been filled with Cr^{3+}:MgO light emitters. Excited emitters in the nanohole (red dots) generate SPPs that undergo in-plane reflection by the nano-grooves. The fluorescence intensity detected in the far field is the result of the interference between the directly emitted field (E_0, vertical solid arrow) and the transmitted field originating from diffractive scattering of round-trip SPPs through the nanoaperture (E_{SPP}, vertical dashed arrows). (**b**) Normalized fluorescence spectra transmitted through the nanohole of a H-3G interferometer with arm length R_G = 4.05 μm (red line) and through an isolated nanohole serving as reference (black line). The inset shows the corresponding normalized intensity ratio *vs.* λ_{em}. (**c**) Fluorescence interferogram obtained by plotting the normalized fluorescence intensity ratio at λ_{em} = 750 nm as a function of arm length R_G. The observed oscillation period ($\lambda_{SPP}/2$) is consistent with round-trip SPP interference. (**d**) Color map reporting spectrally resolved fluorescence interferograms *vs.* emission wavelength λ_{em} (vertical axis) and interferometer arm length R_G (horizontal axis).

RESULTS

Fluorescence Modification Induced by Active Plasmonic Interferometry

Figure 1b shows normalized fluorescence spectra transmitted through the hole of a representative interferometer with $R_G = 4.05\ \mu m$ ($I_{F,n}$, red line) and through a reference isolated hole ($I_{F,0,n}$, black line), evidencing a remarkable spectral modulation due to SPP-mediated interference. The intensity modulation can be quantified by defining a normalized intensity ratio $I_{F,n,r} = I_{F,n}/I_{F,0,n}$ *vs.* emission wavelength λ_{em}, as shown in the inset to Fig. 1b. The normalized intensity modulation at each emission wavelength can also be estimated by measuring the fluorescence spectra through the nanohole of plasmonic interferometers with varying arm length (as reported in Fig. 1c for $\lambda_{em} = 750\,nm$). The fluorescence interferogram oscillates with a period equal to $\lambda_{SPP}/2$, i.e. half the corresponding SPP wavelength, suggesting that the fluorescence modulation is the result of interference from SPPs undergoing a round trip, as reported schematically in Fig. 1a. A color map of the normalized intensity ratio as a function of emission wavelength λ_{em} and arm length R_G is constructed in Fig. 1d by stacking up all of the measured spectra. The color of each pixel in this map corresponds to constructive (red) or destructive (blue) interference, while white pixels correspond to null interference. The 3-groove grating pitch was designed to be half of the SPP wavelength in the three-layer system (silver/Cr^{3+}:MgO/air) to keep the reflection by different grooves in phase, thus enhancing the fluorescence modulation up to a factor of ~2.

The fluorescence modulation can be totally suppressed if the emitters in the hole are purposely removed, suggesting that the emitters on the planar metal surface play no significant role. Therefore, the working hypothesis of this Article is that the fluorescence modification originates solely from interference effects mediated by SPPs that are generated by excited emitters within the nanohole. For this reason, the fluorescence modulation should also be independent of the specific excitation conditions, being only affected by the structural parameters of the interferometer and by changes in the refractive index of the dielectric materials.

Plasmonic Interferometry with Incoherent Light

To demonstrate that coherent excitation is not required to achieve fluorescence modulation, a broadband thermal light source was used to excite the emitters in both a H-3G plasmonic interferometer and a reference isolated hole, using Köhler illumination to tune the degree of spatial coherence26 (see Fig. 2a). Figure 2b reports normalized fluorescence transmission intensity ratios for various subtended angles $\Delta\theta = 9°$ (blue), 18° (green) and 36° (red), showing no appreciable changes in the observed modulation. Although the emitters are excited by an external light source with low spatial coherence, they retain temporal coherence since the emission occurs from a subwavelength cavity. Indeed, any given photon detected in the far field has to be transmitted through the nanohole and it can originate from two main optical paths, i.e. (1) direct emission from an excited emitter or (2) SPP excitation, round-trip propagation and diffractive scattering followed by transmission through the same nanoaperture. Since the two paths are indistinguishable due to the subwavelength nature of the aperture, their complex amplitude probabilities (corresponding to the complex field amplitudes E_0 and E_{SPP}) can be summed up in order to calculate the total far-field fluorescence intensity (as reported in Fig. 1a and Methods). This leads to a coherent interference process that is independent of how the emitter is excited (i.e. either coherently or incoherently) provided the temporal coherence of the emitter is sufficiently high.

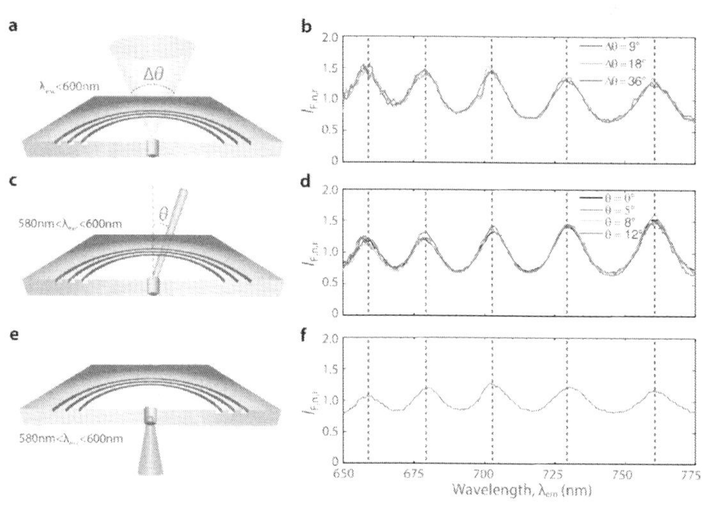

Figure 2. Fluorescence modulation induced by plasmonic interferometry: role of coherence, spectral bandwidth and incidence angle of the external light source. (**a**) Schematic of a plasmonic interferometer with embedded emitters illuminated by a broadband thermal light source (with excitation wavelengths <600 nm and variable transverse coherence length, achieved by changing the

subtended angle $\Delta\theta$, see Methods). (**b**) Corresponding normalized fluorescence intensity ratios measured at $\Delta\theta=9°$ (blue), 18° (green) and 36° (red). (**c**) Quasi-monochromatic (580–600 nm) top illumination with varying angle of incidence θ. (**d**) Corresponding normalized fluorescence intensity ratios with $\theta=0°$ (normal incidence, black), 5° (blue), 8° (green) and 12° (red). (**e**) Quasi-monochromatic (580–600 nm) bottom illumination. (**f**) Corresponding normalized fluorescence intensity ratio. The fluorescence modulation is independent of the specific excitation conditions (coherence, bandwidth, or angle of incidence), as evidenced by the vertical dashed lines in (**b**,**d**,**f**). The arm length of the measured interferometer is 6 μm, the grating pitch is 320 nm.

To prove that the fluorescence modulation is also independent of precise alignment of the external source, a quasi-monochromatic source (580–600 nm) was used to illuminate the same active plasmonic interferometer with varying angles of incidence (Fig. 2c). Figure 2d shows that the measured normalized fluorescence intensity ratios are also invariant for different incident angles: $\theta=0°$ (normal incidence, black), 5° (blue), 8° (green) and 12° (red).

In addition, the measured optical response of the sensor under bottom illumination (Fig. 2e) is shown in Fig. 2f. This configuration is especially attractive since for conventional passive plasmonic interferometry the light source and the detector generally need to be placed on opposite sides of the sample[16,17,18,19,27]. Moreover, to measure biological samples or liquids, a microfluidic channel is typically required to ensure uniform sample delivery and illumination, thus adding complexity to the sensing architecture. Here we show freedom from that constraint in the novel design, whereby pump and detection can be achieved from the same (i.e. the substrate) side of the sensor. This arrangement does not require a microfluidic channel and solves the common problem arising from potential biological sample damage occurring under direct illumination.

Despite the significantly different illumination conditions, the experimental results in Fig. 2 demonstrate that the optical response of plasmonic interferometers with embedded light emitters is no longer affected by coherence, spectral bandwidth and incidence angle of the excitation source, thus providing for a more reliable sensing architecture compared to passive interferometry. The spectroscopic capabilities can still be retained by employing a broadband light emitter such as Cr^{3+}:MgO.

Biochemical Sensing Based on Active Plasmonic Interferometry

Figure 3a shows a schematic of a sensing experiment performed under top illumination. Two representative liquids with slightly different refractive index ($\Delta n \approx 0.05$) were delivered onto the metal surface of the active plasmonic interferometers using a microfluidic channel, resulting in a change of the relative SPP optical path length equal to $2n_{SPP}R_G$, see equation (2) in Methods. The pitch of the 3-groove grating in the H-3G plasmonic interferometer was designed to be 230 nm to keep the SPPs reflected by different grooves in phase while flowing the liquid dielectric materials (with $n \approx 1.35$). An additional 10 nm-thick Al_2O_3 film was deposited by atomic layer deposition (ALD) to protect the surface of the emitting layer from potential chemical contamination. The inset to Fig. 3b shows the normalized spectra through an isolated single hole in the presence of water (blue line, $n \approx 1.33$) and isopropanol (red line, $n \approx 1.38$). No appreciable difference is observed, suggesting that the change of local density of optical states has negligible effects on light emission. In striking contrast, SPP-mediated interference in a H-3G interferometer can determine a dramatic spectral change in the normalized fluorescence spectra and corresponding intensity ratios, as evidenced in Fig. 3b,c, respectively. A figure of merit for this sensing scheme can be defined as follows:

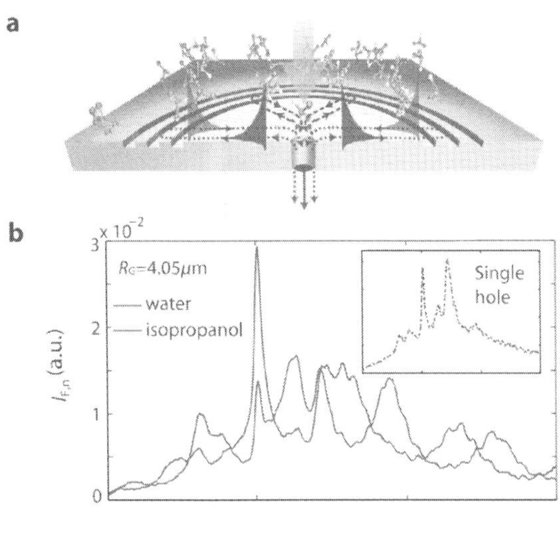

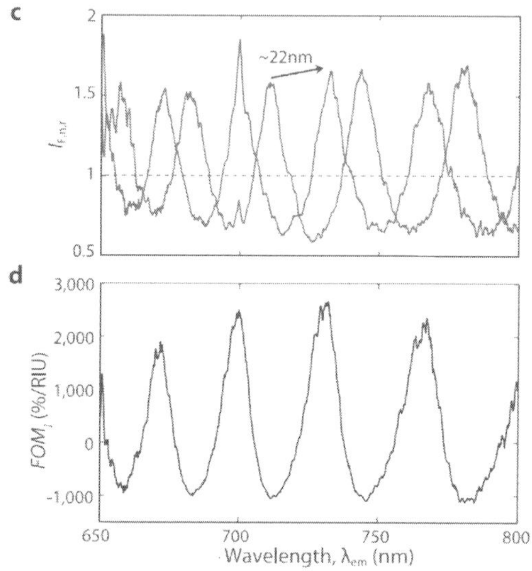

Figure 3. Biochemical sensing experiment in a microfluidic channel: top excitation and bottom detection of fluorescence modulation by plasmonic interferometry. (**a**) Schematic of a circular H-3G plasmonic interferometer with embedded Cr^{3+}:MgO emitters under top illumination. Transmitted fluorescence is detected from the bottom side. Molecules of isopropanol are also shown (not to scale). (**b**) Normalized fluorescence spectra transmitted through the nanohole of a circular H-3G plasmonic interferometer with arm length $R_G = 4.05\,\mu m$ when deionized water (blue) or isopropanol (red) are flown in the microfluidic channel (not shown in **a**). Inset reports the respective reference spectra transmitted through an isolated hole in water (dashed blue) or isopropanol (dashed red). (**c**) Corresponding normalized fluorescence intensity ratios. A clear wavelength shift is observed as a result of the refractive index change. (**d**) Representative figure of merit FOM_I for this specific plasmonic interferometer, calculated according to data in (**c**) and using equation (1).

$$FOM_I = \frac{I_{F,n,r,\text{isop.}} - I_{F,n,r,\text{water}}}{I_{F,n,r,\text{water}}} \times \frac{100\%}{\Delta n}, \tag{1}$$

describing the relative intensity change per refractive index unit (RIU). Figure 3d shows the estimated FOM_I, with values up to 2,600% RIU^{-1}. A ~22 nm red shift is also observed when going from water to isopropanol, with a ~4% refractive index change sufficient to turn constructive interference into destructive and vice versa (see Fig. 3c). The spectral figure of merit of this sensor can therefore be calculated as $FOM_{\lambda_{em}} = \Delta\lambda_{em}/\Delta n = 440\text{ nm RIU}^{-1}$.

Open-Surface Biochemical Sensing on Micro-Droplets

To illustrate the advantage of the same-side illumination and fluorescence detection scheme as previously discussed in Fig. 2e,f, we performed sensing experiments by delivering a single micro-droplet of liquid with different concentrations of isopropanol dissolved in water directly on top of an individual H-3G interferometer (with $R_G = 5\,\mu m$ and 230 nm grating pitch), as schematically shown in Fig. 4a,b. The embedded emitters were illuminated from the bottom side and fluorescence emission was detected from the same side. Normalized intensity ratios are displayed in Fig. 4c for individual water/isopropanol micro-droplets with varying isopropanol concentrations, from 0 to 50%, in steps of 10% corresponding to a refractive index change of ~0.4%. Clear wavelength shifts are observed, with a red shift of $\Delta\lambda_{em} \approx 11$ nm determined by a refractive index change $\Delta n \approx 0.025$, in agreement with the results reported in Fig. 3. The estimated sensitivity to refractive index change of the proposed sensing scheme (limited by the resolution of our spectrometer) is $<5 \times 10^{-4}$, using <1 pL sample volumes.

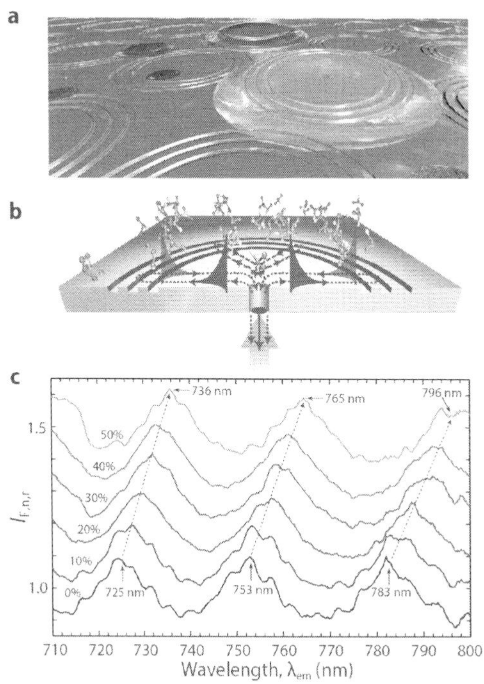

Figure 4. Biochemical sensing experiment on a micro-droplet: same-side (i.e. bottom) excitation and detection of fluorescence modulation by plasmonic interferometry. (**a**) Schematic of a micro-droplet on an array of H-3G plasmonic

interferometers with embedded light emitters. Selective detection of the droplet is accomplished by bottom illumination of a specific plasmonic interferometer and detection of the modulated fluorescence (red) transmitted through the nanohole. (**b**) Schematic showing bottom illumination and fluorescence detection for an active H-3G plasmonic interferometer. (**c**) Normalized fluorescence intensity ratios *vs.* λ_{em} measured through the nanohole of a H-3G plasmonic interferometer with $R_G = 5\,\mu m$, sequentially covered by single micro-droplets with different concentrations of isopropanol: 0% (deionized water), 10%, 20%, 30%, 40% and 50%. Red shifts of the constructive interference peaks are clearly observed (the dashed lines are guides to the eye). For clarity, the curves corresponding to 10–50% isopropanol have been vertically shifted by a constant offset value, in multiples of 0.1.

DISCUSSION

The performance of the proposed sensing scheme may be limited by the low fluorescence intensity compared with the external pump light and the requirement for spectral analysis as described above. However, such shortcomings are not intrinsic to the proposed platform. For example, by using more efficient light emitters, the fluorescence signal can be greatly enhanced. The light intensity throughput can also be enhanced by integrating a large number of identical interferometers as well as using different configurations (e.g. circular slit-groove plasmonic interferometers that are characterized by a polarization-independent response and higher throughput[27] compared with the bullseye structure employed in this work). Moreover, by employing shorter wavelength excitation (e.g. ultraviolet light sources), together with bandpass filters or emitters with narrow fluorescence wavelength range, the spectral analysis required here due to the small pump effect can be entirely eliminated.

In summary, we have realized optical interference at the nanoscale with incoherent external light sources. We have accomplished this by implementing solid-state light emitters directly into subwavelength cavities of an array of plasmonic interferometers. The spatial confinement of the emitters results in the generation of coherent SPPs regardless of the coherence state of the external excitation source. The proposed platform enables a novel sensor design that allows measurements in a broader set of environments than previously thought possible. We have demonstrated the efficacy and potential of this alternative sensing scheme in four ways. One is that the fluorescence spectra transmitted through the nanoaperture of each interferometer can be significantly modulated (up to a factor of two) by SPP-mediated interference and used to detect small refractive index changes. Second, in contrast to conventional plasmonic interferometry, the proposed sensing architecture eliminates the need for a coherent, broadband and precisely aligned external light source, without sacrificing the figures of merit (2,600% RIU^{-1} and 440 nm RIU^{-1}). Third, such a sensing scheme enables same-side excitation and detection of fluorescence modulation which allows for direct exposure of the top surface to the analyte of interest. This eliminates the need for a microfluidic channel to control sample flow and addresses the problem of misalignment induced by lensing effects arising from the droplet surface curvature. And fourth, the proposed sensing technology can be further extended to include various light emitters other than Cr^{3+}:MgO, such as semiconductor quantum dots[28,29,30], as well as electrically excited emitters31·32·33 to realize fully integrated, electrically-driven biochemical sensors.

METHODS

Fabrication

A 300 nm-thick silver film was deposited on a quartz substrate previously coated with a 4 nm-thick titanium adhesion layer using electron-beam evaporation. Then, circular hole-3grooves (H-3G) plasmonic interferometers were fabricated on the silver film using focused ion beam (FIB) milling, together with isolated holes as references. Following the FIB milling, a ~40 nm-thick Cr^{3+}:MgO (1 wt%) emitter layer was deposited on the chip surface using electron-beam evaporation. Each H-3G interferometer consists of a ~300 nm-diameter hole and three ~200 nm-wide, ~40 nm-deep concentric grooves. The radius of the inner groove was varied from 0.25 μm to 6 μm in steps of 25 nm (231 distinct devices in total) and the pitch for the grooves was designed to be 320 nm to keep the reflected SPP in phase for propagation at the silver/air interface. All the isolated holes are identical to the corresponding holes in H-3G interferometers. For the biochemical sensing experiments, another chip containing active plasmonic interferometers was fabricated with an additional thin (10 nm) Al_2O_3 layer deposited on top of Cr^{3+}:MgO using atomic layer deposition (ALD) at low temperature (75 °C) to protect the sensor surface and prevent direct contact of the emitting layer with the liquid samples, thus eliminating chemical contamination. Note that in this case (i.e. silver/40 nm Cr^{3+}:MgO/10 nm Al_2O_3/liquid) the groove pitch was designed to be 230 nm to keep the reflected SPP in phase while propagating at the silver/liquid interface ($n \approx 1.35$).

Experimental Setup

Although in this Article we demonstrate that the sensing scheme is unaffected by the specific light source and illumination conditions, to perform the experiments we used a conventional Xenon arc lamp and a supercontinuum laser, coupled to an optical microscope and spectrograph to detect fluorescence spectra. In particular, the data shown in Figs. 1, 2d and 3 were acquired by illuminating the plasmonic interferometers with a quasi-monochromatic (590 ± 10 nm) light beam from a supercontinuum laser (NKT Photonics, SuperK Extreme) using a wavelength selector (NKT Photonics, SuperK Varia) and then focused on the fabricated plasmonic interferometers by the microscope condenser lens to excite the emitters (top illumination). An inverted microscope (Nikon, Eclipse Ti) with a 40× and 0.6 NA objective was used to collect the fluorescence spectrum transmitted through each nanohole (either individual, or as part of the plasmonic interferometer), which was then projected onto the entrance slit of a spectrograph (Princeton Instruments, Acton SpectraPro SP-2300) and dispersed on a CCD camera (Princeton Instruments, Pixis 100) by a diffraction grating optimized for the specific detection wavelength range. An automated stage was used to scan and acquire the corresponding spectra from all the H-3G/hole nanostructures. Note that active fluorescence under bottom illumination (shown in Figs. 2e,f and 4) was acquired through the same 40×, 0.6 NA objective, used for both excitation and collection of the fluorescence emission. The broadband thermal light source used in Fig. 2a,b was a Xenon arc lamp combined with a 600 nm short pass filter. The transverse spatial coherence was tuned by employing Köhler illumination[26]. Specifically, by increasing the diameter of the microscope condenser aperture diaphragm, the transverse wavevector of the incident light can be made to span a broader range, which in turn determines a decrease in the spatial coherence at the hole location[26]. A microfluidic channel made by polydimethylsiloxane (PDMS) was used to deliver and control the flow of water and isopropanol onto the plasmonic interferometer to obtain the results shown in Fig. 3. For the sensing experiments reported in Fig. 4, liquid droplets containing various concentrations of water and isopropanol were generated using a micro-pipette dispenser.

Data Analysis

The fundamental optical properties of plasmonic interferometers can be captured by the ratio ($I_{F,r}$) between the measured fluorescence transmission intensity through the nanoaperture of a plasmonic interferometer (I_F) and the reference fluorescence intensity transmitted through an isolated identical nanoaperture ($I_{F,0}$):

$$I_{F,r}(\lambda_{em}, R_G) = \frac{I_F(\lambda_{em}, R_G)}{I_{F,0}(\lambda_{em})} = \frac{|E_0 + E_{SPP}|^2}{|E_0|^2} = \left|1 + \beta e^{i(2k_0 n_{SPP} R_G + \phi_G)}\right|^2, \tag{2}$$

where the subscript F indicates active fluorescence transmission, E_0 and E_{SPP} are the complex electric field amplitudes of the direct emission and SPP-induced fluorescence transmitted through the nanoaperture, respectively. $k_0=2\pi/\lambda_{em}$ is the amplitude of the free-space wave vector, λ_{em} is the free-space emission wavelength, R_G is the interferometer arm length, defined as the inner circular groove radius (see Fig. 1a), β is a wavelength-dependent effective SPP scattering coefficient defined with respect to the directly downward emission, which also includes the in-plane reflection coefficient of the groove, $n_{SPP}=\lambda_{em}/\lambda_{SPP}$ is the complex, wavelength-dependent SPP refractive index in the three-layer system, ϕ_G is the wavelength-dependent phase change induced by the scattering processes and the constant 2 in the argument accounts for the SPP round trip.

In order to eliminate any additional (although weak) excitation effects due to the direct modulation of the pump intensity caused by the plasmonic interferometer, the fluorescence spectra were first normalized to their integrated intensity. The ratio of the normalized spectra can then be calculated as:

$$I_{F,n,r}(\lambda_{em}, R_G) = \frac{I_{F,n}(\lambda_{em}, R_G)}{I_{F,0,n}(\lambda_{em})} = \frac{I_F(\lambda_{em}, R_G)/\int I_F(\lambda_{em}, R_G)d\lambda_{em}}{I_{F,0}(\lambda_{em})/\int I_{F,0}(\lambda_{em})d\lambda_{em}}, \tag{3}$$

where $I_{F,n}$ and $I_{F,0,n}$ are the normalized fluorescence spectra transmitted through the nanoaperture of an interferometer and through a reference isolated aperture, respectively. Combining equations (2) and (3), $I_{F,n,r}$ can also be expressed as:

$$I_{F,n,r}(\lambda_{em}, R_G) = I_{F,r}(\lambda_{em}, R_G)\frac{\int I_{F,0}(\lambda_{em})d\lambda_{em}}{\int I_{F,r}(\lambda_{em}, R_G)I_{F,0}(\lambda_{em})d\lambda_{em}}. \tag{4}$$

Equation (4) was used to model the normalized fluorescence intensity ratio color map. Note that equation (2) is valid only when the SPP generation process represents a coherent alternative path for the excited emitters in the nanoaperture. The agreement between simulations and experiments suggests that the degree of coherence of the emitters in the nanoaperture is sufficiently high to justify the use of equation (2). This is further supported by the fact that no appreciable attenuation is observed in the fluorescence intensity amplitude modulation within the range of interferometer arm lengths used in this experiment.

ACKNOWLEDGEMENTS

The authors thank R. Zia, E. Marcora, D. Morrill, P. Liu and A. Nurmikko for fruitful discussions and suggestions. This material is based upon work supported by the National Science Foundation under Grant No. CBET-1159255.

AUTHOR CONTRIBUTIONS

D.P. and D.L. planned and designed the experiment. All authors performed the experimental measurements, discussed the results, and helped with the data analysis and manuscript editing.

REFERENCES

1. Barnes, W. L., Dereux, A. & Ebbesen, T. W. Surface plasmon subwavelength optics. Nature 424, 824–830 (2003).
2. Zia, R. & Brongersma, M. L. Surface plasmon polariton analogue to Young's double-slit experiment. Nature Nanotech. 2, 426–429 (2007).
3. Anker, J. N. et al. Biosensing with plasmonic nanosensors. Nature Mater. 7, 442–453 (2008).
4. Pacifici, D., Lezec, H. J., Sweatlock, L. A., Walters, R. J. & Atwater, H. A. Universal optical transmission features in periodic and quasiperiodic hole arrays. Opt. Express 16, 9222–9238 (2008).
5. Schuller, J. A. et al. Plasmonics for extreme light concentration and manipulation. Nature Mater. 9, 193–204 (2010).
6. Atwater, H. A. & Polman, A. Plasmonics for improved photovoltaic devices. Nature Mater. 9, 205–213 (2010).
7. Ebbesen, T. W., Lezec, H., Ghaemi, H., Thio, T. & Wolff, P. Extraordinary optical transmission through sub-wavelength hole arrays. Nature 391, 667–669 (1998).
8. Pacifici, D., Lezec, H. J., Atwater, H. A. & Weiner, J. Quantitative determination of optical transmission through subwavelength slit arrays in Ag films: Role of surface wave interference and local coupling between adjacent slits. Phys. Rev. B 77, 115411 (2008).
9. Pacifici, D., Lezec, H. J. & Atwater, H. A. All-optical modulation by plasmonic excitation of CdSe quantum dots. Nature Photon. 1, 402–406 (2007).
10. Lezec, H. J. et al. Beaming light from a subwavelength aperture. Science 297, 820–822 (2002).
11. Yu, N. et al. Small-divergence semiconductor lasers by plasmonic collimation. Nature Photon. 2, 564–570 (2008).
12. Jun, Y. C., Huang, K. C. & Brongersma, M. L. Plasmonic beaming and active control over fluorescent emission. Nature Commun. 2, 283 (2011).
13. Aouani, H. et al. Bright unidirectional fluorescence emission of molecules in a nanoaperture with plasmonic corrugations. Nano Lett. 11, 637–644 (2011).

14. Harats, M. G. et al. Full spectral and angular characterization of highly directional emission from nanocrystal quantum dots positioned on circular plasmonic lenses. Nano Lett. 14, 5766–5771 (2014).
15. Jain, P. K., Huang, X., El-Sayed, I. H. & El-Sayed, M. A. Noble metals on the nanoscale: optical and photothermal properties and some applications in imaging, sensing, biology and medicine. Accounts of Chemical Research 41, 1578–1586 (2008).
16. Feng, J. et al. Nanoscale plasmonic interferometers for multispectral, high-throughput biochemical sensing. Nano Lett. 12, 602–609 (2012).
17. Gao, Y. et al. Plasmonic interferometric sensor arrays for high-performance label-free biomolecular detection. Lab on a Chip 13, 4755–4764 (2013).
18. Siu, V. S., Feng, J., Flanigan, P. W., Palmore, G. T. R. & Pacifici, D. A plasmonic cuvette: dye chemistry coupled to plasmonic interferometry for glucose sensing. Nanophotonics 3, 125–140 (2014).
19. Cetin, A. E. et al. Handheld high-throughput plasmonic biosensor using computational on-chip imaging. Light: Science & Applications 3, e122 (2014).
20. Kuck, S. et al. Investigation of Cr-doped MgO and Sc2O3 as potential laser sources for the near infrared spectral range. Laser Phys. 10, 411–416 (2000).
21. Karaveli, S., Wang, S., Xiao, G. & Zia, R. Time-resolved energy-momentum spectroscopy of electric and magnetic dipole transitions in Cr^{3+}: MgO. ACS Nano 7, 7165–7172 (2013).
22. Ford, G. W. & Weber, W. H. Electromagnetic interactions of molecules with metal surfaces. Phys. Reports 113, 195–287 (1984).
23. Lakowicz, J. R. Radiative decay engineering 5: metal-enhanced fluorescence and plasmon emission. Analytical Biochemistry 337, 171–194 (2005).
24. Ditlbacher, H., Krenn, J., Schider, G., Leitner, A. & Aussenegg, F. Two-dimensional optics with surface plasmon polaritons. Appl. Phys. Lett. 81, 1762–1764 (2002).
25. Kuttge, M., Garca de Abajo, F. J. & Polman, A. How grooves reflect and confine surface plasmon polaritons. Opt. Express 17, 10385–10392 (2009).

26. Born, M. & Wolf, E. Principles of optics: Electromagnetic theory of propagation, interference and diffraction of light (Cambridge university press, 1999).
27. Feng, J., Li, D. & Pacifici, D. Circular slit-groove plasmonic interferometers: a generalized approach to high-throughput biochemical sensing. Opt. Mat. Express 5, 2742–2753 (2015).
28. Espiau de Lamaëstre, R., Bernas, H., Pacifici, D., Franzo, G. & Priolo, F. Evidence for a "dark exciton" state of PbS nanocrystals in a silicate glass. Appl. Phys. Lett. 88, 1115 (2006).
29. Dang, C. et al. Red, green and blue lasing enabled by single-exciton gain in colloidal quantum dot films. Nature Nanotech. 7, 335–339 (2012).
30. Roh, K. et al. Surface-emitting red, green and blue colloidal quantum dot distributed feedback lasers. Opt. Express 22, 18800–18806 (2014).
31. Dabbousi, B., Bawendi, M., Onitsuka, O. & Rubner, M. Electroluminescence from CdSe quantum-dot/polymer composites. Appl. Phys. Lett. 66, 1316–1318 (1995).
32. Irrera, A. et al. Excitation and de-excitation properties of silicon quantum dots under electrical pumping. Appl. Phys. Lett. 81, 1866–1868 (2002).
33. Krames, M. R. et al. Status and future of high-power light-emitting diodes for solid-state lighting. J. Disp. Technol. 3, 160–175 (2007).

CHAPTER 6

QUANTUM ENHANCEMENT OF ACCURACY AND PRECISION IN OPTICAL INTERFEROMETRY

Florian Kaiser[1,2,3], Panagiotis Vergyris[1], Djeylan Aktas[1], Charles Babin[1,4], Laurent Labonté[1], and Sébastien Tanzilli[1]

[1]Université Côte d'Azur, Institut de Physique de Nice (INPHYNI), Nice 06108, France
[2]Now at 3. Physikalisches Institut, Universität Stuttgart, Stuttgart 70569, Germany
[3]Now at Center for Integrated Quantum Science and Technology (IQST), Stuttgart 70569, Germany
[4]École Normale Supérieure de Lyon, Lyon 69364, France

ABSTRACT

White-light interferometry is one of today's most precise tools for determining the properties of optical materials. Its achievable precision and accuracy are typically limited by systematic errors due to a high number of interdependent data-fitting parameters. Here, we introduce spectrally resolved quantum white-light interferometry as a novel tool for optical

Citation: (APA): Kaiser, F., Vergyris, P., Aktas, D., Babin, C., Labonté, L., & Tanzilli, S. (2018). Quantum enhancement of accuracy and precision in optical interferometry. Light: Science & Applications, 7(3), 17163-17163. (5 pages)

property measurements, notably, chromatic dispersion in optical fibres. By exploiting both spectral and photon-number correlations of energy-time entangled photon pairs, the number of fitting parameters is significantly reduced, which eliminates systematic errors and leads to an absolute determination of the material parameter. By comparing the quantum method to state-of-the-art approaches, we demonstrate the quantum advantage of 2.4 times better measurement precision, despite requiring 62 times fewer photons. The improved results are due to conceptual advantages enabled by quantum optics, which are likely to define new standards in experimental methods for characterising optical materials.

Keywords: chromatic dispersion, interferometry, quantum optics, quantum metrology

INTRODUCTION

Quantum technologies have received substantial attention as a means to improve the resolution and precision of metrological tasks by reducing statistical errors due to quantum noise[1, 2, 3, 4, 5, 6, 7, 8]. Far less attention has been given to their ability to reduce systematic errors. However, statistical and systematic errors are of equal importance in any measurement, and the latter are typically more difficult to characterise. Notable examples of quantum-improved measurements are the combination of multiple fundamental electronic quantum effects for a more accurate definition of the ampere[9] and quantum-correlated 'twin photon beams' in establishing absolute and universal optical power standards[10]. In this letter we demonstrate a new use of quantum optics to reduce systematic errors in the technologically prominent application of spectrally resolved white-light interferometry (WLI). WLI is used for precise measurements of chromatic dispersion, that is, the second derivative of the wavelength-dependent optical phase. Classical WLI, however, requires precise interferometer equalization[11, 12] and is influenced by third-order dispersion[13, 14]. This leads to systematic errors that are difficult to account for.

We eliminate these drawbacks by inferring chromatic dispersion using energy–time entangled photon pairs and coincidence counting to measure spectral correlation functions. In addition, we exploit photon–number correlations to achieve a twofold resolution enhancement. Our results demonstrate that this new strategy outperforms the precision and accuracy of previous quantum[15, 16] and state-of-the-art techniques[11, 12]. Moreover, because our approach is essentially alignment-free, it enables the use of the same interferometer in a user-friendly manner for analysing a wide variety of different optical materials in terms of type, optical properties, length, etc.

Standard WLI

The standard scheme for WLI is shown in Figure 1a. The emission of a white-light source is directed to an interferometer in which the reference arm is free-space (with well-known optical properties) and the other arm comprises the sample under test (SUT). Recombining both arms at the output beam splitter leads to an interference pattern for which the intensity follows $I \propto 1+\cos(\varphi(\lambda))$, with $\phi(\lambda) = \frac{2\pi}{\lambda}(n(\lambda)L_s - L_r)$. Here, λ represents the wavelength, L_r and L_s are the physical lengths of the reference arm and the SUT, respectively, and $n(\lambda)$ is the effective refractive index of the SUT. It is worth noting that interference is observed only when the interferometer is precisely balanced to within the larger of the coherence length of the white-light source and the coherence length imposed by the resolution of the spectrometer, which is typically on the order of microns to millimetres[11, 12]. In this case, the phase term reads:

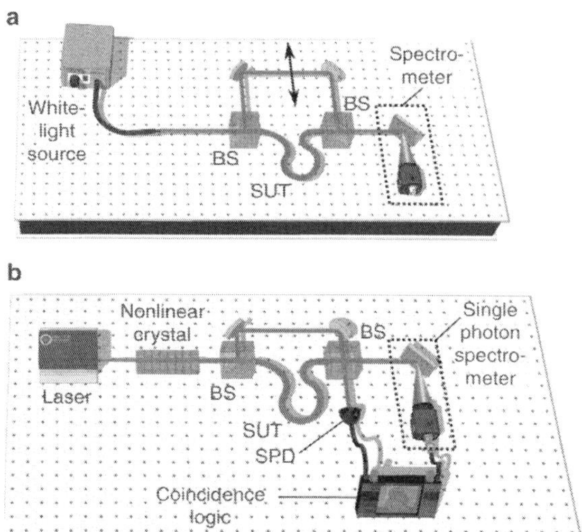

Figure 1. Typical experimental set-ups. (**a**) Standard spectrally resolved WLI. (**b**) Quantum WLI. BS, beam splitter.

$$\phi(\lambda) \approx 2\pi L_s \left(\frac{1 \mathrm{d}^2 n}{2 \mathrm{d}\lambda^2} \bigg|_{\lambda_0} \cdot \frac{(\Delta\lambda)^2}{\lambda_0 + \Delta\lambda} + \frac{1 \mathrm{d}^3 n}{6 \mathrm{d}\lambda^3} \bigg|_{\lambda_0} \cdot \frac{(\Delta\lambda)^3}{\lambda_0 + \Delta\lambda} \right) + \phi_{\mathrm{off}} \quad (1)$$

Here, λ_0 represents the stationary phase point, that is, the wavelength at which the absolute phase difference between the two interferometer arms is exactly zero. In standard WLI, λ_0 is extracted experimentally by identifying the symmetry point of the observed interferogram[11, 12]. Additionally, $\Delta\lambda=\lambda-\lambda_0$, and φ_{off} is a constant offset phase. Provided that L_s is precisely known, the optical material parameters $\frac{d^2n}{d\lambda^2}\big|_{\lambda_0}$ and $\frac{\mathrm{d}^3n}{\mathrm{d}\lambda^3}\big|_{\lambda_0}$ can be extracted from a fit to the data as a function of $\Delta\lambda$. It is noteworthy that the three free parameters, that is, λ_0, $\frac{\mathrm{d}^2n}{\mathrm{d}\lambda^2}\big|_{\lambda_0}$ and $\frac{\mathrm{d}^3n}{\mathrm{d}\lambda^3}\big|_{\lambda_0}$, are usually all interdependent in a non-trivial manner such that uncertainties in one parameter systematically induce uncertainties in the others. In fact, the high number of fitting parameters required and the necessity to re-equilibrate the interferometer for every new SUT are the main limiting factors of this technique[13, 14].

However, more accurate optical measurements are eagerly demanded in almost all fields involving optics. A special focus is made on the optical parameter $\frac{\mathrm{d}^2n}{\mathrm{d}\lambda^2}\big|_{\lambda_0}$, as it is directly related to the chromatic dispersion coefficient $D = -\frac{\lambda_0}{c} \cdot \frac{\mathrm{d}^2n}{\mathrm{d}\lambda^2}\big|_{\lambda_0}$, where c is the speed of light[13, 17, 18, 19, 20, 21, 22, 23]. Chromatic dispersion causes optical pulse broadening, and more accurate measurements of D would have significant repercussions for optimising today's telecommunication networks, developing new-generation pulsed lasers and amplifiers, and designing novel linear and nonlinear optical components and circuits, as well as for assessing the properties of biological tissues.

MATERIALS AND METHODS

Quantum WLI

Figure 1b depicts the new experimental schematic dedicated to spectrally resolved quantum WLI (Q-WLI) intended to overcome the above issues. The quantum white-light source is composed of a continuous-wave pump laser and a nonlinear crystal in which energy–time entangled photon pairs are generated through spontaneous parametric downconversion[24, 25]. This process obeys the conservation of the energy, that is, $\frac{1}{\lambda_p} = \frac{1}{\lambda_1} + \frac{1}{\lambda_2}$. Here, $\lambda_{p,1,2}$ respectively represent the wavelengths in vacuum of the pump laser photons and the individual photons for each generated pair. Another implication of the conservation of the energy is that the degenerate vacuum wavelength of the emission spectrum is $\lambda^*=2\lambda_p$. We send the paired photons to the interferometer; however, as opposed to standard WLI, we now intentionally unbalance it. This provides us with two advantages: first, we avoid single-photon interference, and second, we obtain a means to distinguish events in which the two photons take opposite paths (strongly delayed arrival times at the interferometer's outputs) or the same path (near-zero arrival time difference)[24]. We postselect the latter events by considering only two-photon coincidence detection events in which both the single-photon detector (SPD) and the single-photon-sensitive spectrometer fire simultaneously. Our goal is now to observe quantum interference between these two-photon contributions, which necessitates that they be coherent and indistinguishable. Coherence is ensured by operating the interferometer at a path-length difference that is shorter than the coherence length of the pump laser (~100 m) such that the photon pair contributions are in phase[26]. Indistinguishability concerns mainly the temporal envelope of the photon pair wave packet, which is distorted from its original shape by the dispersion-induced temporal walk-off between the individual photons in the SUT. For standard fibres, this means that path-length differences up to ~10 m are acceptable[27].

Thus, provided that the interferometer is operated in these conditions, near-zero arrival time coincidence detection results in a two-photon $N00N$ state:

$$|\psi\rangle = \frac{|2\rangle_r|0\rangle_s + e^{i\phi_{N00N}}|0\rangle_r|2\rangle_s}{\sqrt{2}} \tag{2}$$

Here, the ket vectors, indexed by s and r, indicate the number of photons in the reference and SUT arms, respectively, and $\varphi_{N00N}=\varphi(\lambda_1)+\varphi(\lambda_2)$. We obtain the spectral dependence of φ_{N00N} by computing $\varphi(\lambda_1)$ and $\varphi(\lambda_2)$ according to Equation (1) and respecting the conservation of the energy during the down conversion process:

$$\phi_{N00N} \approx \left.\frac{d^2 n}{d\lambda^2}\right|_{\lambda^*} \cdot \frac{\pi L_S \cdot (\Delta\lambda)^2}{\frac{\lambda^*}{2} + \Delta\lambda} + \phi_{\text{off}} \tag{3}$$

Here, $\phi_{\text{off}} = \frac{4\pi(n(\lambda^*)L_S - L_r)}{\lambda^*}$ is an offset term, and $\Delta\lambda=\lambda-\lambda^*$. The phase-dependent two-photon coincidence rate R is then $R \propto 1+\cos(\varphi_{N00N})$. In the past, numerous studies have investigated the term φ_{off}, as it allows measuring optical phase shifts at constant wavelengths with double resolution compared to the standard approach[28, 29, 30].

We access here, for the first time, the wavelength-dependent term in Equation (3) by recording R as a function of $\Delta\lambda$; that is, the two-photon coincidence rate is measured as a function of the paired-photons' wavelengths. This leads to several pertinent purely quantum-enabled features. Due to the use of an energy–time entangled two-photon $N00N$ state, the required precision of equilibrating the interferometer is ~10 m instead of microns to millimetres in standard WLI[11, 12, 13, 14]. This is particularly interesting for improving the ease of use, as no realignment is necessary when changing the SUT; compared to Equation (1), the third-order term $\frac{d^3 n}{d\lambda^3}\Big|_{\lambda^*}$ in Equation (3) is cancelled owing to energy–time correlations[16]. Furthermore, the wavelength at which chromatic dispersion is measured, λ^*, need not be extracted from the data, as it is exactly twice the wavelength of the continuous-wave pump laser, λ_p, and can therefore be known with extremely high accuracy. This means that the quantum strategy allows data fitting using exactly one free parameter, namely, $\frac{d^2 n}{d\lambda^2}\Big|_{\lambda^*}$, which is an essential step towards absolute optical-property determination with high precision without systematic errors. Finally, due to the use of a two-photon $N00N$ state, double resolution of $\frac{d^2 n}{d\lambda^2}\Big|_{\lambda^*}$ is achieved, enabling measurements on shorter samples and components compared to standard WLI, that is, down to the technologically interesting mm to cm scale.

Detailed Optical Setup and Data Acquisition

To benchmark standard and quantum approaches, we used a 1-m long SMF28e fibre from Corning as the SUT. We used the same interferometer for all measurements and actively stabilised it using a reference laser and a piezoelectric transducer on one mirror in the reference arm (additional details are provided in the methods section). This ensured that φ_{off} remained constant.

For chromatic dispersion measurements using classical WLI, we used a state-of-the-art superluminescent diode. At the output of the interferometer we measured an average spectral intensity of ~125 pW nm^{-1} from 1450 to 1650 nm. Interferograms were recorded using a spectrometer from Anritsu (model MS9710B, Atsugi-shi, Japan) with 0.1 s integration time and 0.5 nm resolution, which are standard parameters for this kind of measurement[11, 12].

For the Q-WLI approach, the light source was a 780.246 nm laser pumping a type-0 periodically poled lithium niobate waveguide. We stabilised the laser wavelength against the $F = 2 \rightarrow F' = 2 \times 3$ hyperfine crossover transition in atomic ^{87}Rb such that λ_p and λ^* were known with a precision of the order of 1 fm. The quasi-phase matching in the periodically poled lithium niobate waveguide was engineered to generate energy–time entangled photon pairs around the degenerate wavelength of λ^*=1560.493 nm with a bandwidth of ~140 nm[25]. To detect the paired photons, we used an InGaAs SPD (IDQ 220) at one interferometer output. The single-photon spectrometer at the other output was made of a wavelength-tunable 0.5 nm bandpass filter followed by another InGaAs SPD (IDQ 230). To avoid saturation of these detectors, the spectral intensity at the interferometer output was reduced to ~25 fW nm^{-1}, which was partially compensated by increasing the integration time to 8 s.

All measurements were repeated 100 times on the same SUT to infer the statistical accuracy of both WLI and Q-WLI approaches.

RESULTS AND DISCUSSION

Statistical Analysis for Comparing Measurement Precision

Typical interference patterns for chromatic dispersion measurements using both methods are shown in Figure 2a and 2b. With the Q-WLI setup, we found twice as many interference fringes for the same spectral bandwidth, which is a direct consequence of the doubled phase sensitivity of the two-photon *N*00*N* state. After acquiring 2 × 100 measurements on the same SUT, we inferred the precisions of both approaches. The results of the statistical data analysis are shown in Figure 3. For standard WLI, we obtained, on average, $D = 17.047\ \text{ps}\ (\text{nm}\,\text{km})^{-1}$ at λ_0≈1560.5 nm with a standard deviation of $\sigma_{\text{classical}} = 0.051\ \text{ps}\ (\text{nm}\,\text{km})^{-1}$. This result is among the most precise reported to date in the literature[13, 17, 18, 19, 20, 21, 22]. For Q-WLI, we measured, on average, $D = 17.035\ \text{ps}\ (\text{nm}\,\text{km})^{-1}$ at λ^*=1560.493 nm with a significantly smaller standard deviation of $\sigma_{N00N} = 0.021\ \text{ps}\ (\text{nm}\,\text{km})^{-1}$.

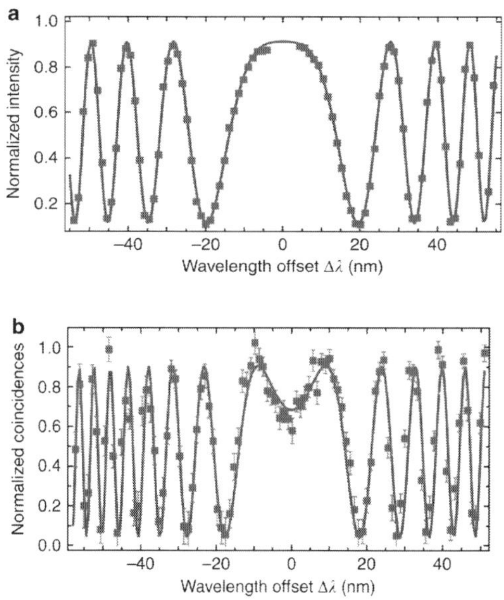

Figure 2. Typical measurements acquired for inferring chromatic dispersion in a 1-m-long standard single-mode fibre. (**a**) Results obtained with standard WLI, and (**b**) using Q-WLI. Red dots are data points; blue curves are appropriate fits to the data following Equations (1) and (3), from which D is extracted. Error bars assume Poissonian photon number statistics. For standard WLI, normalization was obtained by measuring two reference spectra. For Q-WLI, normalization was performed on the fly by counting non-zero arrival time difference coincidences.

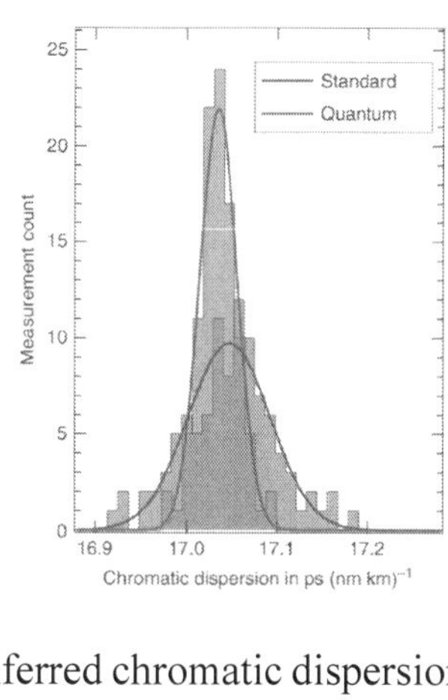

Figure 3. Histogram of inferred chromatic dispersion coefficients after 100 repetitions with the same SUT for both standard (blue) and quantum-enhanced (red) measurements. Fits to the data assumed a normal distribution.

In our two sets of data, we observed a difference of $0.012\,\mathrm{ps}\,(\mathrm{nm\,km})^{-1}$ between the central values, which is larger than the deviation expected from statistical uncertainties $(0.006\,\mathrm{ps}\,(\mathrm{nm\,km})^{-1})$. Polarization mode dispersion can be excluded as it would introduce at most an offset of $0.003\,\mathrm{ps}\,(\mathrm{nm\,km})^{-1}$. Consequently, the difference in central values must originate from systematic errors. In this sense, we compute that, for standard WLI, the difference can be explained by either a slight wavelength offset of the spectrometer (<0.2 nm) or by a slightly unbalanced interferometer (~1.5 μm). Both types of errors induce an error of the fitting parameter λ_0 that translates to an error in $\frac{d^2n}{d\lambda^2}|_{\lambda_0}$ (Refs. 11, 12). At this point, we emphasise again that in our Q-WLI approach, λ^* is known with essentially absolute accuracy, and an unbalanced interferometer does not influence the measurement. Because Q-WLI presents fewer sources of systematic errors, it is therefore natural to conclude that Q-WLI determines chromatic dispersion with absolute accuracy.

We further emphasise that our measurements performed with Q-WLI involve ~62 times fewer photons transmitted through the SUT compared to standard WLI. It is therefore interesting to compare the achievable precision normalised to the number of transmitted photons. For each standard and quantum interferogram, the number of photons reaching the interferometer outputs was $N_{std} \approx 2.0 \times 10^{10}$ and $N_{quant} \approx 3.1 \times 10^{8}$, respectively. Consequently, the standard and quantum methods achieve precisions of $(\Delta D)_{std} = 7146\,\mathrm{ps}\,(\mathrm{nm\,km})^{-1}\,\sqrt{N_{std}}^{-1}$ and $(\Delta D)_{quant} = 372\,\mathrm{ps}\,(\mathrm{nm\,km})^{-1}\,\sqrt{N_{quant}}^{-1}$, respectively. In other words, in addition to being more prone to systematic errors, the standard measurement requires 369 times more photons to achieve the same precision as Q-WLI.

Device Calibration using Q-WLI

Another advantage provided by Q-WLI lies in its straightforward device calibration. All of the optical components in the interferometer actually show small residual chromatic dispersion, and this undesired offset needs to be evaluated and subtracted from the data to avoid systematic errors. In both cases, this implies performing a measurement without any SUT.

Note that in standard WLI, removing the SUT significantly unbalances the interferometer, and to observe interference, the length of the reference arm must be reduced accordingly (typically on the order of 1 m). This procedure is technically challenging, time-consuming, and might lead to additional systematic errors.

At this point, Q-WLI demonstrates its ability for user-friendly operation. Even after removing the SUT, interference is observed without any interferometer realignment. Figure 4 shows the experimental results that we have obtained when measuring chromatic dispersion in our bare interferometer, that is, without the SUT. It turned out that in our interferometer, residual chromatic dispersion amounted to ~10% of the measured values on the 1 m SUT. For all of the data discussed above, except for the raw data in Figure 2a and 2b, we have subtracted the residual chromatic dispersion.

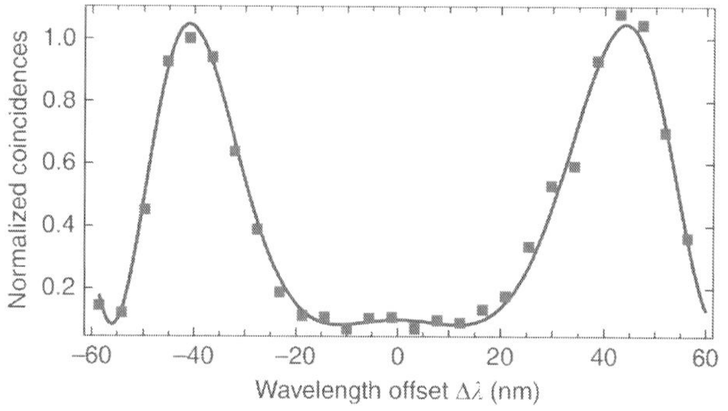

Figure 4. Experimental results when using Q-WLI for inferring residual chromatic dispersion in our interferometer without the SUT. Red dots, data points; the blue curve is an appropriate fit to the data following Equation (3).

CONCLUSIONS

We have introduced and demonstrated the concept of spectrally resolved Q-WLI, exploiting energy–time entangled two-photon *N*00*N* states. Compared to standard measurements, the *N*00*N* state permits achieving a phase sensitivity higher by a factor of two. More strikingly, this use of such quantum states of light reduces the number of free parameters for fitting experimental data from three to one, representing a major advantage for determining optical properties with high precision and absolute accuracy. In addition, our setup does not require a balanced interferometer for performing the measurement, which represents a significant time-saving advantage compared to standard WLI. This is of particular interest for device calibration and when measuring a large set of samples.

As an exemplary demonstration, we have applied our method to infer chromatic dispersion in a standard single-mode fibre, obtaining 2.4 times more precise results compared to state-of-the-art realizations, despite using ~62 times fewer photons.

We note that the sensitivity of our approach could be further doubled by using a double-pass configuration[18]; this could achieve measurements on short samples, such as optical components and waveguide structures (mm to cm length scale). Such measurements would also be of interest for medical applications where precise knowledge of chromatic dispersion in tissues is required to yield optimal image quality in optical coherence tomography[31]. From this perspective, the reduced number of photons required for quantum WLI is also highly interesting for measurements performed on photosensitive biological samples[32, 33, 34]. In optical telecommunication systems, by rotating the polarizations of the entangled photon pairs, our setup could be used for measuring polarization mode dispersion in optical components, which would lead to refinement of manufacturing processes.

In addition, total measurement times could be reduced far below 1 s by using high-speed superconducting detectors with ~3 orders of magnitude higher saturation levels compared to the InGaAs SPDs used here[35]. Alternatively, quantum-inspired strategies may also prove to be suitable[36, 37].

In summary, we believe that combining the fundamental and conceptual advantages enabled by quantum light is a very promising approach for the future development and improvement of applications requiring absolute and high-precision measurements of optical properties.

ACKNOWLEDGEMENTS

We acknowledge financial support from the Foundation Simone & Cino Del Duca, the European Commission for the FP7-ITN PICQUE project (Grant Agreement No. 608062), l'Agence Nationale de la Recherche (ANR) for the CONNEQT, SPOCQ and SITQOM projects (Grants ANR-EMMA-002-01, ANR-14-CE32-0019 and ANR-15-CE24-0005, respectively) and the iXCore Research Foundation. We also thank MT Pérez Zaballos for assistance with statistical data treatment, as well as M Mitchell, T Debuisschert, A Levenson and P Neumann for fruitful discussions.

REFERENCES

1. Giovannetti V, Lloyd S, Maccone L . Quantum-enhanced measurements: beating the standard quantum limit. *Science* 2004; **306**: 1330–1336.
2. Walther P, Pan J-W, Aspelmeyer M, Ursin R, Gasparoni S *et al.* De Broglie wavelength of a non-local four-photon state. *Nature* 2004; **429**: 158–161.
3. Mitchell MW, Lundeen JS, Steinberg AM . Super-resolving phase measurements with a multiphoton entangled state. *Nature* 2004; **429**: 161–164.
4. Nagata T, Okamoto R, O'Brien JL, Sasaki K, Takeuchi S . Beating the standard quantum limit with four-entangled photons. *Science* 2007; **316**: 726–729.
5. Higgins BL, Berry DW, Bartlett SD, Wiseman HM, Pryde GJ . Entanglement-free Heisenberg-limited phase estimation. *Nature* 2007; **450**: 393–396.
6. Matthews JCF, Politi A, Stefanov A, O'Brien JL . Manipulation of multiphoton entanglement in waveguide quantum circuits. *Nat Photonics* 2009; **3**: 346–350.
7. Afek I, Ambar O, Silberberg Y . High-NOON states by mixing quantum and classical light. *Science* 2010; **328**: 879–881.
8. Giovannetti V, Lloyd S, Maccone L . Advances in quantum metrology. *Nat Photonics* 2011; **5**: 222–229.
9. Brun-Picard J, Djordjevic S, Leprat D, Schopfer F, Poirier W . Practical quantum realization of the ampere from the elementary charge. *Phys Rev X* 2016; **6**: 041051.
10. Migdall AL, Datla RU, Sergienko A, Orszak JS, Shih YH . Absolute detector quantum-efficiency measurements using correlated photons. *Metrologia* 1995; **32**: 479–484.
11. Naganuma K, Mogi K, Yamada H . Group-delay measurement using the Fourier transform of an interferometric cross correlation generated by white light. *Opt Lett* 1990; **15**: 393–395.
12. Diddams S, Diels J-C . Dispersion measurements with white-light interferometry. *J Opt Soc Am B* 1996; **13**: 1120–1129.
13. Grósz T, Kovács AP, Kiss M, Szipöcs R . Measurement of higher order chromatic dispersion in a photonic bandgap fiber: comparative study of spectral interferometric methods. *Appl Opt* 2014; **53**: 1929–1937.

14. Galle MA, Saini SS, Mohammed WS, Qian L . Virtual reference interferometry: theory and experiment. *J Opt Soc Am B* 2014; **29**: 3201–3210.

15. Brendel J, Zbinden H, Gisin N . Measurement of chromatic dispersion in optical fibers using pairs of correlated photons. *Opt Commun* 1998; **151**: 35–39.

16. Nasr MB, Saleh BEA, Sergienko AV, Teich MC . Dispersion-cancelled and dispersion-sensitive quantum optical coherence tomography. *Opt Express* 2004; **12**: 1353–1362.

17. Hlubina P, Szpulak M, Ciprian D, Martynkien T, Urbańczyk W . Measurement of the group dispersion of the fundamental mode of holey fiber by white-light spectral interferometry. *Opt Express* 2007; **15**: 11073–11081.

18. Labonté L, Roy P, Pagnoux D, Louradour F, Restoin C *et al.* Experimental and numerical analysis of the chromatic dispersion dependence upon the actual profile of small core microstructured fibres. *J Opt A* 2006; **8**: 933–938.

19. Kardaś TM, Radzewicz C . Broadband near-infrared fibers dispersion measurement using white-light spectral interferometry. *Opt Commun* 2009; **282**: 4361–4365.

20. Hlubina P, Kadulová M, Mergo P . Chromatic dispersion measurement of holey fibres using a supercontinuum source and a dispersion balanced interferometer. *Opt Laser Eng* 2013; **51**: 421–425.

21. Ye QH, Xu C, Liu X, Knox WH, Yan MF *et al.* Dispersion measurement of tapered air-silica microstructure fiber by white-light interferometry. *Appl Opt* 2002; **41**: 4467–4470.

22. Galle MA, Mohammed WS, Qian L, Smith PWE . Single-arm three-wave interferometer for measuring dispersion of short lengths of fiber. *Opt Express* 2007; **15**: 16896–16908.

23. Kovács AP, Osvay K, Bor Z, Szipöcs R . Group-delay measurement on laser mirrors by spectrally resolved white-light interferometry. *Opt Lett* 1995; **20**: 788–790.

24. Kaiser F, Issautier A, Ngah LA, Alibart O, Martin A *et al.* A versatile source of polarization entangled photons for quantum network applications. *Laser Phys Lett* 2013; **10**: 045202.

25. Alibart O, D'Auria V, De Micheli M, Doutre F, Kaiser F *et al.* Quantum photonics at telecom wavelengths based on lithium niobate waveguides. *J Opt* 2016; **18**: 104001.
26. Franson JD . Bell inequality for position and time. *Phys Rev Lett* 1989; **62**: 2205–2208.
27. Vergyris P, Kaiser F, Gouzien E, Sauder G, Lunghi T *et al.* Fully guided-wave photon pair source for quantum applications. *Quantum Sci Technol* 2017; **2**: 024007.
28. Crespi A, Lobino M, Matthews JCF, Politi A, Neal CR *et al.* Measuring protein concentration with entangled photons. *Appl Phys Lett* 2012; **100**: 233704.
29. Ono T, Okamoto R, Takeuchi S . An entanglement-enhanced microscope. *Nat Commun* 2013; **4**: 2426.
30. Israel Y, Rosen S, Silberberg Y . Supersensitive polarization microscopy using NOON states of light. *Phys Rev Lett* 2014; **112**: 103604.
31. Drexler W, Morgner U, Kärtner FX, Pitris C, Boppart SA *et al. In vivo* ultrahigh-resolution optical coherence tomography. *Opt Lett* 1999; **24**: 1221–1223.
32. Frigault MM, Lacoste J, Swift JL, Brown CM . Live-cell microscopy—tips and tools. *J Cell Sci* 2009; **122**: 753–767.
33. Celebrano M, Kukura P, Renn A, Sandoghdar V . Single-molecule imaging by optical absorption. *Nat Photonics* 2011; **5**: 95–98.
34. Piliarik M, Sandoghdar V . Direct optical sensing of single unlabelled proteins and super-resolution imaging of their binding sites. *Nat Commun* 2014; **5**: 4495.
35. Zadeh IE, Los JWN, Gourgues RBn, Steinmetz V, Bulgarini G *et al.* Single-photon detectors combining high efficiency, high detection rates, and ultra-high timing resolution. *APL Photonics* 2017; **2**: 111301.
36. Mazurek MD, Schreiter KM, Prevedel R, Kaltenbaek R, Resch KJ . Dispersion-cancelled biological imaging with quantum-inspired interferometry. *Sci Rep* 2013; **3**: 1582.
37. Manceau M, Khalili F, Chekhova M . Improving the phase super-sensitivity of squeezing-assisted interferometers by squeeze factor unbalancing. *N J Phys* 2017; **19**: 013014.

CHAPTER 7

THERMOMECHANICAL BEHAVIOUR OF A PWB BY SPECKLE INTERFEROMETRY TECHNIQUE

Bartolomeo Trentadue and Giuseppe Illuzzi

Dipartimento di Meccanica, Matematica e Management, Politecnico di Bari, Viale Japigia 182, 70126 Bari, Italy

ABSTRACT

The speckle interferometry technique has been used in this work in order to determine the thermomechanical behaviour of Printed Wiring Board (PWB) (circuits) of a radio integrated with tape player and speakers. A preliminary experiment of such technique has been carried out on a single electronic component (silicon transistor), during the thermal transient and at the steady state. The thermal deformation and stresses on PWB have been obtained through related experimental analyses on both cases. The results showed a very good applicability of speckle technique on the irregular object surface as PWB.

Citation: (APA): Trentadue, B., & Illuzzi, G. (2015). Thermomechanical Behaviour of a PWB by Speckle Interferometry Technique. Active and Passive Electronic Components, 2015. (7 pages)

INTRODUCTION

Stresses in electronic packages can cause premature mechanical failures of their components: fracture of the die, severing of connections, die bond failure, solder and components fatigue failure, and so forth. These stresses are often thermally induced and result from uneven expansions and contractions of various assembly materials due to mismatches in the coefficients of thermal expansion [1–3].

Many experimental and numerical analyses have been carried out on electronic packages. In particular, the research has been concentrated on the local strains and stresses of solder joints. This research has been aimed at achieving the optical measurement on the Printed Wiring Board of a radio integrated with tape player and speakers, considering that PWB can deform under the moments and forces anchoring the thermally stressed solder joints. On the other hand, even though the overstress of boards is rare, the circuits lines running on the surface should be stress checked.

Many available full-field and noncontact optical techniques, such as Moiré and its variations, Moiré Interferometry, have been used for this kind of problems [3]. However, most of these methods need a good flatness of object surface for the replication of gratings. Considering the applicability limits for each technique in relation to the model being analysed, the speckle techniques have been shown undoubtedly to be the unique method for our purpose, which has been also verified by the authors' experience on a silicon transistor during the thermal transient and the steady state [4].

SPECKLE INTERFEROMETRY

Speckle interferometry relies on the cyclic correlation analyses of speckle patterns for different loading states. It can be carried out in real time as electronic speckle pattern interferometry (ESPI) or digital speckle pattern interferometry (DSPI), respectively, if speckle patterns of the object surface obtained through a CCD-camera at different loads are analogically or digitally subtracted [5–10].

By means of the speckle technique shown in Figure 1, it was possible to obtain the in-plane displacement fields of the specimen subject to mechanical and thermal loads. Displacement components (x, y) and $V(x, y)$ are measured at each pixel (x, y). Such displacement fields are represented by the speckle fringes with bright and dark zones.

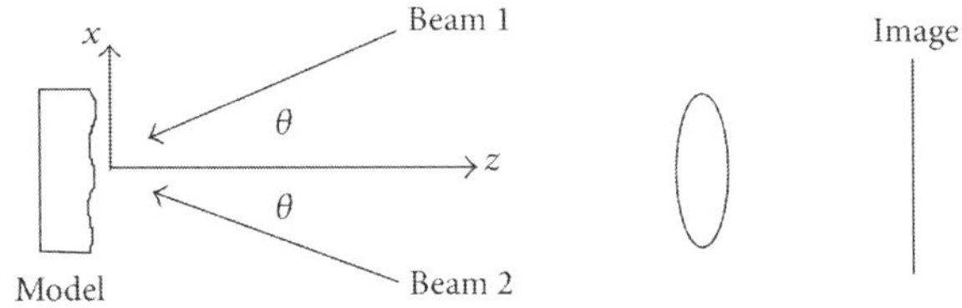

Figure 1. In-plane speckle interferometry.

For the bright zones, the following holds:

$$u = \frac{(2N+1)\lambda}{4\sin\theta}, \tag{1}$$

where N is the fringe order, θ is the illumination angle, and λ is the wavelength of laser beam.

For the dark zones (which is also called the correlated zones), the following holds:

$$u = \frac{N\lambda}{2\sin\theta}. \tag{2}$$

By the displacement obtained, the strains of each point in the measuring field can be determined by

$$\varepsilon_{xx} = \frac{\partial u(x,y)}{\partial x},$$
$$\varepsilon_{yy} = \frac{\partial v(x,y)}{\partial y},$$
$$\gamma_{xy} = \frac{\partial v(x,y)}{\partial x} + \frac{\partial u(x,y)}{\partial y}. \tag{3}$$

By the Hooke law, the stresses have the following calculations:

$$\sigma_{xx} = \frac{E}{(1+\upsilon)(1-2\upsilon)}\left[(1-\upsilon)\varepsilon_{xx} + \upsilon\varepsilon_{yy}\right];$$
$$\sigma_{yy} = \frac{E}{(1+\upsilon)(1-2\upsilon)}\left[(1-\upsilon)\varepsilon_{yy} + \upsilon\varepsilon_{xx}\right];$$
$$\tau_{xy} = G\gamma_{xy} = \frac{E}{2(1+\upsilon)}\gamma_{xy}, \tag{4}$$

where E is Young's modulus, G is shear modulus, and υ is Poisson ratio.

The whole field fringe analyses can be accomplished by simple manually curve drawing or by recently developed automatic fringe analysis methods such as Fourier transform and phase-shifting.

The benefits in using speckle interferometry in the PWB analyses are

- the possibility of carrying out the test in real time and following thermal transient processes;
- nonmandatory requirement of surface flatness under measurement as in the other methods, for example, Moiré Interferometry, where a good plane surface is necessary for the replication of grating;
- solder joints in the backside of PWB that may also have the speckle distribution; however in other experimental techniques they are considered pure defects.

Automatic fringe analysis methods [10–14] have enforced greatly the application of speckle methods overcoming its shortcoming of few fringes and low contrast.

EXPERIMENTAL SET-UP

The optical set-up is shown in Figure 2. A 35 mW He-Ne laser of wavelength (λ) 632.8 nm was utilized for this application. A parabolic mirror M1 is used as reflector and collimator.

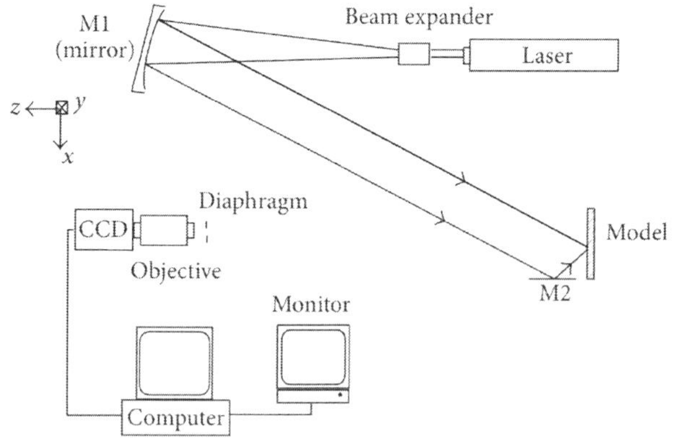

Figure 2. Experimental set-up.

A spatial filter (pinhole) of 10 μm was placed after the beam expander in order to ensure a good quality of speckle intensity pattern. The mirror M2 was positioned perpendicular to the specimen surface to obtain a symmetric incident beam with that coming directly from the mirror M1 at angle θ. The objective diaphragm was used to control the dimension and form of speckle patterns. The speckle patterns were recorded and processed through CCD-camera to an "Image Processing System."

EXPERIMENTS ON A SILICON TRANSISTOR

A transistor (Figure 3) was tested first in order to analyze the effect of the constraint points on the wiring board. The transistor was considered to be subjected to a plane strain field (shown in Figure 4(a)). Each point has two displacement components in the plane: (x, y) in the x direction and $v(x, y)$ in the y direction.

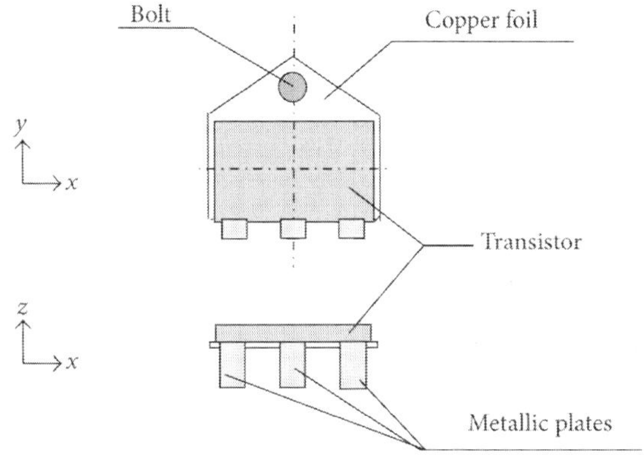

Figure 3. Transistor scheme.

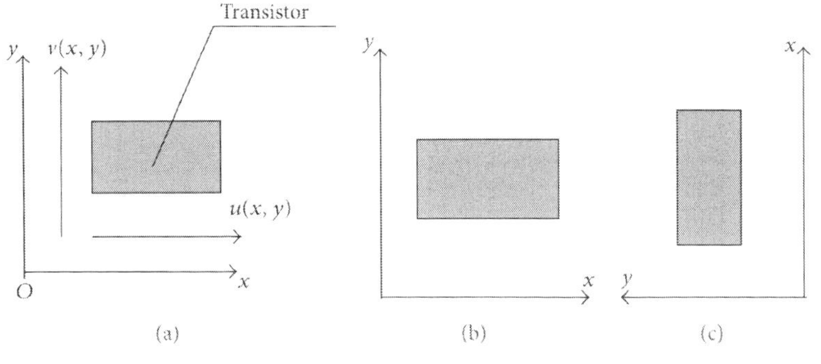

Figure 4. Displacements and positions of the transistor in the test: (a) displacements; (b) horizontal test; (c) vertical test.

The fringe analysis gives directly the displacements, so that it is possible to calculate the strains without considering the analytical solution of (x, y)

and $v(x, y)$. The two displacement components were figured out analyzing two different tests: one with the specimen in horizontal position (Figure 4(b)) and the other, after a rotation of 90°, in vertical position (Figure 4(c)). Several measurements were carried out for both tests during the transient state and at the steady state. Figure 5 shows the speckle fringe patterns of horizontal and vertical displacements, respectively. The transistor has the dimensions of $L_x = 14.7$ mm and $L_y = 10$ mm. Considering the He-Ne laser wavelength $\lambda = 632.8$ nm and the beam incident angle $\theta = 40°$, it is possible to determine the average elongation (ΔL_x, ΔL_y) for the whole transistor surface due to the thermal load:

$$\Delta L_x = \frac{N_x \lambda}{2 \sin \vartheta};$$

$$\Delta L_y = \frac{N_y \lambda}{2 \sin \vartheta}, \tag{5}$$

where N_x and N_y represent the fringe number in the x and y directions, respectively. Thus, the average strains can also be calculated by

$$\varepsilon_{xx} = \frac{\Delta L_x}{L_x};$$

$$\varepsilon_{yy} = \frac{\Delta L_y}{L_y}. \tag{6}$$

Figure 6 gives the average strain in the function of the time for both directions. The average ε_{xx} and ε_{yy} at steady state were calculated as 2.97×10^{-4} and 3.83×10^{-4}, respectively. The strain in the y direction is considerably higher due to the different level of kinematic constraints deriving from the soldering process.

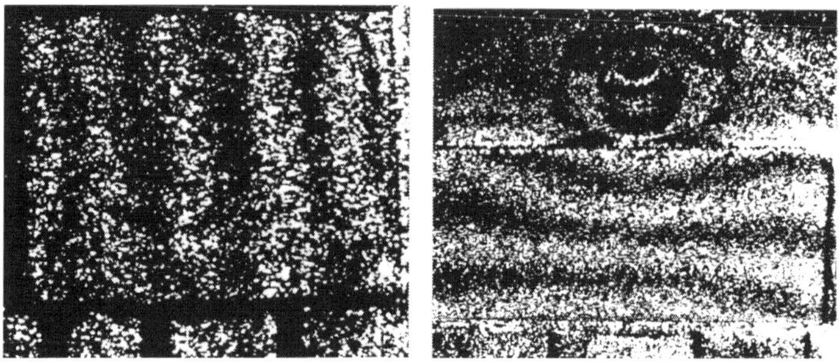

(a) (b)

Figure 5. Correlation fringes: (a) horizontal displacement; (b) vertical displacement.

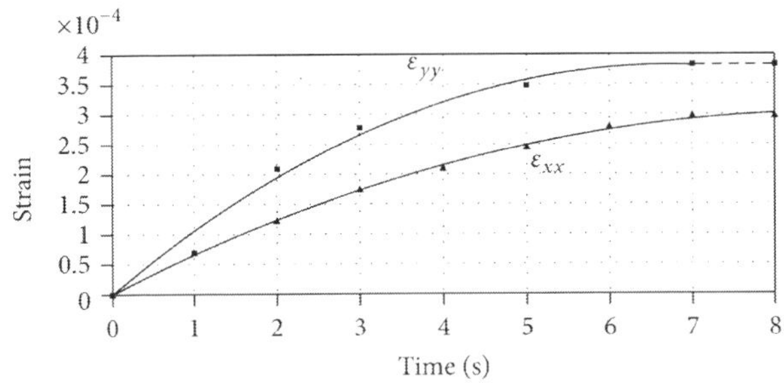

Figure 6. Average strain developed in the transistor with the time.

The local strains $\varepsilon_{xx} = \partial u/\partial x$ and $\varepsilon_{yy} = \partial v/\partial y$ in the sections (a-a) and (b-b) shown in Figure 7 were determined. The strain and stress distribution [2] along section (a-a) are given in Figure 8.

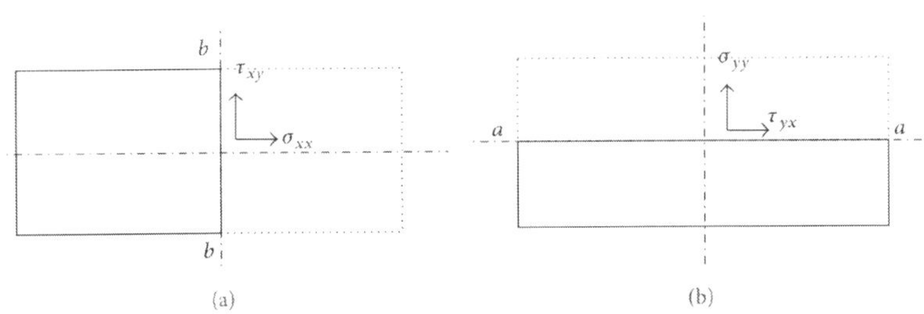

Figure 7. Transistor sections analyzed in the ESPI measurements.

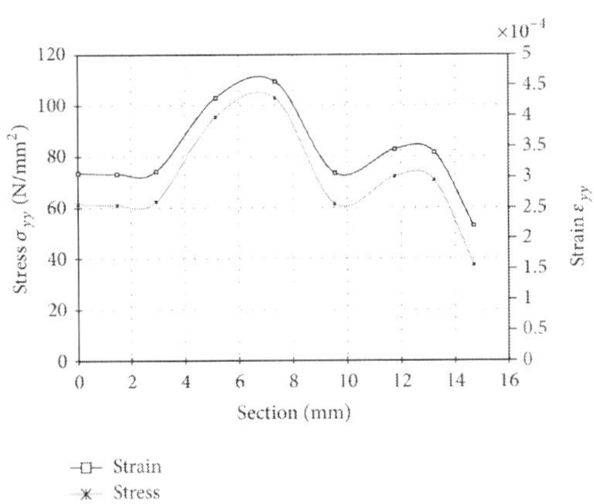

Figure 8. Strain ε_{yy} and stress σ_{yy} distribution along (a-a) section.

The σ_{yy} value decreases while moving to the edges of the transistor. In the diagram shown in Figure 8, a small increment was observed at the location close to the other two constraints (metallic welded plates). The maximum σ_{yy} value was found to be 109.4 MPa.

MEASUREMENTS ON A PRINTED WIRING BOARD

A Printed Wiring Board (PWB) of a radio integrated with tape player and speakers has been taken under experiments (Figure 9). On the backside of the board, the speckle interferometry technique has been applied. The natural surface of PWB consisted of a substrate of polyimide material covered with printed circuit lines and solder joints pattern. In order to obtain a good quality of speckle pattern, the surface was coated with an insulated silver spray, which can be seen in Figure 10.

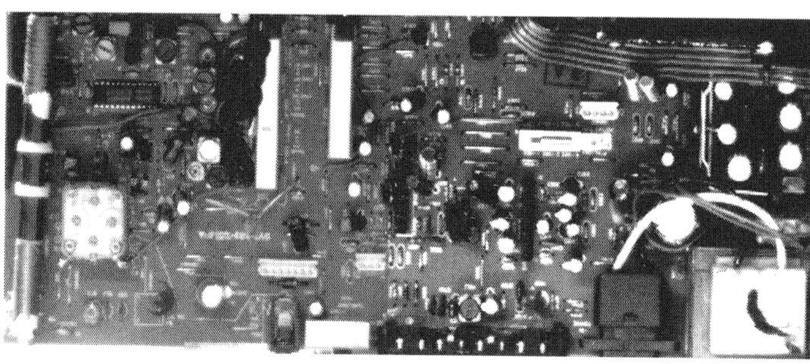

Figure 9. The PWB of a radio and its supported components.

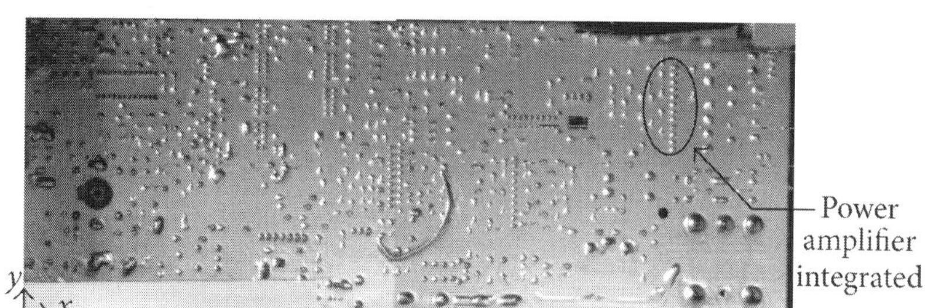

Figure 10. The coated backside PWB (the black ellipse indicates the power amplifier integrated).

During the experiments, the electronic connections among different components remain unchanged, but the board has been taken out of the radio frame to expose its surface to the laser light, which has obviously underestimated the thermoelastic behavior in the real operation condition having less cooling transfer than that in the experiments. The board however has been mounted at the same constraint condition as in its operation.

A series of repeating experiments revealed the thermal steady state of the PWB reached in 15 minutes after operation started with about 90% of the maximum power available. Thus, in order to achieve a measurement of the whole field, the fringe patterns at 15 minutes after the starting point for each measuring zone in the PWB have been obtained as the steady state.

Figure 11 shows the speckle fringe pattern of the whole *u*-displacement field at the steady state from which the stress distribution is outlined. Images were processed with the state-of-the-art image processing software, Holo Moiré Strain Analyzer (HMSA), developed by C. A. Sciammarella and his collaborators [15]. It is clear to see that the stress distribution is not very uniform over the board. The stress concentration on the upper right hand side is due to the settlement of the power amplifier integrated at that zone as shown in Figure 10 and due to the fact that the moments and forces anchoring these high thermally stressed solder joints act on the board.

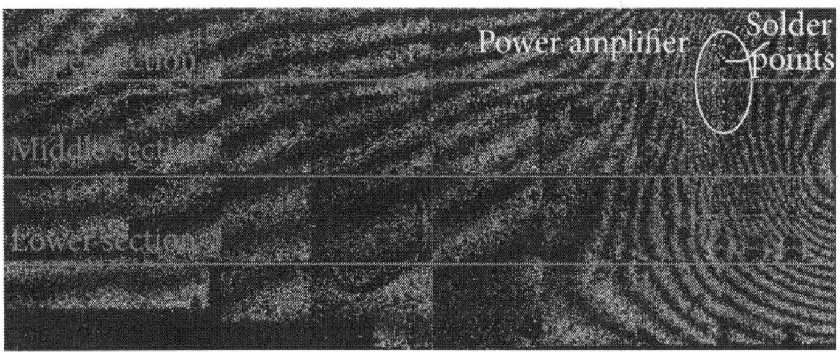

Figure 11. Whole *u*-displacement map of the PWB.

This stress concentration has a clear tension/compression characteristic since the fringes are nearly vertical while in other locations the fringes are almost horizontal which indicates the shear stress predomination. Such nonuniform stress distribution gives a very inconvenient failure time due to the breaking of local printed wires and solder joints.

The elastic modulus of common polyimide boards is about 2.7 GPa and the Poisson ration is between 0.23 and 0.35, while the elastic modulus of copper is about 124 GPa [1]. The sensitivity of speckle interferometer was about 0.48 μm. Distributions of *u*-displacements along three control sections (namely, "upper," "middle," and "lower") are reported in Figure 12. It can be seen that the displacement gradient increases sharply near the location of the power amplifier.

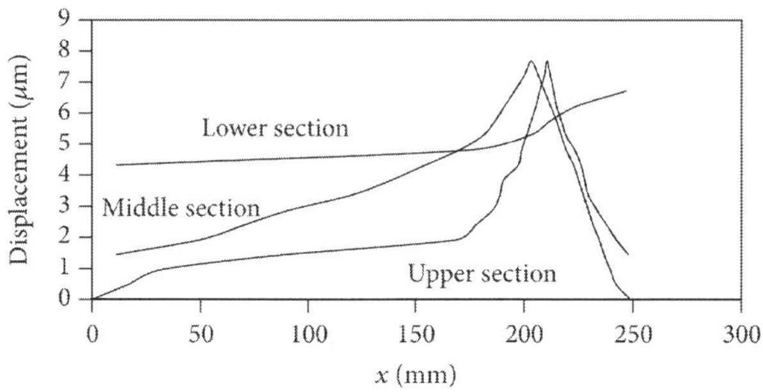

Figure 12. Displacement distribution in x direction measured for the PWB.

Based on the above measurement, a simplified x direction stress calculation is given here for the stress concentration zone. As correlation fringes are mainly oriented in the vertical direction, it can be supposed that strains in the x direction predominate. Therefore, (4) can be simplified as follows:

$$\sigma_{xx} = \frac{E(1-\upsilon)\,\varepsilon_{xx}}{(1+\upsilon)(1-2\upsilon)}. \tag{7}$$

For the board material, the strain ε_{xx} is calculated by (7) as 0.45×10^{-3}. This value represents the frequency of the correlation fringes in correspondence with the solder points of the power amplifier (see dotted/etched lines in Figures 11 and 12). Taking $E = 2.7$ GPa and $\upsilon = 0.3$ for the board and $E = 124$ GPa and $\upsilon = 0.3$ for the printed copper wires in the region of interest (i.e., where the power amplifier is soldered onto the board), the x direction stresses are 3.4 MPa for the board and 75 MPa for the wires. These stresses are lower than those measured in the y direction as the soldering develops right in the vertical direction, thus limiting the deformation and increasing the reaction forces.

CONCLUSIONS

From the experimental results, the very good applicability of the speckle technique for the problems studied has been demonstrated. The displacement map of the whole field has been obtained and can be taken as a verification during the electronic packaging design stages. High stress values are evident in the constraint points of the single components and in the location of the power component on the PWB. The stress state is strongly connected to the constraint reactions and also to the thermal disposal. For these reasons, in order to have a good design, it is very important to consider these two above-mentioned factors, so as to avoid probable and dangerous cracks in the structures. Considering that the wires are subjected to cyclic thermal loading, the design of the board should be also done very carefully to take into account possible fatigue failure.

Besides classical information on displacements and strains available from speckle measurements, we also determined stress concentrations at critical locations. This is an interesting improvement with respect to other studies on thermomechanical behavior of electronic packaging [16–19].

REFERENCES

1. P. A. Engel, *Structural Analysis of Printed Circuit Board Systems*, Springer, Berlin, Germany, 1993.
2. T. P. Ma, *Stresses in SiO_2-on-Si Structures*, vol. 4 of *EMIS Datareviews*, 1987.
3. B. Han, D. Barker, and R. Mahaian, "Experimental/numerical mechanics in electronic packaging," in *Proceedings of the 8th International Congress on Experimental Mechanics*, vol. 1, Nashville,Tenn, USA, June 1996.
4. C. Pappalettere, G. Rizzo, W. M. Sun, and B. Trentadue, "Thermal stress determination in a transistor by speckle interferometry," in *Proceedings of the 8th International Congress on Experimental Mechanics*, vol. 1, Nashville, Tenn, USA, June 1996.
5. J. C. Dainty and J. W. Goodman, *Laser Speckle*, Springer, New York, NY, USA, 1984.
6. R. Jones and C. Wykes, *Holographic and Speckle Interferometry*, Cambridge University Press, Cambridge, UK, 1989.
7. M. Francon, *Laser Speckle and Application in Optics*, Academic Press, New York, NY, USA, 1979.
8. L. Pagnotta and A. Poggialini, "Speckle pattern correlation methods: performance analysis by numerical simulation," in *Proceedings of the International Conference on Materials Engineering*, Gallipoli, Italy, 1996.
9. P. R. Slangen, L. Berwart, C. de Veuster, J.-C. Golinval, and Y. F. Lion, "Digital speckle pattern interferometry (DSPI): a fast procedure to detect and measure vibration mode shapes," in *First International Conference on Vibration Measurements by Laser Techniques: Advances and Applications*, vol. 2358 of *Proceedings of SPIE*, Ancona, Italy, October 1994.
10. C. A. Sciammarella and F. M. Sciammarella, *Experimental Mechanics of Solids*, John Wiley & Sons, 2012.
11. D. W. Robinson and D. C. Williams, "Digital phase stepping speckle interferometry," *Optics Communications*, vol. 57, no. 1, pp. 26–30, 1986.
12. K. Creath, "Phase-shifting speckle interferometry," *Applied Optics*, vol. 24, no. 18, pp. 3053–3058, 1985.

13. M. Kujawinska, *Interferogram Analysis*, Institute of Physics Publishing, Philadelphia, Pa, USA, 1993.
14. P. Jacquot, "Speckle interferometry: a review of the principal methods in use for experimental mechanics applications," *Strain*, vol. 44, no. 1, pp. 57–69, 2008.
15. General Stress Optics, *Holo-Moiré Strain Analyzer (HoloStrain) Version 2.0*, General Stress Optics, Chicago, Ill, USA, 2008.
16. K. Genovese, L. Lamberti, and C. Pappalettere, "A comprehensive ESPI based system for combined measurement of shape and deformation of electronic components," *Optics and Lasers in Engineering*, vol. 42, no. 5, pp. 543–562, 2004.
17. F. Su, L. Liu, and T. Wang, "Evaluation of residual stress in flip chip using 3-D optical interferometry/FEM hybrid technique," *Strain*, vol. 43, no. 4, pp. 289–298, 2007.
18. H. R. Schubach, "Investigation of the thermal deformation of electronic packages," in *Experimental Analysis of Nano and Engineering Materials and Structures*, E. E. Gdoutos, Ed., Springer, Dordrecht, The Netherlands, 2007.
19. C. Casavola, L. Lamberti, V. Moramarco, G. Pappalettera, and C. Pappalettere, "Experimental analysis of thermo-mechanical behaviour of electronic components with speckle interferometry," *Strain*, vol. 49, no. 6, pp. 497–506, 2013.

CHAPTER 8

PHASE-SENSITIVE INTERFEROMETRY OF DECORRELATED SPECKLE PATTERNS

Hendrik Spahr[1,2], Clara Pfäfe[1,2], Sazan Burhan[1], Lisa Kutzner[1,2], Felix Hilge[1,2], Gereon Hüttmann[1,2,3], and Dierck Hillmann[1,4]

[1]Institute of Biomedical Optics, University of Lübeck, Peter-Monnik-Weg 4, 23562, Lübeck, Germany.
[2]Medical Laser Centre Lübeck GmbH, Peter-Monnik-Weg 4, 23562, Lübeck, Germany.
[3]Airway Research Center North (ARCN), Member of the German Center for Lung Research (DZL), 22927 Großhansdorf, Germany.
[4]Thorlabs GmbH, MariaGoeppert-Straße 9, 23562, Lübeck, Germany.

ABSTRACT

Phase-sensitive coherent imaging exploits changes in the phases of backscattered light to observe tiny alterations of scattering structures or variations of the refractive index. But moving scatterers or a fluctuating refractive index decorrelate the phases and speckle patterns in the images. It

Citation: (APA): Spahr, H., Pfäffle, C., Burhan, S., Kutzner, L., Hilge, F., Hüttmann, G., & Hillmann, D. (2019). Phase-sensitive interferometry of decorrelated speckle patterns. Scientific reports, 9(1), 1-15. (15 pages)

is generally believed that once the speckle pattern has changed, the phases are scrambled and any meaningful phase difference to the original pattern is removed. As a consequence, diffusion and tissue motion that cannot be resolved, prevent phase-sensitive imaging of biological specimens. Here, we show that a phase comparison between decorrelated speckle patterns is still possible by utilizing a series of images acquired during decorrelation. The resulting evaluation scheme is mathematically equivalent to methods for astronomic imaging through the turbulent sky by speckle interferometry. We thus adopt the idea of speckle interferometry to phase-sensitive imaging in biological tissues and demonstrate its efficacy for simulated data and imaging of photoreceptor activity with phase-sensitive optical coherence tomography. We believe the described methods can be applied to many imaging modalities that use phase values for interferometry.

INTRODUCTION

Phase-sensitive, interferometric imaging measures small changes in the time-of-flight of a light wave by detecting changes in its phase. But in many applications, statistically varying optical properties of the scattering structure randomize the phase of the backscattered light, resulting in a speckle pattern with random intensity and phase[1]. As a consequence, we can only extract meaningful phase differences from images with identical or at least almost identical speckle patterns[2]. However, if the detected wave's speckle pattern changes over time, it inevitably impedes phase sensitive imaging[3,4]. For example, holographic interferometry or electronic speckle pattern interferometry (ESPI) compare at least two states of backscattered light acquired at different times, and it can only be applied if the respective speckle patterns are still correlated[2,3,4,5]. While some techniques, such as dynamic light scattering[6,7] or optical coherence angiography[8] make use of these decorrelations, other imaging techniques are severely hindered by it.

Among the most important effects that decorrelate speckles with time, are random motions and changes of the optical path length on a scale below the resolution, e.g., by diffusion[6,7]. These effects are impossible to prevent. But also bulk sample motion can cause the phase evaluation to compare different parts of the same speckle pattern. While in 3D imaging this effect of bulk motion can often be corrected to some extent by co-registration and suitable algorithms, in 2D sectional imaging we lack the data to correct this as the acquired slices of the specimen change if the specimen moves perpendicularly to the imaging plane.

We recently demonstrated imaging of the activity of photoreceptors and neurons by phase-sensitive full-field swept-source optical coherence tomography (FF-SS-OCT)[9,10]. The optical path length of neurons and photoreceptors in the human retina changes by few nanometers upon activation, e.g., after white light stimulation, which can be used as functional contrast for those cellular structures[11,12,13]. While most OCT systems lack the axial resolution required to resolve this, phases of the OCT signal encode small changes of the optical path length. Using the phases, FF-SS-OCT successfully measured nanometer changes in the optical path length of neurons and receptors due to activation over a few seconds. However, we could neither increase the measurement time nor determine tissue changes in single cross-sectional scans (B-scans). In both cases speckle patterns changed after a few seconds due to diffusion, bulk tissue motion, or tissue deformations and the phase information was lost, ultimately limiting the applicability of the method. Apart from this application, phase evaluation beyond the speckle decorrelation time can be of importance in other fields, such as optical coherence elastography (OCE)[14], for which Chin *et al.* noticed the importance of phase correlation[15]. Finally, countless applications of OCT make use of the phase[16,17,18,19] and might benefit from better phase evaluation schemes.

In this paper, we reconstruct phase changes over times significantly longer than the decorrelation time of the corresponding speckle patterns. The idea is to calculate phase differences in a series of consecutive images over small time differences (short-time phase differences) that still show sufficient correlation. If this is done for several measurements with different speckle patterns, averaging of these independent measurements will cancel the contributions of the disturbing phase. Integrating all these phase differences then yields the real time evolution of the phase. By averaging the phase in multiple speckles to obtain a single phase value, the real phase change can be obtained beyond the correlation time of the speckles. In essence, we combine the information from multiple speckles, each of which carries information on phase changes over a certain time (see Fig. 1a).

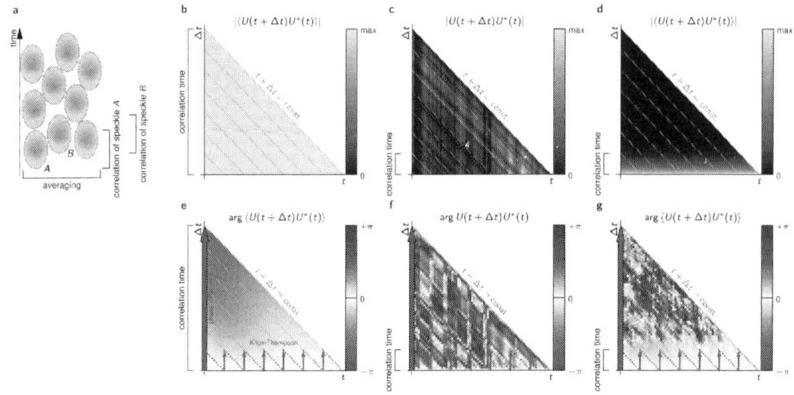

Figure 1. (**a**) In average, each speckle (schematically represented by the blue ellipses) carries valid phase information only for the correlation time of the speckle pattern. Assuming that all speckle in the averaged area are subject to a common phase change in addition to random uncorrelated phase changes, one can use the phase of speckle *A* as long as it is valid and then continue with the phase of speckle *B*. Using multiple speckle, phase information for times significantly exceeding the correlation time can be extracted. (**b**,**e**) Exemplary cross-spectrum magnitude and phase assuming infinite correlation time. Total phase changes can either be computed directly (blue arrow) or using the Knox-Thompson path (green arrows). (**c**,**f**) Magnitude and phase of the cross-spectrum without averaging; neither method can extract the phase beyond the correlation time. (**d**,**g**) Magnitude and phase of ensemble averaged cross-spectrum. Direct phase differences (blue) cannot be used for phase extraction, but Knox-Thompson (green) method can be applied since phase values for small Δt are valid.

Just integrating the phase difference of successive frames followed by a summation, as it is commonly used for phase unwrapping[16,17,18,19,20], accumulates the phase noise; except for phase unwrapping it is mathematically equivalent to the direct computation of phase differences. Without averaging, calculating short-time phase differences will not yield any information about phase differences of uncorrelated speckle patterns. The averaging of short-time phase differences before integration is the essential step as it effectively cancels the phase noise.

In the 1970s and 1980s the same idea, known as speckle interferometry, was originally used in astronomy for successful imaging through the turbulent atmosphere with diffraction limited resolution[1,21,22,23,24,25,26,27,28]. Short time exposures were used to reconstruct the missing phase of the Fourier transform of diffraction limited undisturbed images: By utilizing a large number of short-exposure images with different disturbances, true and undisturbed phase differences for small distances in the aperture plane could be computed. From these phase differences, the complete phase was reconstructed by integration. We will frame the similarities and the mathematical analogy between the two methods, which allows us to benefit from advances in astronomy for interferometric measurements during coherent imaging.

Methods that allow phase evaluation beyond the correlation time have also been developed for synthetic aperture radar (SAR) interferometry (InSAR)[29,30]. In InSAR, phase evaluation is used to monitor ground elevation, but small changes on the ground or turbulence in the sky interfere with an evaluation of the phase[29] similar to interferometric measurements in biological imaging. For InSAR, two techniques have been developed which selectively evaluate the phase from only minimally affected images or structures. The first technique is referred to as small baseline subset technique (SBAS)[31]. It selects optimal images from a time series to compare phase values based on good correlation. The other method only compares phases of single selected scatterers that maintain good correlation over long times, so called permanent scatterers[32].

In the first part of this paper, we lay down the basics and establish the commonality between coherent phase-sensitive imaging and astronomic speckle interferometry. Afterwards, we demonstrate the efficacy of the resulting methods by simulating simple phase resolved images in backscattering geometry and evaluating these with the proposed approach. Finally, we apply our method to *in* vivo data of phase-sensitive optical coherence tomography.

THEORY

Mathematical Formulation of the Problem

We assume coherent imaging of the physical system, in which a deterministic change of the optical path length, e.g., swelling or shrinking of cells, is superimposed by random changes. These may be caused by diffusion, fluctuations of the refractive index, or rapid uncorrelated micro motion. The complex amplitude in each pixel of a coherently acquired and focused image can be represented as a sum of (random) phasors (complex values), where each summand comprises amplitude and phase. Such a coherently acquired field $U(\boldsymbol{x},t)$ has contributions from the random phases $\phi_i(\boldsymbol{x},t)$ and a systematic phase $\phi(\boldsymbol{x},t)$. $U(\boldsymbol{x},t)$ may then be written as

$$U(\boldsymbol{x}, t) = \sum_i A_i(\boldsymbol{x}, t) e^{i\phi_i(\boldsymbol{x},t)} e^{i\phi(\boldsymbol{x},t)}, \tag{1}$$

where A_i represents the unknown amplitude of each scatterer, and $\boldsymbol{x}$ and t denote location and time, respectively. Depending on the imaging scenario, U might not depend on the location $\boldsymbol{x}$ at all or $\boldsymbol{x}$ might be up to three-dimensional as in tomographic imaging.

We can now separate U into its systematic phase contribution U_0 and a random modifier H, which yields

$$\begin{aligned} U(\boldsymbol{x}, t) &= \underbrace{\left(\sum_i A_i(\boldsymbol{x}, t) e^{i\phi_i(\boldsymbol{x},t)} \right)}_{H(\mathbf{x},t)} \underbrace{e^{i\phi(\boldsymbol{x},t)}}_{U_0(\boldsymbol{x},t)} \\ &= H(\boldsymbol{x}, t) U_0(\boldsymbol{x}, t). \end{aligned} \tag{2}$$

The modifier H covers all effects that alter the speckle patterns over time, e.g., changing phase, changing amplitude, and scatterers moving out of or into the detection area. In this form, we have no way to distinguish H from U_0. We can, however, compute the ensemble average by

$$\langle U(\boldsymbol{x}, t) \rangle = \langle H(\boldsymbol{x}, t) \rangle U_0(\boldsymbol{x}, t), \tag{3}$$

when assuming that U_0 (though not H) is constant over the averaged ensemble. The ensemble could consist of repeated measurements or certain pixels from a volume, in which U_0 is constant. In this averaged expression, we can now obtain the time evolution of U_0 for times where $\langle H \rangle$ remains approximately constant, i.e., where the contribution of the random phases is small. For t exceeding the correlation time of $\langle H \rangle$, however, the decorrelation of $\langle H \rangle$ still prevents us from computing changes in the systematic phase U_0.

Solving the Phase Problem in Astronomic Speckle Interferometry

The resolution of large ground-based telescopes is limited by atmospheric turbulences, which modify the phase of the optical transfer function (OTF) and cause time varying speckles in the image. During a long exposure these speckle patterns average out to a blurred image. Speckle interferometry is a way to obtain one diffraction limited image from several short exposure images. In each of these images the random atmospheric turbulences are constant and each image essentially contains one speckle pattern1. If the actual diffraction limited image of the object is $I_0(x)$ and has a Fourier spectrum $\tilde{I}_0(\boldsymbol{k})$, the Fourier spectra of the short exposure images are given by

$$\tilde{I}(\boldsymbol{k}) = \mathcal{H}_S(\boldsymbol{k})\tilde{I}_0(\boldsymbol{k}), \tag{4}$$

where $\mathcal{H}_S$ is the short exposure OTF[1]. $\mathcal{H}_S$ represents the disturbing effect of the atmosphere, changes randomly in each acquired image, and causes speckle noise in each image $I(\boldsymbol{x})$. A single short exposure image $I(\boldsymbol{x})$ will thus not yield the object information because phase and magnitude of $\mathcal{H}_S$ are not known. However, the magnitude of the spectrum $\tilde{I}_0(\boldsymbol{k})$ of the diffraction limited images is obtained by averaging the squared magnitude of all short exposure spectra $\tilde{I}(\boldsymbol{k})$,

$$\langle |\tilde{I}(\boldsymbol{k})| \rangle^2 = \langle |\mathcal{H}_S(\boldsymbol{k})|^2 \rangle |\tilde{I}_0(\boldsymbol{k})|^2, \tag{5}$$

since the squared short exposure OTF averaged over all measurements $\langle |\mathcal{H}_S(\boldsymbol{k})|^2 \rangle$ is larger than zero and can be computed on the basis of known statistical properties of $\mathcal{H}_S$[1]. But the magnitude $|\tilde{I}_0|$ of the Fourier spectrum only allows computation of the autocorrelation of I_0. To obtain the full diffraction limited image the phase of $\tilde{I}_0(\boldsymbol{k})$ is required too. This phase is contained in the Fourier transforms $\tilde{I}(\boldsymbol{k})$ of the short time exposures, but it is corrupted by a random phase contribution of the atmospheric turbulences in the same way as the systematic phase in coherent imaging is corrupted by random phase noise.

Hence, retrieving diffraction limited images by speckle interferometry faces the same problem as retrieving the temporal phase evolution in coherent imaging. Only over small spatial frequency distances in the Fourier plane the phase difference in a spectrum $\tilde{I}(\boldsymbol{k})$ corresponds to the phase difference in the spectrum $\tilde{I}_0(\boldsymbol{k})$ of the diffraction limited image. Over larger distances, the phase information is lost due to atmospheric turbulence. The similarity of the problem is seen in the analogy of Eqs (4) with (2), where $\mathcal{H}_S$ and I correspond to H and U, respectively. However, while the astronomic problem has a two-dimensional vector $\boldsymbol{k}$ of spatial frequencies as the dependent variable, the phase-sensitive imaging problem has only the time t. Both use a series of measurements, in which only phase information over small $\Delta\boldsymbol{k}$ or

Δt is contained, respectively. We will show that by analyzing a large number of measurements, in which the corrupting phase contribution changes, the full phase can be reconstructed.

Phase Information in the Cross-Spectrum

In astronomy, it was Knox and Thompson who solved the recovery of phases from multiple, statistically varying images[21] with short exposure times using the so-called cross-spectrum. Later, another approach, called the triple-correlation or bispectrum technique, even improved the achieved results[22]. One major advantage of the bispectrum over the cross-spectrum technique for astronomic imaging is that it is only sensitive to the non-linear phase contributions of the transfer function $\mathcal{H}_S$, since shifted images yield to the same bispectrum and do not pollute the obtained phases (see, for example[23], for details). However, for phase sensitive imaging, the linear part of the phase evolution of U_0 is generally of interest and thus the bispectrum technique is not applicable here.

Following we will apply the algorithm of Knox and Thompson to recover the phase beyond the speckle correlation time in coherent imaging. For convenience, we will retain the term cross-spectrum, even though in our scenario it is evaluated in time-domain as a function of t rather than in (spatial) frequency-domain as a function of $\boldsymbol{k}$.

We define the cross-spectrum C_H of the phase disturbing speckle modulation H by

$$C_H(\boldsymbol{x}, t, \Delta t) = H(\boldsymbol{x}, t)H^*(\boldsymbol{x}, t + \Delta t).$$

It is subject to speckle noise since H itself merely contains speckle (Fig. 1c,f); but its ensemble average

$$\langle C_H(\boldsymbol{x}, t, \Delta t)\rangle = \langle H(\boldsymbol{x}, t)H^*(\boldsymbol{x}, t + \Delta t)\rangle \tag{6}$$

has in general a magnitude larger than 0 and speckles of H are averaged out, at least for small Δt. Most importantly, it is to a good approximation real-valued, i.e., its phase is zero, as long as Δt is within the correlation time of H. This real-valuedness was previously shown for the corresponding cross-spectrum $\langle C_{\mathcal{H}_S}\rangle$ in astronomic imaging[1,27]. For phase imaging, we show the real-valuedness of C_H for small Δt for which the autocorrelation is still strong in the Methods section. If Δt is larger than the correlation time, $H(\boldsymbol{x},t)$

and $H^{*}(\boldsymbol{x},t+\Delta t)$ become statistically independent, the phase of $\langle C_H \rangle$ gets scrambled and C_H follows speckle statistics.

Now, if we again assume that U_0 and thus C_{U_0} are constant over the ensemble, we compute the averaged cross-spectrum of the measurements of U to

$$\begin{aligned}\langle C_U(\boldsymbol{x}, t, \Delta t)\rangle &= \langle U(\boldsymbol{x}, t)U^{*}(\boldsymbol{x}, t + \Delta t)\rangle \\ &= \langle C_H(\boldsymbol{x}, t, \Delta t)\rangle C_{U_0}(\boldsymbol{x}, t, \Delta t).\end{aligned} \tag{7}$$

Knowing that $\langle C_H \rangle$ is real-valued for small Δt, the phases of $\langle C_U \rangle$ are determined only by the phases of C_{U_0}. If we find phases of U_0 that yield the phases of $\langle C_U \rangle$ for these small Δt, we obtain the systematic phase $\phi(\boldsymbol{x},t)$ we are looking for as introduced in (1).

Note that $\langle C_U(\boldsymbol{x},t,\Delta t)\rangle$ is related to the time-autocorrelation of U, except averaging being performed over the ensemble instead of the time t and thus being a function not only of Δt bus also of t. Due to this relation, the cross-spectrum maintains comparably large magnitudes for time differences Δt with large autocorrelation values. The magnitudes and phases of exemplary cross-spectra, in the decorrelation-free scenario and with strong decorrelation, with and without the ensemble averaging (obtained from simulations, see Results and Methods) is shown in Fig. 1b–g. The figures illustrate that $\langle C_U \rangle$ yields a deterministic increase of the phases for small Δt (Fig. 1g) as long as its magnitude (Fig. 1d, related to the autocorrelation) remains large, but only if the ensemble averaging is performed. It can further be seen, that the valid phase differences visible in Fig. 1g are small compared to the phases for large t in Fig. 1e since they represent merely phase difference for small Δt. Nevertheless, when reconstructing the phase of U_0 from the values of $\langle C_U \rangle$, we need to ensure that only values for (small) Δt with strong autocorrelation are taken into account. This condition coincides with the real-valuedness of $\langle C_H \rangle$.

Ensemble averaging needs multiple independent measurements, which cannot be performed easily in a real experiment. However, averaging over multiple lateral pixels or averaging over multiple detection apertures can be done from a single experiment and approximates the ensemble average. In our case, we either use an area over which we strive to obtain one mean phase-curve, or we use a Gaussian filter with a width determined as a compromise between spatial resolution and sufficient averaging statistics to obtain images of the phase evolution (see Methods).

Phase Retrieval from the Cross-Spectrum

Having obtained the cross-spectrum $\langle C_U \rangle$, the systematic phase function U_0 needs to be extracted. We assume a given initial phase $\phi(\boldsymbol{x},t=t_0)$ and evaluate methods to extract the systematic phase evolution $\phi(\boldsymbol{x},t)$ from the cross-spectrum: An approach equivalent to computing phase differences directly is given by

$$\phi(\boldsymbol{x}, t_0 + t) = \phi(\boldsymbol{x}, t_0) + \arg\langle C_U(\boldsymbol{x}, t_0, t)\rangle, \tag{8}$$

for $t \geq t_0$ (blue arrows in Fig. 1b–g). However, it only yields phases within the correlation time. In astronomy, this approach never works, unless the image was not disturbed by turbulence in the first place. Instead, Knox and Thompson originally demonstrated diffraction limited imaging through the turbulent sky[21] by iteratively walking in small increments of Δt (e.g., one time step) through the cross-spectrum, i.e.,

$$\phi(\boldsymbol{x}, t_0 + n\Delta t) = \phi(\boldsymbol{x}, t_0 + (n-1)\Delta t) + \arg\langle C_U(\boldsymbol{x}, t_0 + (n-1)\Delta t, \Delta t)\rangle, \tag{9}$$

for integer $n \geq 1$ with increasing iteration number n (green arrows in Fig. 1b–g). This method uses only phase values well within the correlation time. If $\langle C_U \rangle$ is valid for the single time steps Δt, i.e., $\langle C_H \rangle$ is real valued, the iterative formula will yield valid results, even for larger t. However, a single outlier at a time t_{err}, i.e., one false step, ruins all following values for $t \geq t_{err}$.

The influence of these single events can be reduced by taking multiple ways with different Δt for stepping through the cross-spectrum. In astronomy, this is known as the extended Knox-Thompson method[23,28]. However, most of the algorithms for finding the optimal paths have been developed for use with the bispectrum technique[24,25,26]; nevertheless, in general, they can be easily transferred to the cross-spectrum. Here we minimized the sum of weighted squared differences between the measured cross-spectrum and the cross-spectrum resulting from the systematic phase $\phi(\boldsymbol{x},t)$. The algorithm is described step-by-step in the methods section.

Realization of the Algorithm

To actually implement the presented extended Knox-Thompson algorithm for phase-sensitive OCT imaging we proceeded as follows and as illustrated in Fig. 2. A more detailed description of the required steps is given in the Methods section. We assume the OCT data is reconstructed, co-registered and the layers of interest are segmented and denoted z_1 and z_2, with the volumes being given by $U(\boldsymbol{x},t)$. For example, these layers can be the inner segment/outer segment junction of the retina and the outer segment tips, respectively, in which case phase difference will show elongation of the photoreceptor outer segments. We now compute the cross-spectra for all points in the respective depths of z_1 and z_2 by

$$C_U(\boldsymbol{x}, t, \Delta t) = U(\boldsymbol{x}, t)U^*(\boldsymbol{x}, t + \Delta t).$$

To average signals over multiple depths, the complex cross spectra C_U can also be averaged over multiple z values here. Next the phase difference between these cross-spectra in the respective layers of interest is performed by computing

$$C_U^{z_1-z_2}(x, y, t, \Delta t) = C_U(x, y, z_1, t, \Delta t)C_U^*(x, y, z_2, t, \Delta t).$$

Utilizing the difference of two layers makes the measurement almost independent from the phase stability of the system (see also Fig. S4 in the Supplementary Information). This step is followed by the actual ensemble average yielding $\langle C_U \rangle$ as in Eq. (7). To create images, this can be done by applying a Gaussian filter in x and y direction. In parallel, weights $w(x,y,t,\Delta t)$ are computed that describe the quality of each entry in all cross-spectra and are required to perform a fit afterwards (see Methods section, Eqs (12) and (13)). After phase unwrapping the ensemble averaged cross-spectra $\langle C_U \rangle$, a phase function is fitted to the phase of $\langle C_U \rangle$ for each x and y coordinate, respecting the weights w. This yields the phase $\phi(x,y,t)$. These values then compose the resulting images. In case only a single curve is to be extracted, the dependency of x and y is dropped in the ensemble averaging step and only one curve is fitted.

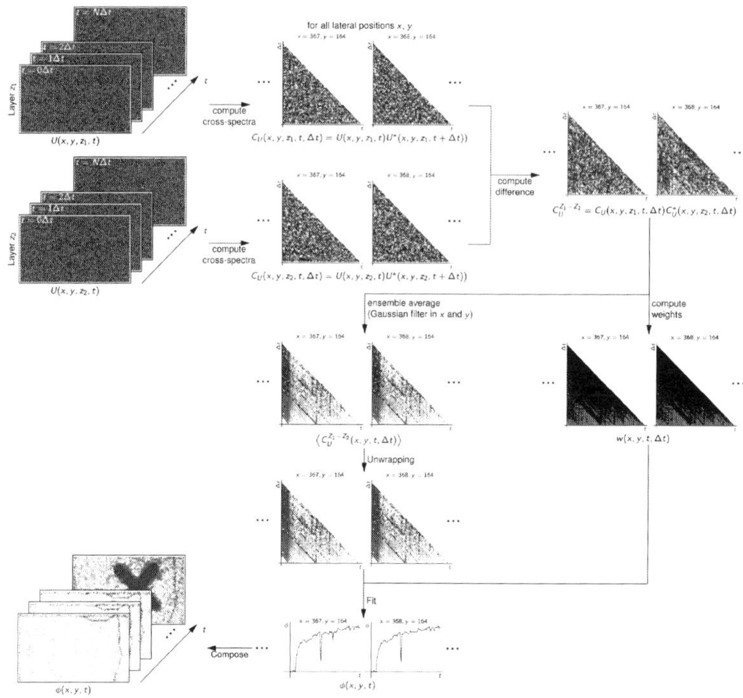

Figure 2. Flowdiagram of the actual implementation of the extended Knox-Thompson algorithm.

Acquisition and Reconstruction Parameters

The algorithm depends on several acquisition and reconstruction parameters. Among the obvious ones are the number of values used for (ensemble) averaging and the trade-off between obtainable signal-to-noise ratio, e.g., by increasing the integration time of the data acquisition, and the time distance between time-adjacent images. While it is difficult to give definite answers on how to choose these parameters, rough guide-lines can be provided: Accuracy of obtained phases will improve if the averaging size is increased, provided that all averaged data points exhibit the same or at least similar systematic phase changes. In many cases, the practical consideration here will be related to the resolution loss that is accompanied by more averaging or larger filter sizes.

The choice of integration time is ultimately limited by the requirement, that at least temporally adjacent images need to be well correlated. If this is not the case, however, an increased integration time also no longer yields increased signal-to-noise ratio, but rather phase wash-out diminishes the signals. In our cases, we were always limited by technical implications preventing us from decreasing the time between successive measurements and thereby limiting the temporal sampling frequency. However, we are fairly certain that our phase signals would still benefit from more frequent measurements. We believe this holds for most applications since some inherent averaging is done when fitting data to the cross-spectrum in the extended Knox-Thompson method. Exceptions might occur when strong phase noise dominates the cross-spectra.

Implementation

We implemented the respective algorithms in C++ using the Clang compiler. Vectorization and OpenMP were used to increase performance where possible. Most importantly, to achieve good performance, all cross-spectra in single-precision complex floating point numbers had to fit in memory at the same time, requiring in our case about 10 GB of RAM. Parts requiring linear algebra were implemented using the Armadillo library[33,34].

RESULTS

Simulation

We simulated images of a large number of point scatterers that exhibit a random Gaussian-distributed 3-dimensional motion (with variance σ^2) between frames in addition to a common axial motion in a specific "x"-shaped area. A simulation has the advantage that the actual phase change (ground-truth) is well known and that the speckle decorrelation can be tightly controlled. While both methods show indistinguishable results for the noise-free scenario ($\sigma=0$), the simulation in the case of degrading speckle demonstrates the power of the extended Knox-Thompson method. For $\sigma=\lambda/32$, the phase evaluation with a simple phase difference to the first image and the extended Knox-Thompson method yield almost indistinguishable results for the first 10 steps (Fig. 3b,f). After 60 time steps, a simple phase difference to the first image yields only noise (Fig. 3c), while the "x"-shaped pattern is still well visible (Fig. 3g) using the extended Knox-Thompson method. The speckle patterns appear correlated after 10 steps but are completely changed after 69 steps (Fig. 3i–k). Doubling the scatterer motion to $\sigma=\lambda/16$, the "x"-shaped pattern begins to deteriorate after 69 steps even with the Knox-Thompson method. However, the "x" remains visible (Fig. 3h), whereas the phase difference again yields only phase noise (Fig. 3d).

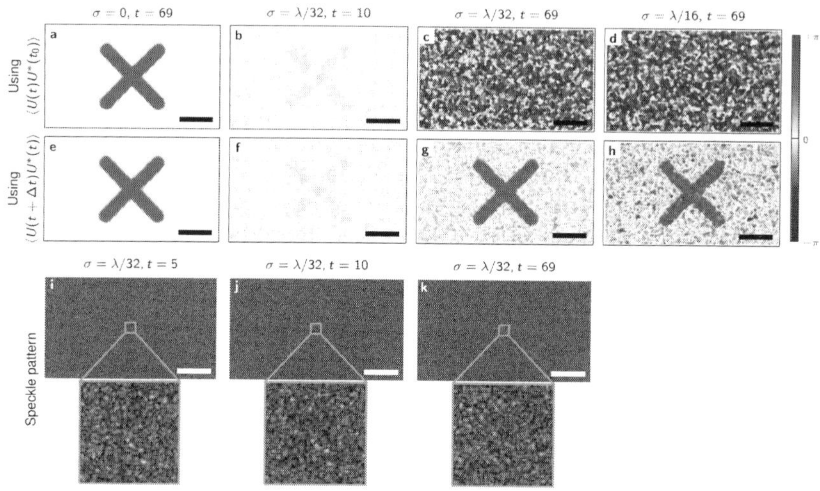

Figure 3. Phase images and speckle patterns obtained from the simulation. (**a–d**) Phase evaluated by phase differences for different σ and time steps. (**e–h**) Phase evaluated using the extended Knox-Thompson methods for the same σ and time steps as shown in (**a–d**). (**i–k**) Speckle patterns for σ=λ/32 after a different number of time steps. Scale bars are 500λ.

The curves extracted in multiple simulations from the "x" confirm the results of the images and demonstrate the reproducibility of the method. Without random scatterer motion (σ=0, Figs 3a,e and 4a,e), the phase can be calculated by the phase difference to the first frame yielding the exact curve that was supplied to the simulation. But introducing random Gaussian-distributed motion of the scatterers with as little as σ=λ/32 between frames accumulates huge errors in the phase of the directed motion after 70 frames (Fig. 4b); the mean autocorrelation of the complex fields (Fig. 4i) in this case drops to one half in less than ten frames. Increasing motion amplitude (Fig. 4c,d) has devastating effects on the calculated phase differences. Retrieving the phase with the cross-spectrum based Knox-Thompson method yields the directed motion up to σ=λ/16 (Fig. 4f–h) even though the autocorrelation halves after a few frames (Fig. 4k).

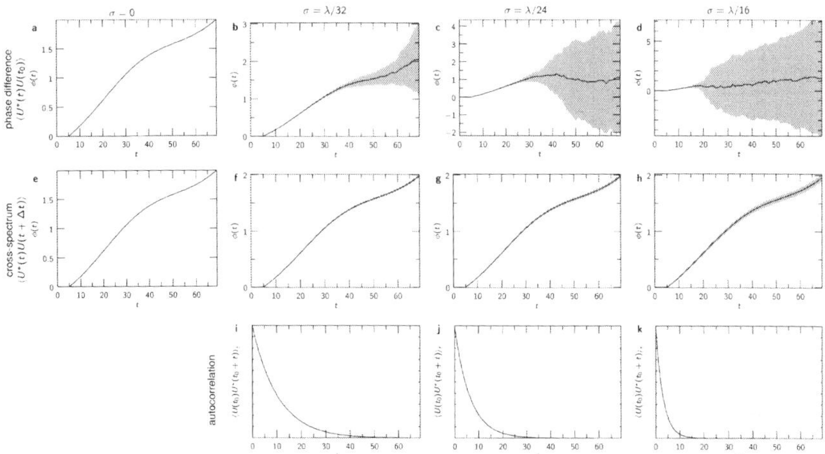

Figure 4. Extracting a single phase curve from simulated images of point scatterers that move between successive measurements by subtracting the phase of the first image (**a–d**) and extended Knox-Thompson evaluation of phases (**e–h**) while random 3D motion of the simulated scatterers is increased from 0 to λ/16. (**i–k**) Autocorrelation of temporal changes of the wave field. Each curve was simulated 100×; the grey areas indicate the standard deviation of the obtained values.

Finally, the dependency on the number of independent samples that are averaged is shown in the Supplementary Information. Fig. S1 shows that smaller averaging numbers increase the fluctuations of the results but do not lead to a general bias. However, for small averaging numbers, the fluctuations increase dramatically. In this simulation, the statistical fluctuations of these averaged and laterally adjacent points should be fully independent. Consequently, these points only share the systematic phase change that we aim to extract.

The resulting curves also show that deviations from the ground truth add up over time, i.e., they increase with time t. Although one might expect that, in particular, the extended Knox-Thompson method would be subject to this, it is also clear from Fig. 4 that the effect with the extended Knox-Thompson method is lower than when computing phase differences. The strength of this overall error depends on multiple factors, such as statistical and independent phase noise, how many values are averaged, and on the strength of the (simulated) Brownian motion. Two kinds of effects might contribute to these phase deviations with increasing time t. Uncorrelated phase noise that is statistically independent in all acquired or simulated images and decorrelation effects that alter the phases and the speckle with time, but where a correlation remains in successive images. The strong discrepancies observed in Fig. 4 result from decorrelating effects instead of uncorrelated phase noise. The extended Knox-Thompson method cannot improve results if uncorrelated phase noise dominates the images or the data and should in many cases be more vulnerable to this phase noise than the standard phase differences.

To further quantify these results we looked at the average sum of all squared differences of the theoretical curves and the actual curves for all simulated datasets, as well as the difference of the final time-point to the ground-truth curve, which might show accumulated errors (Table 1). As soon as some degree of speckle decorrelation is introduced, the extended Knox-Thompson method outperformed the simple phase differences. For the extended Knox-Thompson method, deviations from the ground truth are orders of magnitude lower than for simple phase differences.

Table 1. Quantitative comparison of the simulated results (compare Fig. 4). Values indicate the mean squared errors over the simulated datasets taken all values (mean over all *t*) or only the final values in a time series (final *t*) into account.

	Mean over all *t*		**Final *t***	
	Phase differences	**Extended Knox-Thompson**	**Phase differences**	**Extended Knox-Thompson**
$\sigma=0$	0.0±0.0	0.0±0.0	0.0 ±0.0	0.0 ±0.0
$\sigma=\lambda/32$	0.17±0.38	0.00038±0.00042	2.8±6.6	0.0010±0.0012
$\sigma=\lambda/24$	2.1±1.3	0.00068±0.00082	7.3±8.1	0.0015±0.0020
$\sigma=\lambda/16$	3.7±1.3	0.0024±0.0031	6.8±7.8	0.0054±0.0073

In Vivo Experiments

The *in vivo* experiments confirm the simulation results. Here, we imaged human retina with full-field swept-source optical coherence tomography (FF-SS-OCT, see Methods), which provides three-dimensional tomographic data comprising amplitude and phase, to show the elongation of photoreceptor cells in the living human retina. Phase differences between both ends of the photoreceptor outer segments were evaluated before and during stimulating the photoreceptors with an "x"-shaped light-stimulus, similar to the simulation (see Methods). Figure 5a–g show the extracted phase difference of the 5th frame (the beginning of the stimulus) to various other frames of the data set. The last frame was acquired 14.6 seconds later. After 5 seconds the phase differences are dominated by random phases and the speckle patterns of the compared images changed considerably after this time (Fig. 5o–u). Diffusion and uncorrected tissue motion are likely to be the dominant factors. The phase evaluation using the cross-spectrum is nevertheless able to reconstruct the phase difference and show the elongation of the photoreceptor outer segments (Fig. 5h–n). Even in the non-stimulated area of Fig. 5g, it is clearly seen that the phase difference gives random results, whereas the extended Knox-Thompson approach shows the expected small changes (Fig. 5n). It is also worth mentioning that the phase differences wrap causing the blue colour within the stimulated area in Fig. 5c–g. In contrast to this, the values of the extended Knox-Thompson do not wrap but are clipped at $+\varpi$ in the false-colour representation of Fig. 5j–n.

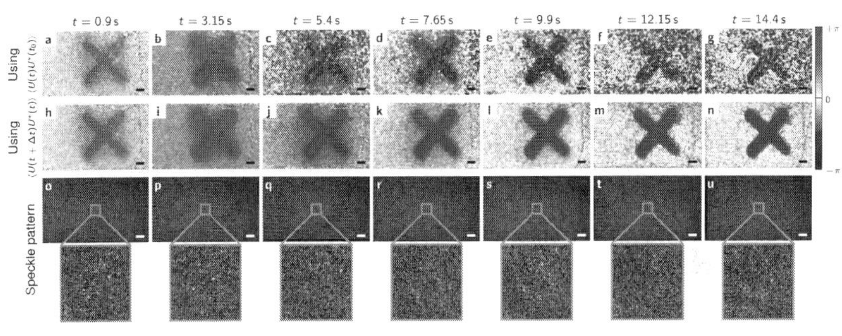

Figure 5. Phase differences between the ends of the photoreceptor outer segment in the living human eye at different times compared to the initial phase at $t=0$. The entire measurement ended at about $t=14.6$ s after initiating a light stimulus. (**a–g**) Phase difference obtained by directly comparing to pre-stimulus phase with phase after different times. (**h–n**) Phase difference obtained from the cross-spectrum as described in the Methods section (by minimizing Eq. (14)). (**o–u**) Speckle pattern of one of the involved layers before the stimulus and at the respective times. It can clearly be seen, that that the speckle pattern changes with time. Scale bars are 200 μm.

A second measurement with a brief stimulation lasting 200 ms that was taken over 8 seconds is shown in Fig. S3. This measurement demonstrates the relaxation of the signals. It is visible that both methods, using phase differences and the extended Knox-Thompson method, yield comparable results here, albeit the extended Knox-Thompson method shows reduced noise in the resulting phase images. Small residual signals remain in either method.

Finally, we also compared phase results for different averaging parameters (Supplementary Document Fig. S2), represented by the size of the Gaussian filter in Eq. (11). The results show that even for small averaging of laterally adjacent pixels the extended Knox-Thompson method outperforms simple phase differences. Contrary to the simulated case, lateral pixels might still show some correlation in their phase fluctuations since the actual motion is unknown and there might be systematic motion that is correlated in regions smaller than the filter size. The results indicate, however, that this does not result in major problems for the extended Knox-Thompson algorithm.

While uncorrelated phase noise is inevitable using FF-SS-OCT, it is not dominating the image degradation in Fig. 5a–g. Rather, major improvements in the *in vivo* experiments are visible when using the extended Knox-Thompson method, which indicates that decorrelating speckle patterns and phases dominate the imaging scenario and these originate from the specimen, i.e., the retina in these images.

Processing Times

Obviously, the extended Knox-Thompson method has additional computational cost compared to the simple phase differences. Table 2 lists the required execution times for a single dataset as shown in Figs 3–5 as measured on an Intel Core i7 8700 with 32 GB of RAM. Although the extended Knox-Thompson method requires a processing time exceeding the phase differences by up to 2 orders of magnitude, processing times remain manageable with under 2 minutes in all tested cases. For our specific imaging scenario, this time is negligible when compared to the required pre-processing time (see Methods), comprising reconstruction, motion correction, registration and segmentation, which takes hours for one dataset.

Table 2. Computational time required to evaluate single datasets of the data shown in Figs 3–5, for both simulated and experimental data.

	Simulated data		**Experimental data**
	Image (Fig. 3)	**Curve (Fig. 4)**	**Image (Fig. 5)**
Standard	0.34 ± 0.08s	0.27 ± 0.06s	2.33 ± 0.06s
Extended Knox-Thompson	109 ± 6s	40 ± 1s	109 ± 2s

CONCLUSION

Building the phase differences of successive frames, and then summing or integrating the differences has been used in phase sensitive imaging[16,17,18,19,20]. It removes the burden of phase unwrapping and thereby improves results. In contrast, including an average over multiple speckles after computing the phase differences and before integrating those differences again, as done by the Knox-Thompson method in astronomic speckle interferometry, allows phase sensitive imaging over more than ten times the speckle decorrelation time. To our knowledge, this important step has never been fully realized in phase sensitive imaging. The resulting method overcomes limitations from decorrelating speckle patterns and recovers phase differences even if speckle are completely decorrelated. Using extended Knox-Thompson methods yields the optimized results presented here and removes the sensitivity to outliers. The method achieves this by attempting to separate random phase and speckle decorrelations (such as Brownian motion) and systematic phase changes. This requires some form of presumptions on the data, e.g., on the spatial dependence of the systematic changes, that will influence the results of the algorithm and inaccurate assumptions introduce some form of error into the results. But, since simple phase differences do not yield any meaningful results after decorrelation of the speckles, the proposed extended Knox-Thompson method is an improved tool for phase sensitive imaging in these scenarios.

Still, measurements and the algorithm can be further improved in several regards. We approximated the ensemble average either by averaging a certain area of lateral pixels or by a Gaussian filter. The former is suitable to obtain the mean expansion in a certain area, the latter is used to obtain images. But to get high-resolution images, using spatial filtering is not ideal. Other speckle averaging techniques, e.g., based on non-local means[35,36], and time-encoded manipulation of the speckle pattern by deliberately manipulating the sample irradiation (similar to[37]) should preserve the full spatial resolution.

Increasing the sampling rate of the t axis of the cross-spectrum, i.e., decreasing the smallest Δt, should also improve results, since then the correlation of compared speckle patterns is improved. The results presented here are limited by the finite sampling interval of the t-axis and not by the overall measurement time; however, at some point, it is expected that errors add up for high frequent sampling of t as data size increases further. Alternatively, one could acquire two immediate frames with a small Δt, followed by a larger time gap to the next two frames (during which speckle patterns can begin to decorrelate). The immediate frames can give the correct current rate of phase changes (the first derivative of the phase change) that can then be extrapolated linearly to the time of the next two frames, etc. Obviously, this can be extended to three or more frames with small Δt giving the second or even higher order local derivative, respectively.

There are also extensions of the presented method possible. For example, the extraction of systematic phase changes in a dynamic, random speckle pattern can be combined with dynamic light scattering (DLS)[6,7] basically separating the random diffusion from systematic particle motion. This approach might give additional contrast compared to either method on its own.

Overall, applications of phase sensitive imaging over long times are manifold. Applications range from biological phase imaging to measure retinal pulse waves[19], detect cellular activity[9,10], all the way to visualize deformations using electronic speckle interferometry (ESPI). In particular, for OCT imaging of neuron function of layers that are more severely decorrelating and have lower signal-to-noise ratio compared to the photoreceptors, the proposed method will be of importance.

METHODS

Simulation

For the simulation we created a complex valued image series with 640×368 pixels and a pixel spacing of 4λ, where λ is the simulated light wavelength. To obtain an image series we created a collection of $50 \times 640 \times 368 = 11{,}776{,}000$ point scatterers each getting a random x, y, and z coordinate, as well as an amplitude A. While the x and y coordinate was distributed entirely random in the image area, the z coordinate was restricted between 0 and 10λ and the amplitude was equally distributed between 0 and 1. To create an image we iterated over all scatterers, and summed the values $A\exp(i2kz)$ for the pixel corresponding to the scatterer's x and y coordinate, where z is the scatterer's z-value, and $k=2\varpi/\lambda$. This basically corresponds to a phase one would obtain in reflection geometry when neglecting any possible defocus. Finally the image was laterally filtered simulating a limited numerical aperture (NA).

For each simulation we created a series of 70 images; starting with the initial scatterers, to create the next frame, we moved each scatterer randomly as specified by a Gaussian distribution with a certain variance σ^2 in x, y, and

z direction. Furthermore, all scatterers currently having x and y-coordinates as found in a pre-created "x"-shaped mask were additionally subjected to the movement

$$\Delta z = \begin{cases} \frac{\lambda}{400} + \frac{\lambda}{800}\sin(0.1\,(t - t_0)) & t \geq t_0 \\ 0 & t < t_0 \end{cases},$$

where t is the frame number and $t_0 = 5$ in each frame.

Experiments

In vivo data was acquired with a Michelson interferometer-based full-field swept-source optical coherence tomography system (FF-SS-OCT) as shown in Fig. 6. The setup is similar to the system previously used for full-field and functional imaging[9,19,38]. Light from a Superlum Broadsweeper BS-840-1 is collimated and split into reference and sample arm. The reference arm light is reflected from a mirror and then directed by a beam splitter onto a high-speed area camera (Photron FASTCAM SA-Z). The sample light is directed in such a way, that it illuminates the retina with a collimated beam; the light backscattered by the retina is imaged through the illumination optics and the beamsplitter onto the camera, where it is superimposed with the reference light.

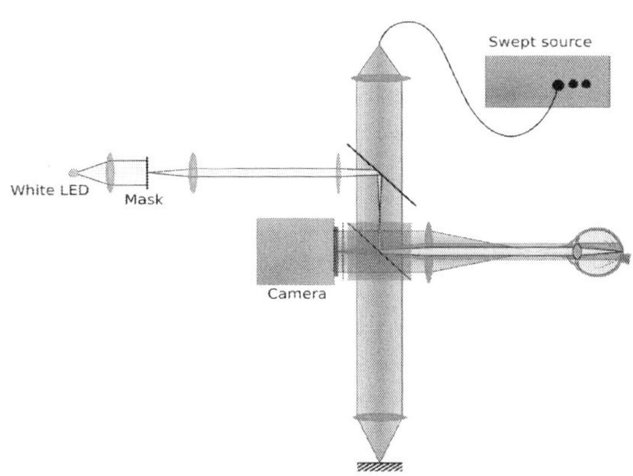

Figure 6. Full-field swept-source optical coherence tomography setup used for acquiring *in vivo* data.

In addition the retina was stimulated with a white LED, where a conjugated plane to the retina contained an "x"-shaped mask. A low-pass optical filter in front of the camera ensured that the stimulation light does not reach the camera.

Swept laser, camera, and stimulation LED were synchronized by an Ardunio Uno microprocessor board. It generates the trigger signals such that the laser sweeps 70× for one dataset. During each sweep the camera acquires 512 images with 640×368 pixels each at a framerate of 60,000 frames/s. Each set of 512 images corresponds to one OCT volume. The stimulation LED is triggered after the first 5 volumes and remains active for the rest of the dataset. The time between volumes determines the overall measurement time. The number of 70 volumes was limited by the camera's internal memory since the acquisition of the camera is too fast to stream the data to a computer in real time.

Measurement light had a centre wavelength of 841 nm and a bandwidth of 51 nm. About 5 mW of this light reached the retina illuminating an area of about 2.6 mm × 1.5 mm, which is the area imaged onto the camera. The "x"-pattern stimulation had significantly less irradiant power, of about 20 μW white light.

Phase Stability

Additionally, we characterized the phase stability of the setup by measuring a cover slip glass that contains two reflecting surfaces over almost 30 seconds. As shown in the Supplementary Information (Fig. S4), the phase observed in a single lateral position in one of the two surfaces of the cover slip fluctuates with time. As a consequence, the autocorrelation of this single layer on its own decreases significantly after a few seconds. Taking the phase difference of both layers shows excellent phase stability and autocorrelation over the entire measurement time. Our measurements are thus limited by phase decorrelation of the specimen and not by the phase stability of the system.

In Vivo Data Acquisition

All investigations were done with healthy volunteers; written informed consent was obtained from all subjects. Compliance with the maximum permissible exposure (MPE) of the retina and all relevant safety rules was confirmed by the responsible safety officer. The study was approved by the ethics board of the University of Lübeck (ethics approval Ethik-Kommission Lübeck 16-080).

Data Reconstruction

Data was reconstructed, registered and segmented as described in previous publications[10]: After a background removal the OCT signal was reconstructed by a fast Fourier transform (FFT) along the spectral axis. Next step was a dispersion correction by multiplication of the complex-valued volumes in the axial Fourier domain (after an FFT along the depth axis) with a correcting phase function that was determined iteratively by an optimization of sharpness metric of the OCT images[38]. Co-registering the volumes aligned the same structures in the same respective voxels of the data series as precisely as the present and changing speckle patterns permitted. This registration also took other layers of the retina into account making use of all retinal structures. Finally, segmenting the average volume allowed aligning of the photoreceptor layers in a certain constant depth for easier phase extraction.

Phase Evaluation

Computing the Cross-Spectrum

To compute the cross-spectrum, we took the series of OCT volumes $U(\boldsymbol{x},t)$ and computed the cross-spectrum by

$$C_U(\boldsymbol{x}, t, \Delta t) = U(\boldsymbol{x}, t)U^*(\boldsymbol{x}, t + \Delta t).$$

Comparing Phases of Two Different Depths

For the *in vivo* experiments we additionally want to compare the phases between different ranges of layers. Let Z_1 and Z_2 denote the sets of layers, then we computed the effective phase difference cross-spectrum by

$$C_U^{Z_1-Z_2}(x, y; t, \Delta t) = \Big(\sum_{z\in Z_1} C_U(x, y, z; t, \Delta t)\Big)\Big(\sum_{z\in Z_2} C_U(x, y, z; t, \Delta t)\Big)^*, \tag{10}$$

where we now represented the vector $\boldsymbol{x}$ by its components (x,y,z). For the simulation, there is only one layer to be compared, and thus we set the phase difference cross-spectrum equal to the cross-spectrum of this one layer, i.e., $C_U^{Z_1-Z_2} = C_U$.

Approximating the Ensemble Average

To extract a curve the phase difference was approximated by averaging the cross-spectrum over the *x* and *y* coordinates belonging to a mask *M*, i.e.,

$$\langle C_U(t, \Delta t)\rangle_M \approx \frac{1}{|M|} \sum_{(x,y)\in M} C_U^{Z_1-Z_2}(x, y; t, \Delta t),$$

where $|M|$ is the number of pixels in the mask *M*.

To obtain images, instead, the ensemble average was approximated by applying a Gaussian filter (convolving with a Gaussian $G_{\sigma^2_{\text{Gauss}}}$ with variance σ^2_{Gauss}):

$$\langle C_U(x, y; t, \Delta t)\rangle \approx G_{\sigma^2_{\text{Gauss}}}(x, y) * C_U^{Z_1-Z_2}(x, y; t, \Delta t), \tag{11}$$

where * is the convolution in *x* and *y* direction. In effect we used a circular convolution by using a fast Fourier transform (FFT) based algorithm. For Figs 3 and 5 a Gaussian filter of width $\sigma_{\text{Gauss}} = 4$ pixels was used. A comparison of different sizes is found in the Supplementary Information in Fig. S2. For Fig. 4 ensemble averaging was performed over the entire '*x*'-shaped pattern, containing 26,437 independent data points. The dependency on the number of data points for curves is shown in the Supplementary Information (Fig. S1).

The Weights

We additionally used weights to specify how reliable the respective value of the cross-spectrum $\langle C_U\rangle$ is. Unfortunately, most algorithms that were used previously to compute weights in astronomy cannot be computed as efficiently as $\langle C_U\rangle$, since we used a fast convolution to apply the Gaussian filter in Eq. (11). We therefore use the weights

$$w(t, \Delta t) = T\Big(\sum_{(x,y)\in M} \frac{C_U^{Z_1-Z_2}(t, \Delta t)}{|C_U^{Z_1-Z_2}(t, \Delta t)|} - b\Big) \tag{12}$$

for phase curve extraction and

$$w(x, y; t, \Delta t) = T\left(\left|\left[G_{\sigma^2_{\text{Gauss}}}(x, y) * \frac{C_U^{Z_1-Z_2}(x, y; t, \Delta t)}{|C_U^{Z_1-Z_2}(x, y; t, \Delta t)|}\right]\right| - b\right) \tag{13}$$

for phase imaging; both could be computed as efficiently as $\langle C_U\rangle$ itself. In these formula *T* is a cut-off function for negative values:

$$T(x) = \begin{cases} x & \text{for } x \geq 0 \\ 0 & \text{for } x < 0 \end{cases}$$

The truncation of negative values and the introduction of b are supposed to keep the expectation value for randomly distributed C_U at 0. The sum in (12) and the convolution term in (13) follow speckle statistics if the phase of C_U is random, since they are a sum of random phasors1. Consequently, a non-zero value is expected. This can be used to determine suitable values for b. The amplitude probability of a speckle pattern is given by a Rayleigh distribution1

$$p(A) = \frac{A}{\sigma^2}\exp\left(-\frac{A^2}{2\sigma^2}\right).$$

Enforcing a certain threshold A_0 gives a total probability

$$P_0(A_0) = \int_0^{A_0} dA'\, p(A') = 1 - e^{-\frac{A_0^2}{2\sigma^2}}$$

and its inverse

$$A_0(P_0) = \sqrt{-2\sigma^2 \ln(1 - P_0)}\,.$$

With this formula, we selected b to be the 99% threshold, i.e., $b = A_0(0.99)$), except for Fig. 3 for which $b = A_0(0.99999)$ was chosen.

For the two scenarios of curve extraction and imaging with a Gaussian convolution, the parameter σ of the Rayleigh distribution remains to be determined. Starting with the derivation of speckle statistics1, it can be computed to yield

$$\sigma = \frac{1}{\sqrt{2|M|}}$$

for curve extraction and

$$\sigma = \frac{1}{\sqrt{8\pi}\,\sigma_{\text{Gauss}}}$$

for imaging, when assuming unit magnitude phasors (as present in (12) and (13)) with completely random phases.

Phase Unwrapping of the Cross-Spectrum

While alternate approaches to extract the phase from the cross-spectrum deal with the phase wrapping problem differently (e.g.[26]), for our scenario phases of the obtained cross-spectrum need to be unwrapped in the Δt axis in order to obtain good results. However, in 1D phase unwrapping, single outliers, i.e., a random phase for single specific Δt, can tremendously degrade results for all following values. We therefore slightly modified the standard 1D-phase unwrapping approach:

We assume the initial D phases to be free of wrapping. Afterwards we moved through the data set from small Δt to larger Δt. We computed the sum of all absolute values of the phase differences to the preceding D phases for each phase value that is to be determined. We increased or decreased the respective phase value by 2ϖ as long as this difference sum kept decreasing. Then me move to the next Δt. This procedure was done once for curve extraction and repeated for each x and y value for imaging. In our scenario, we used $D=10$.

Obtaining the Phase from the Cross Spectrum

Instead of using the approaches described by (8) or (9), we formulate the recovery of the phases from the cross-spectrum as a linear least squares problem. This also served as basis for many of the previously demonstrated approaches in astronomy. Assuming the phase of the cross-spectrum is unwrapped in its Δt axis, we can assume that

$$\sum_{t,\Delta t} w^2(\boldsymbol{x};\, t,\, \Delta t)|\phi(\boldsymbol{x},\, t + \Delta t) - \phi(\boldsymbol{x},\, t) - \langle C_U(\boldsymbol{x};\, t,\, \Delta t)\rangle|^2 \tag{14}$$

minimizes for the desired phase $\phi(\boldsymbol{x},t)$, if the weights w represent the quality of the cross-spectrum for the respective parameters $\boldsymbol{x}$, t, and Δt. However, the solution to this linear least square problem is not unique: since the cross-spectrum only contains phase differences, there will be different solutions for different initial values $\phi(\boldsymbol{x},t=0)$. In addition, all weights might be 0 for one specific t which represents a gap in reliable data. Both problems in the evaluation can be solved by Tikhonov regularization with small parameters. For the former case one should force the resulting phase for one time point t_0 to be small; the corresponding regularization parameter should not influence other results as long as it is chosen sufficiently small to not introduce numerical errors. For the second problem, a difference regularization can be introduced. Again, the parameter can be small; it is only used is to obtain a unique solutions, in case weights approach 0. Consequently, the entire approach can be formulated as a linear least squares problem, and solved by regularized solving of the respective linear equation.

To solve the least squares problem (14) we first need to discretize it properly. To this end, we first create the discrete vectors and matrices **c**ϕ corresponding to the unwrapped cross-spectrum phase arg $\langle(C_U)\rangle$, ϕ corresponding to the phases to be extracted, and a matrix S relating the two. Assume a given systematic phase ϕ′, then the corresponding cross-spectrum $\boldsymbol{c}_\phi'$ would be uniquely given by

$$\mathbf{c}'_\phi = \begin{pmatrix} \phi'_0 - \phi'_1 \\ \phi'_0 - \phi'_2 \\ \vdots \\ \phi'_1 - \phi'_2 \\ \vdots \end{pmatrix} = \underbrace{\begin{pmatrix} 1 & -1 & 0 & \cdots & 0 \\ 1 & 0 & -1 & \cdots & 0 \\ & & \vdots & \ddots & \\ 0 & 1 & -1 & \cdots & 0 \\ & & \vdots & & \end{pmatrix}}_{S} \begin{pmatrix} \phi'_0 \\ \phi'_1 \\ \vdots \\ \phi'_{N-1} \end{pmatrix}.$$

We can write this as

$$\mathbf{c'}_\phi = S\phi'.$$

Given the actually measured cross-spectrum $\boldsymbol{c}_\phi$ and the corresponding phase ϕ to be computed and introducing the diagonal weight matrix $W=\mathrm{diag}(w_0, w_1, \ldots, w_{N-1})$ as computed by (12) or (13) we can write the determination of ϕ as the regularized minimization problem by

$$\left\|W(S\phi - c_\phi)\right\|^2 + \mu_1\|\Gamma_1\phi\|^2 + \mu_2\|\Gamma_2\phi\|^2 + \mu_3\|\Gamma_3\phi\|^2 \rightarrow \min.$$

The corresponding ϕ is found by

$$\phi = (S^T W^2 S + \mu_1{}^2\Gamma_1^T\Gamma_1 + \mu_2{}^2\Gamma_2^T\Gamma_2 + \mu_3{}^2\Gamma_3^T\Gamma_3)^{-1} S^T W^2 \mathbf{c}_\phi,$$

which needs to be performed for each curve or each lateral pixel (x,y) when doing phase imaging. In general, we chose

$$\Gamma_{1,ij} = \begin{cases} 1 & \text{for } i = j = 4 \\ 0 & \text{otherwise} \end{cases},$$

$$\Gamma_{2,ij} = \begin{cases} 1 & \text{if } i = j \neq 0 \\ -1 & \text{if } i = j + 1, \\ 0 & \text{otherwise} \end{cases}$$

and

$$\Gamma_{3,i,j} = \begin{cases} 2 & \text{if } i = j,\ i \neq 0, \text{ and } i \neq N-1 \\ -1 & \text{if } i = j+1 \text{ or } i = j-1 \\ 0 & \text{otherwise} \end{cases},$$

with μ_1=0.02, μ_2=0.0001, and μ_3=0.02. Without the regularization terms, the matrix S is not invertible or not invertible if there are gaps in the data.

Specifying the Initial Value

The regularization term $\mu_1 \|\Gamma_1\|^2$ basically enforces the phase value corresponding to the 5th volume to 0, thereby specifying the initial value. Since in both, simulation and experiment, we only know that no (deliberate) phase change is occurring in the frames <5, we can normalize the final result by

$$\phi_{\mathrm{res},i} = \phi_i - \frac{1}{N_0} \sum_{j=0}^{N_0-1} \phi_j,$$

with N_0=5.

Real-Valuedness of the Modulating Cross-Spectrum

For the entire approach to work it remains to be shown, that the modulation cross spectrum $\langle C_H \rangle$ is real-valued for small Δt. We assume the modulation cross-spectrum is given according to (2) and (6) by

$$\begin{aligned}
\langle C_H \rangle &= \langle H(t) H^*(t+\Delta t) \rangle \\
&= \left\langle \left(\sum_i A_i(t) e^{i\phi_i(t)} \right) \left(\sum_j A_i(t+\Delta t) e^{i\phi_i(t+\Delta t)} \right)^* \right\rangle \\
&= \left\langle \sum_i A_i(t) A_i^*(t+\Delta t) e^{i(\phi_i(t) - \phi_i(t+\Delta t))} + \sum_{i,j,i\neq j} A_i(t) A_j^*(t+\Delta t) e^{i(\phi_i(t) - \phi_j(t+\Delta t))} \right\rangle \\
&= \left\langle \sum_i A_i(t) A_i^*(t+\Delta t) e^{i(\phi_i(t) - \phi_i(t+\Delta t))} \right\rangle \\
&\quad + \left\langle \sum_{i,j,i\neq j} A_i(t) A_j^*(t+\Delta t) e^{i(\phi_i(t) - \phi_j(t+\Delta t))} \right\rangle.
\end{aligned}$$

Now, the second term is small compared to the first term since ϕ_i and ϕ_j are statistically independent making the sum and the average run over random phasors. The phases of the first term can be approximated by $\phi_i(t) - \phi_i(t) + \partial_{t'} \phi_i(t')|_{t'=t} \Delta t = \partial_{t'} \phi_i(t')|_{t'=t} \Delta t$ and thus will be 0 for small Δt that are within good autocorrelation of the modulating function H. Thus $\langle C_H \rangle$ is real for small Δt to a good approximation.

Phase Evaluation by Phase Differences

In the alternative approach we compute merely phase differences. We can still formulate this by using the cross-spectrum of phase difference $C_U^{Z_1-Z_2}$ given by (10). For all values $t \geq t_0$, the phase is obtained by

$$\phi(t) = \arg\{G_{\sigma^2_{\text{Gauss}}} * C_U^{Z_1-Z_2}(x, y; t_0, t)\}$$

for imaging and

$$\phi(t) = \arg \sum_{(x,y)\in M} C_U^{Z_1-Z_2}(x, y; t_0, t)$$

for curve evaluation. For this direct phase computation, the phase-wrapped cross-spectrum was used. For curve extraction the phase was unwrapped in t afterwards.

DATA AVAILABILITY

The datasets generated and/or analysed during the current study are available from the corresponding author on reasonable request.

ACKNOWLEDGEMENTS

This work was funded by the German Research Foundation (DFG), Project Holo-OCT HU 629/6-1.

AUTHOR CONTRIBUTIONS

H.S. and D.H. conceived of the presented method and implemented the algorithm. C.P. and L.K. conducted the experiments. G.H. and D.H. supervised the project. All authors designed the experiments, discussed the results, and contributed to the fnal manuscript.

Supplementary Information

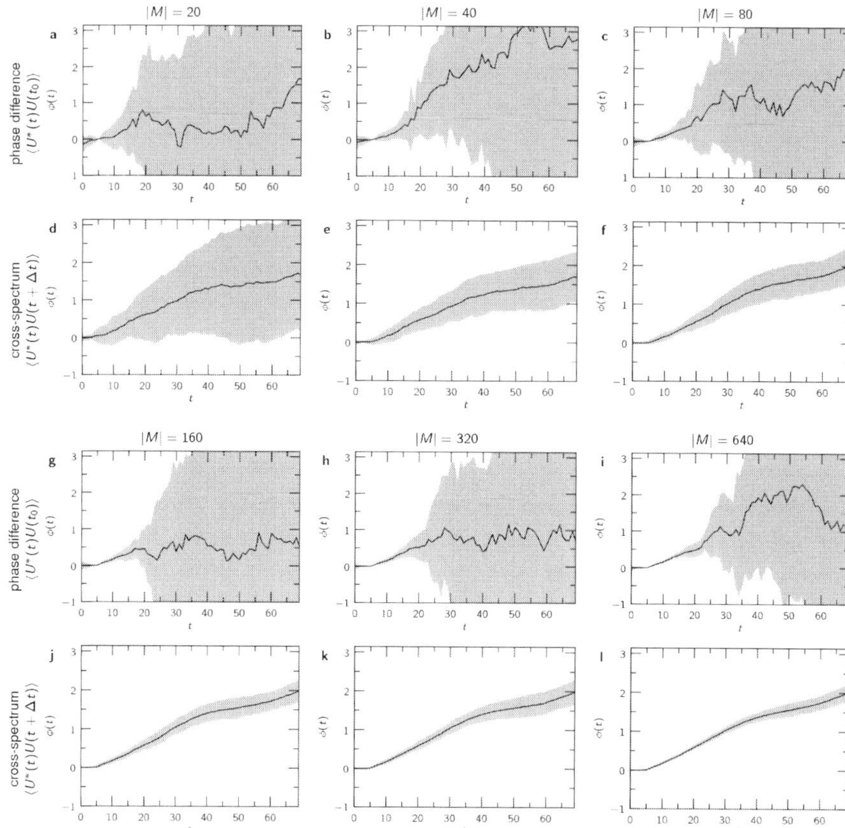

Figure S1. Simulation results (compare Fig. 4 in the main text) for various averaging parameters of sizes |M| between |M| = 20 and |M| = 640 using both standard phase difference method (a-c and g-i) and the extended Knox-Thompson method (d-f and j-l). For all averaging sizes, the extended Knox-Thompson method shows reduced fluctuations and better reproduction of the input curves compared to phase differences. Neither method shows any significant bias in their estimation of the phase response.

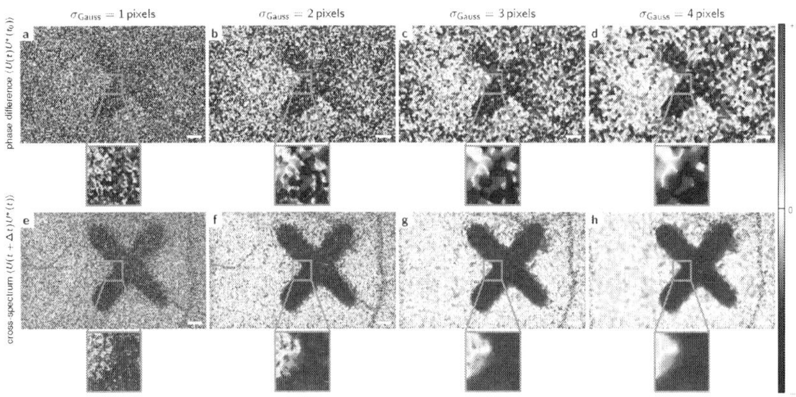

Figure S2. Experimental results (compare Fig. 5 in the main text) for various averaging parameters, here represented by the size of the applied Gaussian filter σ_{Gauss}, using both standard phase difference method (a-d) and the extended Knox-Thompson method (e-h). For all filters, the extended Knox-Thompson method shows reduced fluctuations compared to phase differences.

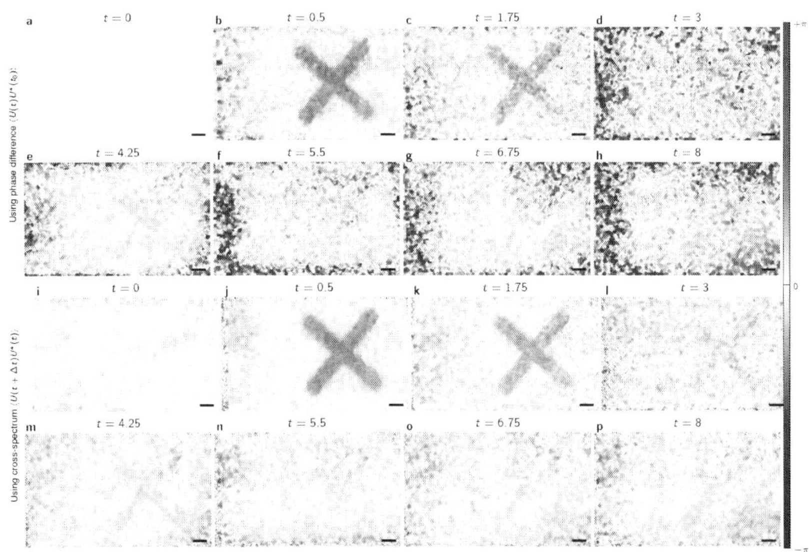

Figure S3. Experimental results (comparable to Fig. 5 in the main text), albeit only with a brief stimulation of 200ms, showing the initial increase and the following partial relaxation of the signal over 8 s. In both methods a slight remaining 'x' is still visible at the end of the measurement indicating that there is remaining expansion in the stimulated area compared to the initial time. The extended Knox-Thompson method shows reduced noise, especially close to the edges at later times (compare f, g, and h with n, o, and p, respectively).

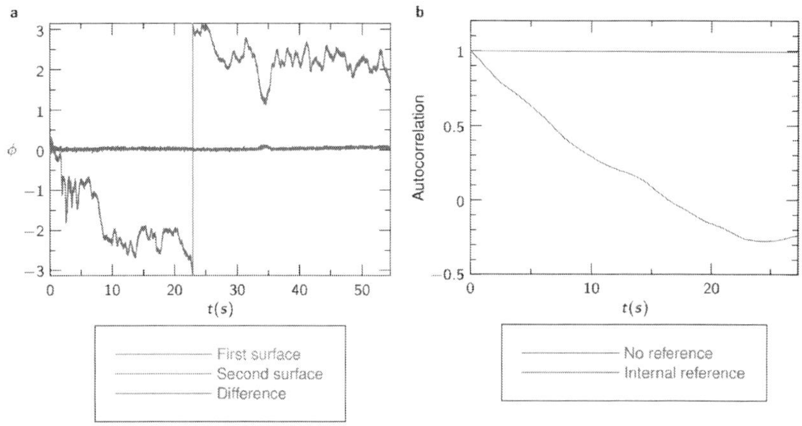

Figure S4. Phase stability of the OCT system used for this study. It was evaluated by measuring phases from a cover slip over more than 50 s and evaluation the autocorrelation over more than 25 s of the complex wave fields for certain depths. a) Phases from two layers of the cover slip (red and green) and their phase difference (blue) for one single lateral position of the coverslip. b) Autocorrelation of the complex field corresponding to one surfaces (red) and to the phase difference of both surfaces (blue). The autocorrelation is averaged over multiple lateral positions. The measurement demonstrates that phases referenced to a different layer from the same measurement can create phase stability exceeding 30 s.

REFERENCES

1. Goodman, J. *Speckle Phenomena in Optics: Theory and Applications* (Roberts & Company, 2007).
2. Creath, K. Phase-shifting speckle interferometry. *Appl. Opt.* **24**, 3053–3058, https://doi.org/10.1364/AO.24.003053 (1985).
3. Joenathan, C., Haible, P. & Tiziani, H. J. Speckle interferometry with temporal phase evaluation: influence of decorrelation, speckle size, and nonlinearity of the camera. *Appl. Opt.* **38**, 1169–1178, https://doi.org/10.1364/AO.38.001169 (1999).
4. Lehmann, M. Decorrelation-induced phase errors in phase-shifting speckle interferometry. *Appl. Opt.* **36**, 3657–3667, https://doi.org/10.1364/AO.36.003657 (1997).
5. Jones, R. & Wykes, C. De-correlation effects in speckle-pattern interferometry. *Opt. Acta* **24**, 533–550, https://doi.org/10.1080/716099421 (1977).
6. Berne, B. & Pecora, R. *Dynamic Light Scattering: With Applications to Chemistry, Biology, and Physics. Dover Books on Physics Series* (Dover Publications, 2000).
7. Goldburg, W. I. Dynamic light scattering. *Am. J. Phys* **67**, 1152–1160, https://doi.org/10.1119/1.19101 (1999).
8. Chen, C.-L. & Wang, R. K. Optical coherence tomography based angiography [invited]. *Biomed. Opt. Express* **8**, 1056–1082, https://doi.org/10.1364/BOE.8.001056 (2017).
9. Hillmann, D. *et al. In vivo* optical imaging of physiological responses to photostimulation in human photoreceptors. *Proc. Natl. Acad. Sci. USA* **113**, 13138–13143, https://doi.org/10.1073/pnas.1606428113 (2016).
10. Pfäffle, C. *et al.* Functional imaging of ganglion and receptor cells in living human retina by osmotic contrast. ArXiv e-prints 1809.02812, https://arxiv.org/abs/1809.02812 (2018).
11. Zhang, P. *et al. In vivo* optophysiology reveals that G-protein activation triggers osmotic swelling and increased light scattering of rod photoreceptors. *Proc. Natl. Acad. Sci.* **114**, E2937–E2946, https://doi.org/10.1073/pnas.1620572114, https://www.pnas.org/content/114/14/E2937.full.pdf (2017).

12. Zhang, F., Kurokawa, K., Lassoued, A., Crowell, J. A. & Miller, D. T. Cone photoreceptor classification in the living human eye from photostimulationinduced phase dynamics. *Proc. Natl. Acad. Sci.* **116**, 7951–7956, https://doi.org/10.1073/pnas.1816360116 (2019).
13. Azimipour, M., Migacz, J. V., Zawadzki, R. J., Werner, J. S. & Jonnal, R. S. Functional retinal imaging using adaptive optics swept-source OCT at 1.6 MHz. *Optica* **6**, 300–303, https://doi.org/10.1364/OPTICA.6.000300 (2019).
14. Kennedy, B. F., Kennedy, K. M. & Sampson, D. D. A review of optical coherence elastography: Fundamentals, techniques and prospects. *IEEE J. Sel. Top. Quantum Electron.* **20**, 272–288, https://doi.org/10.1109/JSTQE.2013.2291445 (2014).
15. Chin, L. *et al.* Analysis of image formation in optical coherence elastography using a multiphysics approach. *Biomed. Opt. Express* **5**, 2913–2930, https://doi.org/10.1364/BOE.5.002913 (2014).
16. Wang, R. K., Kirkpatrick, S. & Hinds, M. Phase-sensitive optical coherence elastography for mapping tissue microstrains in real time. *Appl. Phys. Lett.* **90**, 164105, https://doi.org/10.1063/1.2724920 (2007).
17. An, L., Chao, J., Johnstone, M. & Wang, R. K. Noninvasive imaging of pulsatile movements of the optic nerve head in normal human subjects using phasesensitive spectral domain optical coherence tomography. *Opt. Lett.* **38**, 1512–1514, https://doi.org/10.1364/OL.38.001512 (2013).
18. Li, P. *et al.* Phase-sensitive optical coherence tomography characterization of pulse-induced trabecular meshwork displacement in *ex vivo* nonhuman primate eyes. *J. Biomed. Opt* **17**, 17–17–11, https://doi.org/10.1117/1.JBO.17.7.076026 (2012).
19. Spahr, H. *et al.* Imaging pulse wave propagation in human retinal vessels using full-field swept-source optical coherence tomography. *Opt. Lett.* **40**, 4771–4774, https://doi.org/10.1364/OL.40.004771 (2015).
20. van Brug, H. & Somers, P. A. A. M. Speckle decorrelation: Observed, explained, and tackled. In Jacquot, P. & Fournier, J.-M. (eds) *Interferometry in Speckle Light*, 11–18 (Springer Berlin Heidelberg, Berlin, Heidelberg, 2000).
21. Knox, K. T. & Thompson, B. J. Recovery of images from atmospherically degraded short-exposure photographs. *Astrophys. J.* **193**, L45–L48, https://doi.org/10.1086/181627 (1974).

22. Lohmann, A. W., Weigelt, G. & Wirnitzer, B. Speckle masking in astronomy: triple correlation theory and applications. *Appl. Opt.* **22**, 4028–4037, https://doi.org/10.1364/AO.22.004028 (1983).
23. Ayers, G. R., Northcott, M. J. & Dainty, J. C. Knox–Thompson and triple-correlation imaging through atmospheric turbulence. *J. Opt. Soc. Am. A* **5**, 963–985, https://doi.org/10.1364/JOSAA.5.000963 (1988).
24. Lannes, A. Backprojection mechanisms in phase-closure imaging. Bispectral analysis of the phase-restoration process. *Exp. Astron* **1**, 47–76, https://doi.org/10.1007/BF00414795 (1989).
25. Marron, J. C., Sanchez, P. P. & Sullivan, R. C. Unwrapping algorithm for least-squares phase recovery from the modulo 2π bispectrum phase. *J. Opt. Soc. Am. A* **7**, 14–20, https://doi.org/10.1364/JOSAA.7.000014 (1990).
26. Haniff, C. A. Least-squares Fourier phase estimation from the modulo 2π bispectrum phase. *J. Opt. Soc. Am. A* **8**, 134–140, https://doi.org/10.1364/JOSAA.8.000134 (1991).
27. Roggemann, M., Welsh, B. & Hunt, B. *Imaging Through Turbulence. Laser & Optical Science & Technology* (Taylor & Francis, 1996).
28. Mikurda, K. & Der Lühe, O. V. High resolution solar speckle imaging with the extended Knox–Thompson algorithm. *Sol. Phys.* **235**, 31–53, https://doi.org/10.1007/s11207-006-0069-6 (2006).
29. Bamler, R. & Just, D. Phase statistics and decorrelation in SAR interferograms. In Geoscience and Remote Sensing Symposium, 1993. IGARSS'93. Better Understanding of Earth Environment., International, 980–984 vol.3, https://doi.org/10.1109/IGARSS.1993.322637 (1993).
30. Zebker, H. A. & Villasenor, J. Decorrelation in interferometric radar echoes. *IEEE T. Geosci. Remote* **30**, 950–959, https://doi.org/10.1109/36.175330 (1992).
31. Lanari, R. *et al.* An overview of the Small BAseline Subset algorithm: A DInSAR technique for surface deformation analysis. In Wolf, D. & Fernández, J. (eds) Deformation and Gravity Change: Indicators of Isostasy, Tectonics, Volcanism, and Climate Change, 637–661 (Birkhäuser Basel, Basel, 2007).
32. Ferretti, A., Prati, C. & Rocca, F. Permanent scatterers in SAR interferometry. *IEEE T. Geosci. Remote* **39**, 8–20, https://doi.org/10.1109/36.898661 (2001).

33. Sanderson, C. & Curtin, R. Armadillo: a template-based C++ library for linear algebra. *J. Open Source Softw* **1**, 26, https://doi.org/10.21105/joss.00026 (2016).
34. Sanderson, C. & Curtin, R. A user-friendly hybrid sparse matrix class in C++. *Lect. Notes Comput. Sci* **10931**, 422–430, https://doi.org/10.1007/978-3-319-96418-8_50 (2018).
35. Deledalle, C., Denis, L. & Tupin, F. NL-InSAR: Nonlocal interferogram estimation. *IEEE T. Geosci. Remote* **49**, 1441–1452, https://doi.org/10.1109/TGRS.2010.2076376 (2011).
36. Lucas, A. *et al*. Insights into Titan's geology and hydrology based on enhanced image processing of Cassini RADAR data. *J. Geophys. Res. –Planet* **119**, 2149–2166, https://doi.org/10.1002/2013JE004584 (2014).
37. Liba, O. *et al*. Speckle-modulating optical coherence tomography in living mice and humans. *Nat. Commun.* **8**, 15845, https://doi.org/10.1038/ncomms15845 (2017).
38. Hillmann, D. *et al*. Aberration-free volumetric high-speed imaging of *in vivo* retina. *Sci. Rep* **6**, 35209, https://doi.org/10.1038/srep35209 (2016).

CHAPTER 9

THE APPLICATIONS OF THE HETERODYNE INTERFEROEMETRY

Cheng-Chih Hsu

Department of Photonics Engineering, Yuan Ze University, Yuan-Tung Road, Chung-Li, Taiwan

INTRODUCTION

Optical interferometry is widely used in many precision measurements such as displacement[1, 2], vibration[3, 4], surface roughness[5, 6], and optical properties[7-14] of the object. For example, holographic interferomter [1-3] can be used to measure the surface topography of the rigid object. The emulsion side of the photographic plate faces the object and is illuminated by a plane wave at normal incidence. Therefore, the reflection type hologram is recorded the interference signals between the incident wave and scattered wave from the object within the emulsion layer. Then the hologram is reconstructed with laser light and the information of object

Citation: (APA): Hsu, C. C. (2012). The Applications of the Heterodyne Interferoemetry. In Interferometry-Research and Applications in Science and Technology (pp. 31-64). IntechOpen. (36 pages)

surface can be obtained. The Speckle interferometry [2-4] can be used to measure the motion of the rough surface. To compare the two exposure specklegrams, then the phase difference related to the surface movement can be obtained. Abbe refractometer [7, 8] is an easy method to determine the refractive index of the material based on the total internal reflection (TIR). That means the refractive index of the testing sample will be limit by the hemisphere prism installed in the refractometer. The ellipsometer [9-12] is widely used to measure the thickness and refractive index of film or bulk materials. Typically, the optical components of ellipsometer included polarizer, compensator, sample, and analyzer. Hence, there were many different types of ellipsometer for refractive index and thickness measurement of the sample. Most popular type is rotating polarizer and analyzer ellipsometer which can be divided into rotating polarizer type and rotating analyzer type. Both of them are analysis of the ellipsometric angles (ψ, Δ) which determined directly from the adjustable angular settings of the optical components. The accuracy of the ellipsometric measurement are typically within the range 0.01° and 0.05° in (ψ, Δ) [13, 14].

Compare to previous method, the heterodyne interferometry give much more flexibility of different kinds of the measurement purposes with suitable optical configuration. In this chapter, I will review the heterodyne interferometry and focus on the applications of this kind of interferometer. First of all, I will briefly introduce the history and applications of heterodyne interferometry that will be discussed in this chapter. Before I mention the applications of the heterodyne interferometry, I would like to describe several types of heterodyne interferometry. Then I would like to describe the precision positioning with optical interferometer and focus on the heterodyne grating interferometer. After that, I will review some refractometer using heterodyne interferometer. In this section I would like to quick look some useful methods for measuring the refractive index and thickness of bulk material or thin film structure. In addition, the measurement of the optic axis and birefringence of the birefringent crystal will also be discussed in this section. The final application of the heterodyne interferometry that I would like to talk about is the concentration measurement. In this section, I will roughly classify the method into two categories. One is fiber type sensor; another is a non-fiber type sensor. And I will discuss the surface plasmon resonance (SPR) sensor in fiber-type and non-fiber type sensors. Finally, I would like to give the short conclusion, which summarized the advantages and disadvantages of the heterodyne interferometer.

HETERODYNE INTERFEROMETRY

This section will introduce the development history of the heterodyne interferometry and describe the fundamental theory and basic optical configuration of the heterodyne interferometer.

History of Heterodyne Light Source Development

Hewlett Packard Company (HP) developed the first commercial heterodyne interferometer for precision positioning since 1966. Until now, HP systems have widely used in industry, scientific research, and education. J. A. Dahlquist, D. G. Peterson, and W. Culshaw [15] demonstrated an optical interferometer, which used Zeeman laser properties in 1966. They had the application of an axial magnetic field and resulted in the frequency difference between the right hand and left hand circular polarization states of the He-Ne laser. Because of these two polarization states are affected by equally thermal drift and mechanism vibration of the laser, the frequency difference are extremely stable. Therefore, this light source with different frequency is so called the heterodyne light source. Figure 1 showed that the first heterodyne interferometer which constructed with Zeeman laser. As you can see, the frequency shift coming from the moving mirror will be carried with v_2. Then these two lights with different frequency will be interference at 45 and the distance-varying phase can be detected.

There are many methods can construct the optical frequency shift such as rotation or moving grating method [16, 17], accousto-optical modulator (AOM) [18, 19], electro-optical modulator (EOM) [20, 21], and modulating two slightly different wavelengths of laser diodes [22]. Suzuki and Hioki [16] proposed the idea of moving grating method for constructing the heterodyne light source in 1967. As the grating moves along y-axis with the velocity v, the frequency shift will be introduced into the ±1 order diffracted beam with $\pm\frac{v}{a}$.

By suitable arrangement of the optical configuration, either one of these frequency shifted signals can be selected and to form the heterodyne light source. W. H. Stevenson [17] proposed the rotation radial grating to form the heterodyne light source in which he showed that the frequency shift were linear increased with the rotation rate of the radial grating up to 6k rpm. And the maximum frequency shift in this case was 500 kHz.

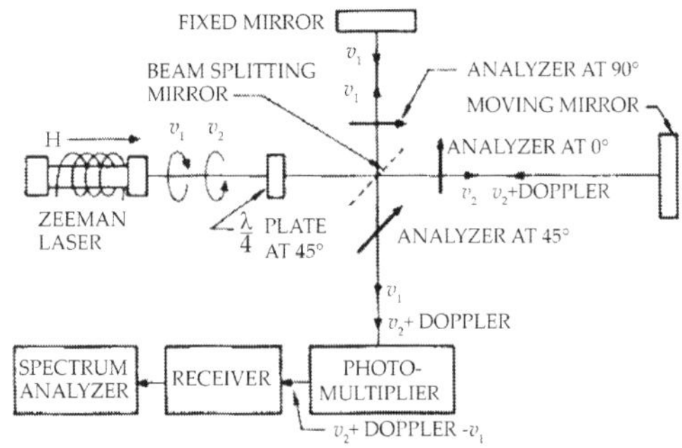

Figure 1. The first heterodyne interferometer constructed by Zeeman laser [15].

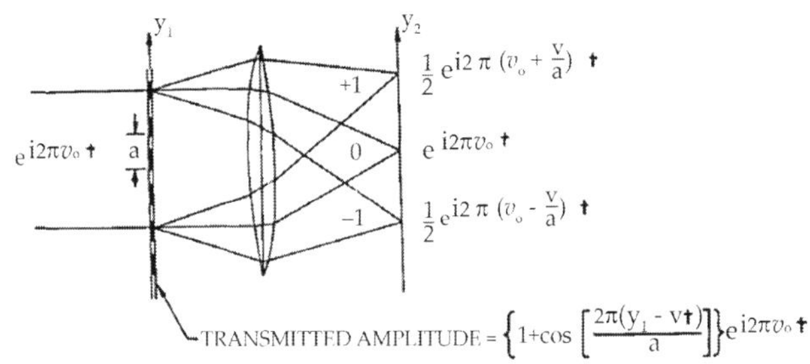

Figure 2. The heterodyne light source constructed with moving grating [17].

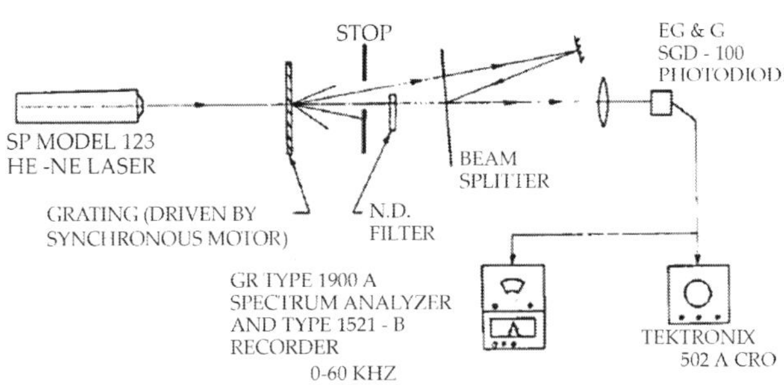

Figure 3. The heterodyne light source constructed with rotation radial grating [17].

An acousto-optic modulator (AOM) uses the acousto-optic effect to diffract and shift the frequency of the light [18, 19]. The piezoelectric transducer attaches to the quartz and the oscillating electric signal drives the transducer to vibrate, which creates sound wave in the quartz and changes the refractive index of the quartz as periodic index modulation. The incoming light diffracts by these moving periodic index modulation planes, which induced the Doppler-shifted by an amount equal to the frequency of the sound wave. That phenomenon is similar to the moving grating method but the fundamental concepts are momentum conservation of the phonon-photon interaction and Bragg diffraction theory. Figure 4 shows the frequency shifted by AOM that proposed in 1988 [18]. A typical frequency shifted varies from 27 MHz to 400 MHz. In the case of M. J. Ehrlich et al. [18], the frequency shifted was 29.7 MHz and the induced phase shifted over 360° by applying the voltage within 15 V.

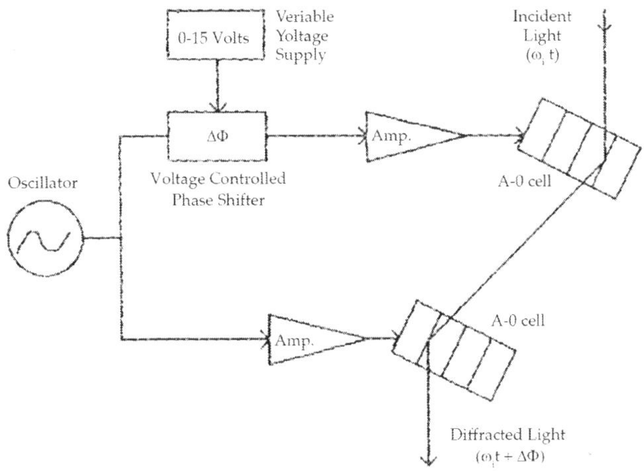

Figure 4. The frequency shifted by AOM [18].

Electro-optic modulator is a signal-controlled optical device that based on the electro-optic effect to modulate a beam of light. The modulation may be imposed the phase, amplitude, or frequency of the modulated beam. Lithium niobate ($LiNbO_3$) is one of the electro-optic crystals that is widely used for integrated optics device because of its large-valued Pockels coefficients. The refractive index of $LiNbO_3$ is a linear function of the strength of the applied electric field, which is called Pockel effect. Figure 5 shows one of the optical configurations of the heterodyne light source constructed by EOM. The linear polarized light into the EOM, which the crystal axis is located at 45° respected to the x-axis and applied half-wave voltage $V_{\lambda/2}$ on it, the outcome light will carry the frequency shifted.

The wavelength of laser diode can be varied as the injection current and temperature of the laser diode. The wavelength increased as the injection current increased. In general, the rate of the increase is about 0.005 nm/mA at 800 nm and that will be different for different types of laser diode [22]. As the wavelength of the laser diode is changed from λ to λ+Δλ periodically, in which the injection current is periodically changed, the frequency shift of the heterodyne signal can be obtained.

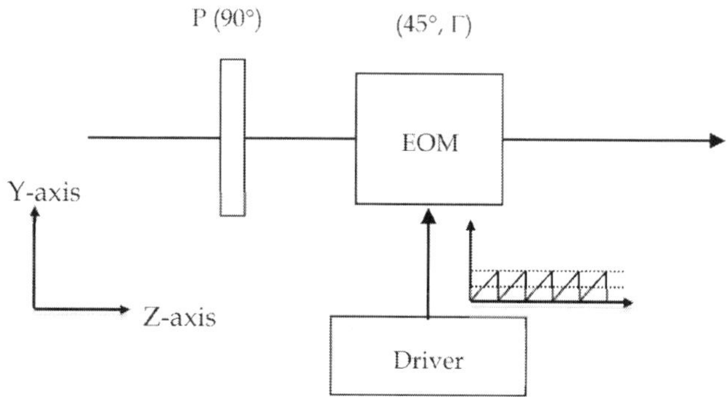

Figure 5. The optical setup of heterodyne light source with EOM.

Type of Heterodyne Interferometer

The heterodyne interferometer can be divided into two categories, one is common-path type and another is non common-path type. The common-path means that the environment influence of the polarization states of the interference signal can be ignored. Of course, one also can divide into linear polarized heterodyne and circular polarized heterodyne interferometers based on the heterodyne light source. In this section, we would like to describe that based on heterodyne light source and focus on the boundary phenomena between the heterodyne light source and testing sample.

The optical configuration of the linear polarized heterodyne light source have described in figure 5. For convenient, assume that the light propagate along z-axis and vertical direction is y-axis. If the fast axis of the EOM is located at 45° respected to the x-axis, the Jones matrix can be described [14, 23]:

$$EO(45°,\Gamma) = \begin{pmatrix} \cos 45^o & -\sin 45^o \\ \sin 45^o & \cos 45^o \end{pmatrix} \begin{pmatrix} e^{i\frac{\Gamma}{2}} & 0 \\ 0 & e^{-i\frac{\Gamma}{2}} \end{pmatrix} \begin{pmatrix} \cos 45^o & \sin 45^o \\ -\sin 45^o & \cos 45^o \end{pmatrix} = \begin{pmatrix} \cos\frac{\Gamma}{2} & i\sin\frac{\Gamma}{2} \\ i\sin\frac{\Gamma}{2} & \cos\frac{\Gamma}{2} \end{pmatrix} \quad (1)$$

where the Γ is the phase retardation of EOM and can be described $\Gamma = \frac{\pi V}{V_{\lambda/2}}$. When we applied half-wave voltage of the EOM with sawtooth electric signal, equation (1) can be approximated as

$$\mathrm{EO}(\omega t) = \begin{pmatrix} e^{-im\pi\frac{i\omega t}{2}} & 0 \\ 0 & e^{im\pi\frac{-i\omega t}{2}} \end{pmatrix} = \begin{pmatrix} e^{i\frac{\omega t}{2}} & 0 \\ 0 & e^{-i\frac{\omega t}{2}} \end{pmatrix} \tag{1}$$

As a linear polarized light with the polarization direction at 45° pass through the EOM, then the E-field can be

$$E = EO(\omega t) \cdot E_{in} = \begin{pmatrix} e^{i\frac{\omega t}{2}} & 0 \\ 0 & e^{-i\frac{\omega t}{2}} \end{pmatrix} \cdot \frac{1}{\sqrt{2}}\binom{1}{1}e^{i\omega_0 t} = \frac{1}{\sqrt{2}}\begin{pmatrix} e^{i\frac{\omega t}{2}} \\ e^{-i\frac{\omega t}{2}} \end{pmatrix} e^{i\omega_0 t} \tag{3}$$

where ω_0 and ω are optical frequency and frequency shifted between two orthogonal polarization state, respectively. Obviously, equation (3) described the linear polarized heterodyne light source.

For a circular polarized heterodyne light source, the optical configuration is showed in figure 6. As a linear polarized light pass through EOM and quarter-wave plate Q with the azimuth angle at 0°, the Jones matrix of the E-field of the outcome light can be described

$$\begin{aligned} E' &= Q(0°) \cdot EO(\omega t) \cdot E_{in} \\ &= \begin{pmatrix} 1 & 0 \\ 0 & i \end{pmatrix} \begin{pmatrix} \cos(\omega t/2) & i\,\sin(\omega t/2) \\ i\,\sin(\omega t/2) & \cos(\omega t/2) \end{pmatrix} \binom{1}{0} = \begin{pmatrix} \cos(\omega t/2) \\ -\sin(\omega t/2) \end{pmatrix} \\ &= \frac{1}{2}\binom{1}{i}e^{\frac{i\omega t}{2}} + \frac{1}{2}\binom{1}{-i}e^{-\frac{i\omega t}{2}}. \end{aligned} \tag{4}$$

Obviously, equation (4) describes the circular heterodyne light source that indicated the frequency shifted ω between left-hand circular polarized light and right-hand circular polarized.

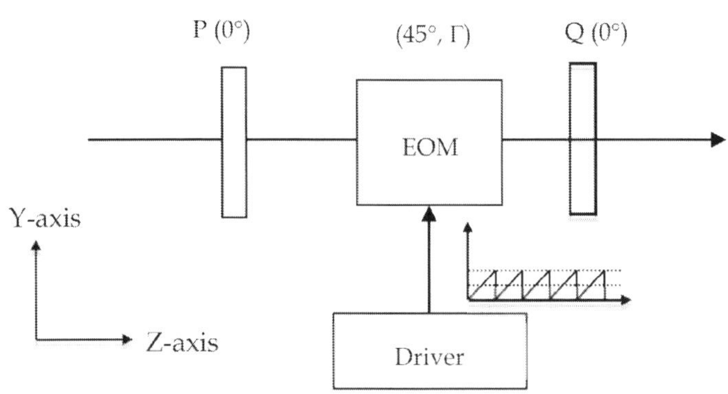

Figure 6. The circular polarized heterodyne light source.

If the optical interferometer is constructed of the circular polarized heterodyne light source, we always call that circular heterodyne interferometer otherwise we call that heterodyne interferometer. For the specific purpose, we will arrange the tested system as transmission type, reflection type, and multi-reflection type according to the optical property of the testing sample. These types are summarized and show in figure 7. It is obvious that the polarization states (p- and s- polarization states or right-hand and left-hand circular polarization states) of the heterodyne light source are propagated at the same optical path, in which we call common-path structure. The advantage of the common-path structure is the influence of the polarization states of the heterodyne light source can be assumed and limited to the acceptable value. In general, we can ignore the error when the measurement system with common-path configuration. In figure 7, the reference signal I_r coming from the function generator can be written as

$$I_r = I'[1 + \cos(\omega t)], \tag{5}$$

and direct into the lock-in amplifier. The heterodyne light source will pass through or reflect from the tested system and then pass through the analyzer AN_t with azimuth angle at α, finally detect by photodetector D_t. The tested system can be divided into three types based on the optical property of the testing sample. There are transmission, reflection, and multi-reflection types.

To consider a heterodyne light source passed through the transmission materials which induced the phase retardation ϕ, the E-field and intensity detected by D_t can be written as

$$E_t = AN_t(\alpha) \cdot W \cdot E_{in} = \frac{1}{2}\begin{pmatrix} 1 & 1 \\ 1 & 1 \end{pmatrix}\begin{pmatrix} e^{\frac{i\varphi}{2}} & 0 \\ 0 & e^{-\frac{i\varphi}{2}} \end{pmatrix}\begin{pmatrix} \cos\frac{\omega t}{2} \\ -\sin\frac{\omega t}{2} \end{pmatrix},$$

$$= \left[\cos\alpha \cos\frac{\omega t}{2} e^{i\frac{\varphi}{2}} - \sin\alpha \sin\frac{\omega t}{2} e^{-i\frac{\varphi}{2}}\right] \cdot \begin{pmatrix} \cos\alpha \\ \sin\alpha \end{pmatrix} \tag{6}$$

and

$$I_t = |E_t|^2 = \frac{1}{2}\left[1 + 2\sqrt{A^2 + B^2}\cos(\omega t + \phi)\right] \tag{7}$$

where W is the Jones matrix of testing sample at transmission condition; A, B, and φ can be written as

$$A = \frac{1}{2}(\cos^2\alpha - \sin^2\alpha), \tag{8a}$$

$$B = \cos\alpha\sin\alpha\cos\varphi, \tag{8b}$$

and

$$\phi = \tan^{-1}\left(\frac{B}{A}\right) = \tan^{-1}\frac{2\cos\alpha\sin\alpha\cos\varphi}{(\cos^2\alpha - \sin^2\alpha)}, \tag{8c}$$

It is obvious that the phase retardation ϕ will be carried by the testing signal It. To compare I_r and I_t with lock-in amplifier, the phase difference φ coming from the testing sample can be obtained. Substitute the phase difference into equation (8c), the phase retardation of the sample can be determined.

Of course, if the testing sample is not transparence, the reflection type or multi-reflection type can be applied to measure the optical property of the testing sample. To consider a circular heterodyne light source is reflected by the testing sample, passed through the analyzer with the azimuth angle α, and finally detected by photodetector. According to Jones calculation, the E-field and intensity can be expressed as

$$E_t = AN(\alpha)\cdot S\cdot E'$$

$$= \begin{pmatrix} \cos^2\alpha & \sin\alpha\cos\alpha \\ \sin\alpha\cos\alpha & \sin^2\alpha \end{pmatrix}\begin{pmatrix} r_p & 0 \\ 0 & r_s \end{pmatrix}\begin{pmatrix} \cos\frac{\omega t}{2} \\ -\sin\frac{\omega t}{2} \end{pmatrix}$$

$$= \left(r_p\cos\alpha\cos\frac{\omega t}{2} - r_s\sin\alpha\sin\frac{\omega t}{2}\right)\begin{pmatrix} \cos\alpha \\ \sin\alpha \end{pmatrix} \tag{9}$$

and

$$I_t = |E_t|^2 = I_0[1 + \cos(\omega t + \phi)] \tag{10}$$

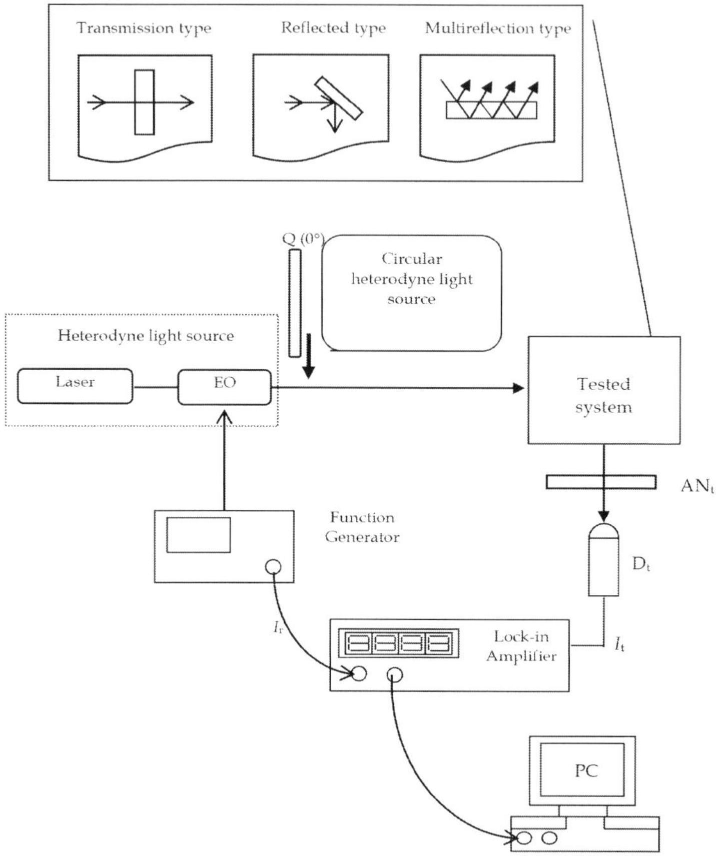

Figure 7. Types of the heterodyne interferometer with common-path structure.

Where **S** is the Jones matrixof testing sample at reflection condition, r_p and r_s are the reflection coefficients, I_0 and φ are the average intensity and phase difference coming from the sample between p- and s- polarizations, which can be written as

$$I_0 = \frac{(r_p{}^2 \cos^2 \alpha + r_s{}^2 \sin^2 \alpha)}{2} \qquad (11a)$$

and

$$\phi = \tan^{-1}\left(\frac{2 \sin \alpha \cos \alpha \cdot r_p r_s}{r_p{}^2 \cos^2 \alpha + r_s{}^2 \sin^2 \alpha}\right) \qquad (11b)$$

The reflection coefficients in the reflection matrix of the sample can be expressed by Fresnel equation that can be divided into single reflection and multi-reflection depended on the testing structure. Hence, the r_p and r_s can be written as [14, 23]

(1) single reflection

$$r_p = \frac{n_2 \cos\theta - n_1 \cos\theta_t}{n_2 \cos\theta + n_1 \cos\theta_t}, \tag{12ab}$$

$$r_s = \frac{n_1 \cos\theta - n_2 \cos\theta_t}{n_1 \cos\theta + n_2 \cos\theta_t}, \tag{12b}$$

(2) multi-reflection

$$r_p(\beta) = \frac{r_{1p} + r_{2p}\exp(i\beta)}{1 + r_{1p}r_{2p}\exp(i\beta)}, \tag{13a}$$

$$r_s(\beta) = \frac{r_{1s} + r_{2s}\exp(i\beta)}{1 + r_{1s}r_{2s}\exp(i\beta)}, \tag{13b}$$

and

$$\beta = \frac{2\pi n_2 d \cos(\theta_t)}{\lambda}. \tag{13c}$$

Where n_1 and n_2 are the refractive indices of air and testing sample, θ and θ_t are the incident angle and refracted angle, β is the phase difference coming from the optical path difference in the testing sample, λ is the wavelength of the heterodyne light source. It is obvious that the optical properties of the testing sample can be obtained by substitute phase difference into the equations (10) ~ (13).

On the other hand, the typical optical configuration of the non-common path is shown in figure 8. It is clear that p- and s- polarizations will be propagated at two different paths when they passed through the polarization beam splitter (PBS). In practice, the environment disturbance will not be neglected in non-common path configuration because of these two orthogonal polarization states will have different influence at different path. Therefore, the non-common path optical interferometry using for precision measurement should be seriously taken consideration of stability of the environment disturbance. Figure 8 shows the optical configuration of the displacement measurement. The p- and s- polarizations will。reflect by mirrors M_1 and M_2, then pass through the analyzer with azimuth angle at

45°. Therefore, the E-field and intensity of the interference signal between two arms can be written as

$$E_t = \begin{pmatrix} e^{\frac{i\omega t}{2} - ik(2d_p)} \\ e^{\frac{i\omega t}{2} - ik(2d_s)} \end{pmatrix} e^{i\omega_0 t}, \tag{14}$$

and

$$I_t = \frac{1}{2}\left\{1 + \cos\left[\omega t - \frac{4\pi}{\lambda}(d_p - d_s)\right]\right\}. \tag{15}$$

If the mirror M_1 is moved with time, the phase difference $\frac{4\pi d_p}{\lambda}$ will be changed and the displacement variation can be measured by comparing the testing signal and reference signal with lock-in amplifier.

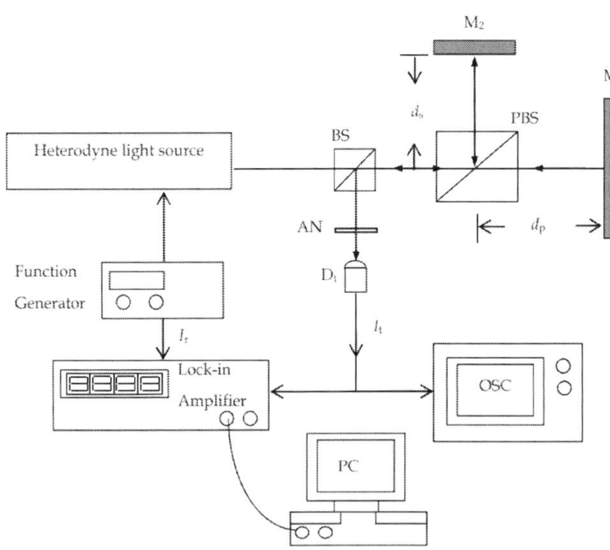

Figure 8. The optical configuration of the non-common path displacement measurement.

ACCURATE POSITIONING WITH HETERODYNE INTERFEROMETER

Nano-scale positioning devices have become a significant requirement in scientific instruments used for nanotechnology applications. These devices can be applied to nano-handling, nanomanipulation, and nanofabrication. In addition, they are an essential part of the scanning probe microscopy (SPM) and widely used in many research fields. The precision positioning devices consist of three principle parts, which are the rolling component, the driving system and the position sensor. Piezoelectric actuator is the most popular method for driving system and commercial products have been on the market for a few decades. Therefore, the piezoelectric actuator and the position sensor will play the role of the positioning and the feedback control of the rolling element. To achieve the high resolution positioning, the sensing methods of position sensor become more important and have attracted great attention over the past two decades. In this section, we will introduce a few of typical precision positioning methods [24-30] which used heterodyne interferometry.

C. C. Hsu [29] proposed the grating heterodyne interferometry (GHI) to measure the in-plane displacement. The schematic diagram of this method is shown in figure 9. The diffracted grating has mounted on the motorized stage and four diffracted lights will diffract and propagate in the x-z and y-z planes which are for measuring the displacement in x- and y- directions respectively. Based on the Jones calculation, the E-field of the ±1 diffracted beams can be expressed

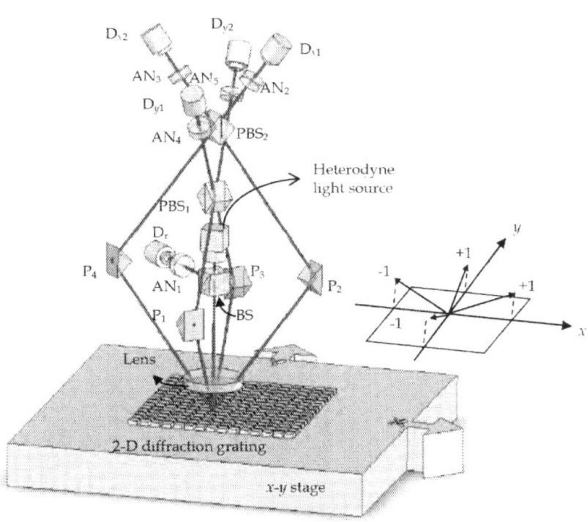

Figure 9. 2-D displacement measurement system with GHI [29].

$$E_{x\pm1} \propto \exp\left(ikl_{x\pm1} \pm i\frac{2\pi}{g_x}d_x\right)\cdot E_H = \exp\left(ikl_{x\pm1} \pm i\frac{2\pi}{g_x}d_x\right)\begin{pmatrix} e^{i\omega t/2} \\ e^{-i\omega t/2}\end{pmatrix}, \text{(x-direction)} \tag{16a}$$

and

$$E_{y\pm1} \propto \exp\left(ikl_{y\pm1} \pm i\frac{2\pi}{g_y}d_y\right)\cdot E_H = \exp\left(ikl_{y\pm1} \pm i\frac{2\pi}{g_y}d_y\right)\begin{pmatrix} e^{i\omega t/2} \\ e^{-i\omega t/2}\end{pmatrix}, \text{(y-direction)} \tag{16b}$$

where g_x and g_y are the grating pitch in x- and y- directions, d_x and d_y are the displacement along the x- and y- directions respectively. To consider x-direction displacement measurement, the ±1st order diffracted lights will be collected by a lens L and propagate into two paths: (1) prism P_2 → polarization beam splitter PBS_2 → analyzer AN_2 (45°) → detector Dx_1, (2) prism P_4 → polarization beam splitter PBS_2 → analyzer AN_3 (45°) → detector Dx_2. It is similar to the y-direction displacement measurement, the ±1st order diffracted lights will be propagated into (3) prism P_1 → polarization beam splitter PBS_2 → analyzer AN_4 (45°) → detector Dy_1, (4) prism P_3 → polarization beam splitter PBS_2 → analyzer AN_5 (45°) → detector Dy_2. After Jones calculation, they can be written as

$$I_{x1} \propto \left|E_{+1s,x} + E_{+1p,x}\right|^2 = \frac{1}{2}\left[1 + \cos\left(\omega t + k\left(l_{-1,x} - l_{+1,x}\right) - \frac{4\pi}{g_x}d_x\right)\right], \tag{17a}$$

$$I_{x2} \propto \left|E_{-1s,x} + E_{+1p,x}\right|^2 = \frac{1}{2}\left[1 + \cos\left(\omega t + k\left(l_{+1,x} - l_{-1,x}\right) - \frac{4\pi}{g_x}d_x\right)\right], \tag{17b}$$

$$I_{y1} \propto \left|E_{+1s,y} + E_{-1p,y}\right|^2 = \frac{1}{2}\left[1 + \cos\left(\omega t + k\left(l_{-1,y} - l_{+1,y}\right) - \frac{4\pi}{g_y}d_y\right)\right], \tag{18a}$$

and

$$I_{y2} \propto \left|E_{-1s,y} + E_{+1p,y}\right|^2 = \frac{1}{2}\left[1 + \cos\left(\omega t + k\left(l_{+1,y} - l_{-1,y}\right) - \frac{4\pi}{g_y}d_y\right)\right]. \tag{18b}$$

To compare the equations (17) and (18), the phase difference coming from the movement in x- and y- directions can be obtained and expressed as

$$\phi_i = \frac{8\pi}{g_i}d_i + 2kl_i \quad (i = x, y), \tag{19}$$

where l_x and l_y are the path difference between grating and PBS_1 and PBS_2 respectively. In practice, the second term in equation (19) can be assumed the initial phase. Therefore, the displacement can be obtained as phase difference is measured and grating pitch is given.

Figure 10 shows that the 2-D displacement measurement with 2-D grating. The movement of the stage is toward to 45° respected to the x-direction and moved 180 nm. The displacement projection in the x- and y-direction are about 120 nm and 140 nm respectively. It is obvious that there are small difference between the results measured by GHI and HP 5529A. Hsu's results can observe that the sensitivity of GHI is higher than

HP 5529A and the smallest displacement variation can be judged is about 6 pm. Besides, GHI can provide the 2-D displacement monitoring with single measurement apparatus which have many advantages such as easy alignment, high cost/preference ratio, and easy integrated to the motorized system.

Recently, J. Y. Lee [30] proposed a novel method to measure the 2-D displacement which have quasi-common optical path (QCOP) configuration. The optical structure is shown in figure 11. Based on the clever arrangement, the expanded heterodyne beam is divided into 4 parts A, B, C and D. According to the Jones calculation, the amplitudes of these 4 parts are given by

$$E_B = E_C = J(180°) \cdot E_H = \begin{pmatrix} e^{-i\omega t/2} \\ e^{i\omega t/2} \end{pmatrix}, E_A = E_D = J(0°) \cdot E_H = \begin{pmatrix} e^{i\omega t/2} \\ e^{-i\omega t/2} \end{pmatrix}. \quad (20)$$

The expanded heterodyne beam will reflect by a mirror and focus by a lens with suitable focal length, in which can make the zero order (m=0) beam overlap with the ±1 order diffracted beams. The beam distribution is shown in detail in the inset. When the grating moves along the x direction, the interference phase changes can be observed from the

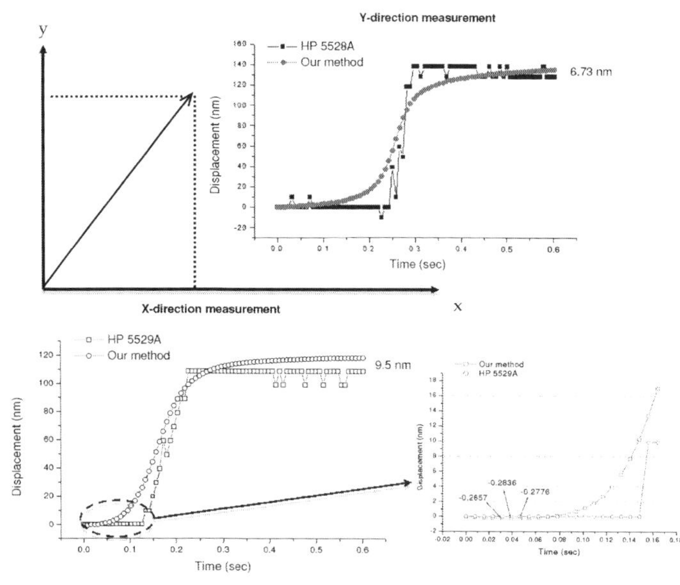

Figure 10. Displacement measurement of 2-D movement of motorized stage with GHI and HP 5529A.

overlapping area O_1 to O_4; when the grating moves along the *y* direction, the interference phase changes can be observed from the overlapping area O_5 to O_8. One can use an iris before the focus lens to control the overlapping area. The overlapping areas (O_1 and O_5) are chosen to pass through two polarizers P_1 and P_2 with transmittancc axes at 0°. The interference of the light is detected using two detectors D_1 and D_2. The interference signal I_1 and *I*2 measured by the detectors D1 and= D_2 can be written as

$$I_1 = 1 + \cos(\omega t - \phi_{x1}), \tag{21a}$$

$$I_2 = 1 + \cos(\omega t - \phi_{y1}). \tag{21b}$$

A polarizer P_3 for which the transmittance axis is at 45° and a detector D_3 are used to measure the intensity of the non-overlapping areas which can be a reference signal I_3

$$I_3 = 1 + \cos(\omega t), \tag{22}$$

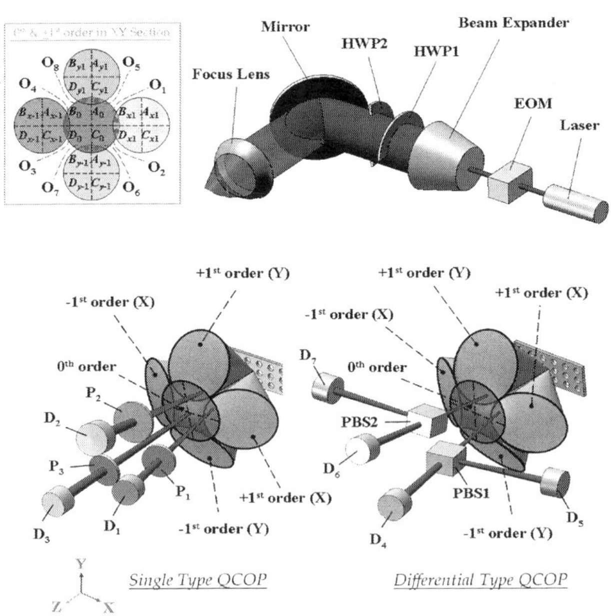

Figure 11. The schematic of the single type and differential type QCOP heterodyne grating interferometer.

These three signals I_1, I_2, and I_3 are sent into the lock-in amplifier and the phase differences $\Phi_x = \varphi_{x1}$ (between I_1 and I_3) and $\Phi_y = \varphi_{y1}$ (between I_2 and I_3) are given by

$$\phi_i = \frac{2\pi}{g_i} d_i \quad (i = x, y), \tag{23}$$

where d_i is the displacement in x- and y- directions; g_i is the grating pitch of 2-D grating in x-and y- directions. It is obvious that the 2-D displacement can be obtained as the phase difference and grating pitch of the 2-D grating are given. In the differential type QCOP method, two polarization beam splitters (PBSs) are used to separate the two overlapping beams into four parts. Therefore, the interference signals detected by D_4, D_5, D_6, and D_7 can

$$I_4 = 1 + \cos(\omega t - \phi_{x1}) \text{ and } I_5 = 1 + \cos(\omega t + \phi_{x1}); \quad \text{(for x- direction)} \tag{24a}$$

$$I_6 = 1 + \cos(\omega t - \phi_{y1}) \text{ and } I_7 = 1 + \cos(\omega t + \phi_{y1}). \quad \text{(for y- direction)} \tag{24b}$$

These two pairs signal are sent into the multi-channel lock-in amplifier, the phase differences $\Phi_x = \varphi_{x1} - (-\varphi_{x1})$ (between I_4 and I_5) and $\Phi_y = \varphi_{y1} - (-\varphi_{y1})$ (between I_6 and I_7) are 4 times of φ_i.

Figure 12 shows a top view of the experimental results in the XY section and the XY stepper moves with a displacement of 1 mm. It is clear that the slight difference between the results measured by the laser encoder and QCOP method. The difference is coming from a tiny angle between the moving direction and the grating, which can be alignment by mounting 2D grating on the rotation stage. In their case, the larger difference was about 12 μm in the y-direction and the smallest difference was about 29 nm in the x- direction for a displacement of 1 mm. Based on the error analysis, if the phase resolution (0.001°) of the lock-in amplifier is considered, the corresponding displacement resolutions of the differential and single type interferometers are estimated to be 9 pm and 4.5 pm for a grating pitch of 3.2 μm, respectively. If only high frequency noise is considered, the measurement resolution of the differential and single type QCOP interferometers can be estimated to be 1.41 nm and 2.52 nm.

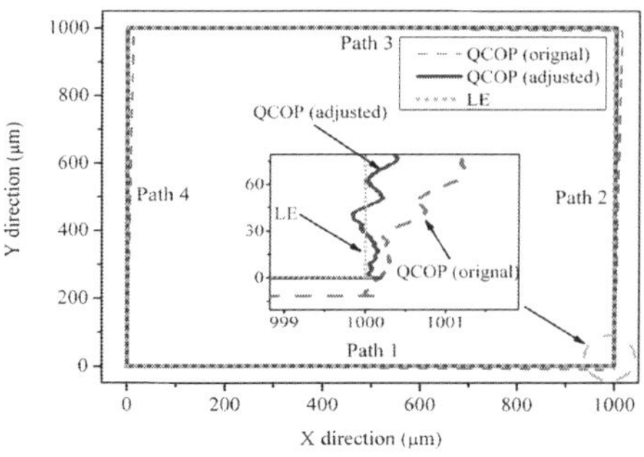

Figure 12. Displacement measurement of the quadrangular motion with the QCOP method [30].

OPTICAL CONSTANTS MEASUREMENT WITH HETERODYNE INTERFEROMETER

Optical constants of the materials such as refractive index, birefringence, optical activity, and thickness are significant parameters in material science. There are many methods [9-12, 31-40] can determine those factors, most popular method is ellipsometer [9-12]. Recently, these factors can be obtained by heterodyne interferometer. In this section, we will review some novel methods [31-40] for optical constants measurement with heterodyne interferometer.

C. C. Hsu [32] proposed a novel method for determine the refractive index of the bulk materials with normal incident circular heterodyne interferometer (NICHI) and the schematic diagram is shown in figure 13. The circular heterodyne light source was incident into the modified Twyman-Green interferometer, in which the testing signal reflected from the sample can be interfered and carried by the circular heterodyne light beams. Based on Jones calculation, the interference signal measured by D can be written as

$$I_t = |E_1 + E_2|^2 = I_0[1 + \gamma \cos(\omega t + \phi)], \tag{25}$$

where E_1 and E_2 are the E-field coming from the optical path 1 and path 2 respectively. I_0, γ, and φ are the mean intensity, the visibility and the phase of the interference signal, respectively. In additions, they can derive from the Jones calculation and written as

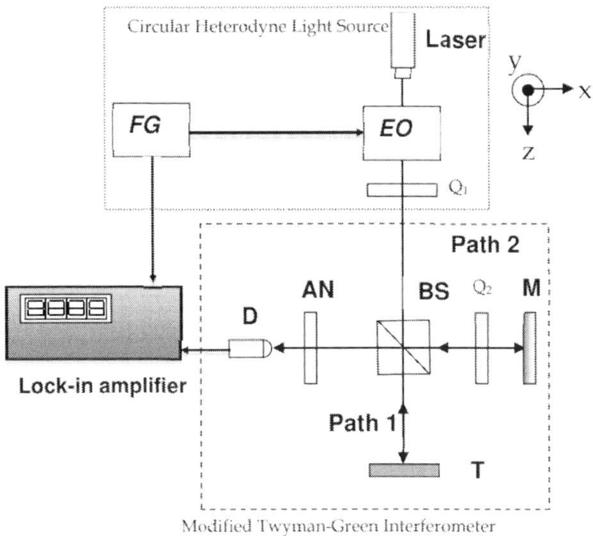

Figure 13. The optical configuration of normal incident circular heterodyne interferometer [32].

$$I_0 = \left[\frac{1}{2}\left(\frac{n-1}{n+1}\right)^2 + 2{r_m}^2 + 2\left(\frac{n-1}{n+1}\right)r_m \sin 2\alpha \sin(\phi_{d1} - \phi_{d2})\right], \tag{26a}$$

$$\gamma = \frac{\sqrt{A^2+B^2}}{\left[\frac{1}{2}\left(\frac{n-1}{n+1}\right)^2+2{r_m}^2+2\left(\frac{n-1}{n+1}\right)r_m \sin 2\alpha \sin(\phi_{d1}-\phi_{d2})\right]}, \tag{26b}$$

$$\phi = \tan^{-1}\left(\frac{B}{A}\right), \tag{26c}$$

where the symbols A and B can be written as

$$A = \frac{1}{2}\left[\left(\frac{n-1}{n+1}\right)^2 - 4{r_m}^2\right] \cdot (\cos^2\alpha - \sin^2\alpha), \tag{27a}$$

$$B = \frac{1}{2}\left[\left(\frac{n-1}{n+1}\right)^2 - 4{r_m}^2\right] \cdot \sin 2\alpha - 2\frac{n-1}{n+1}r_m, \tag{27b}$$

where r_m is the normal reflection coefficients of the test medium. If the phase can be measured and the reflectivity of mirror is given, the refractive index of the testing sample will be obtained. Furthermore, it is clear that the resolution of refractive index is strongly related to the azimuth angle of analyzer and the reflectivity of the mirror. To derive equation (26) to *n*, the resolution of refractive index can be written as

$$\Delta n = \frac{1}{\frac{d\phi}{dn}}|\Delta\phi| = [\frac{ac-b}{cd}]|\Delta\phi|, \tag{28}$$

where a, b, c, and d are

$$a = \frac{2[-\frac{4r_m}{(1+n)^2}+\frac{2(n-1)}{(1+n)^3}\sin 2\alpha]}{[\left(\frac{n-1}{1+n}\right)^2-4{r_m}^2][\cos^2\alpha-\sin^2\alpha]}, \tag{29a}$$

$$b = 2\frac{4(n-1)}{(1+n)^3}\left\{\frac{-2(n-1)r_m}{1+n}+\frac{1}{2}[\left(\frac{n-1}{n+1}\right)^2-4{r_m}^2]\cdot\sin 2\alpha\right\}, \tag{29b}$$

$$c = [\left(\frac{n-1}{n+1}\right)^2-4{r_m}^2]^2(\cos^2\alpha-\sin^2\alpha), \tag{29c}$$

$$d = 1+\frac{4\left\{\frac{-2(n-1)r_m}{n+1}+\frac{1}{2}[\left(\frac{n-1}{n+1}\right)^2-4{r_m}^2]\sin 2\alpha\right\}^2}{[\left(\frac{n-1}{n+1}\right)^2-4{r_m}^2]^2[\cos^2\alpha-\sin^2\alpha]^2}. \tag{29d}$$

The simulation results were shown in figure 14 and resolution of the refractive index can be reached 10^{-5} as the suitable experimental conditions were chosen.

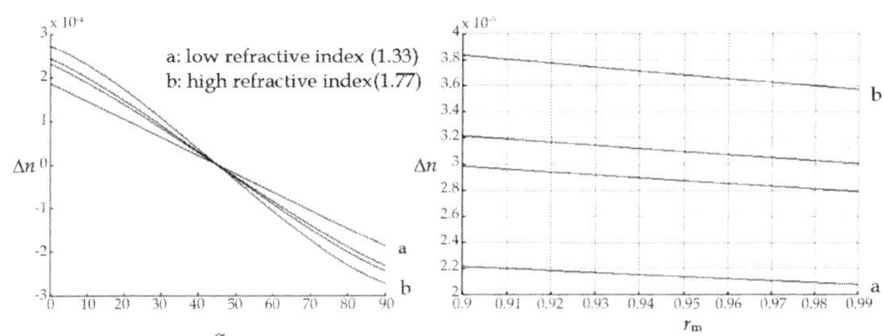

Figure 14. The relationship between the azimuth angle, reflectivity of the mirror, and resolution of the refractive index [32].

In 2010, Y. L. Chen and D. C. Su [38] developed a full-field refractive index measurement of gradient-index lens with normal incident circular heterodyne interferometer (NICHI). They used high speed CMOS camera to record 2D interference signal and the optical configuration was shown in figure 15.

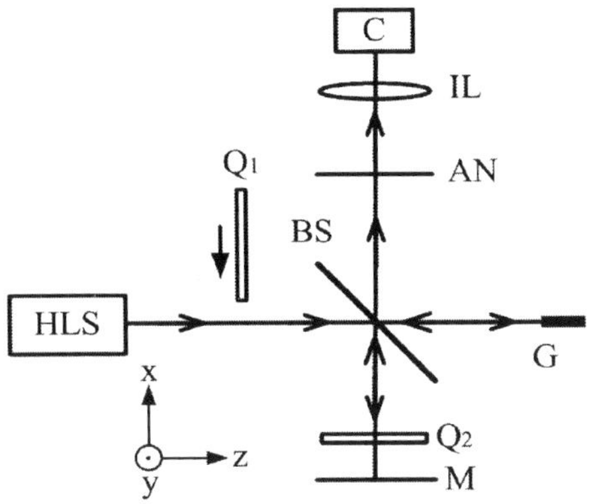

Figure 15. Optical configuration of full-field normal incident circular heterodyne interferometer [38].

In this interferometer, one is for reference beam (BS→Q2→M→Q2→BS→AN→IL→C) and one is for testing beam ((BS→G→BS→AN→IL→C). Here, G means GRIN lens. They were interfere with each other after passing through AN. Before insert the Q_1, the interference signal can be written as

$$I_1 = I_0 + \gamma_1 \cos(2\pi f t + \phi_1)$$

$$= \frac{1}{2}\left\{r^2 + r_m{}^2 - 2 r r_m \cos\left[2\pi f t + \frac{\pi}{2} - (\phi_{d1} - \phi_{d2} + \phi_r)\right]\right\} \tag{30}$$

where I_0 , γ_1 , and φ_1 are the mean intensity, visibility, and phase of the interference signal, respectively. Then insert the Q_1, the interference signal can be written as

$$I_2 = [r r_m \sin(\phi_{d1} - \phi_{d2} + \phi_r)] \cos(2\pi f t) + \left|\frac{1}{2}(r_m{}^2 - r^2)\right| \cos(2\pi f t) + C, \tag{31a}$$

and

$$\phi_2 = \cot^{-1}\left[\frac{2 r r_m \sin(\phi_{d1} - \phi_{d2} + \phi_r)}{(r_m{}^2 - r^2)}\right]. \tag{31b}$$

It is obvious that the refractive index of the GRIN lens G can be the function of φ_1 and φ_2 which expressed as

$$n = \frac{\cot\phi_2 - r_m\cos\phi_1 + r_m\sqrt{\cos^2\phi_1 + \cot^2\phi_2}}{\cot\phi_2 + r_m\cos\phi_1 - r_m\sqrt{\cos^2\phi_1 + \cot^2\phi_2}}. \quad \textbf{(32)}$$

Therefore, for a specificd r_m, φ_1 and φ_2 are given by the measurement, the refractive index of GRIN lens can be obtained.

For full-filed heterodyne phase detection can be realized with three-parameter sine wave fitting method that proposed by IEEE standards 1241-2000. The fitting equation has the form of

$$I(t) = \sqrt{{A_0}^2 + {B_0}^2}\cos(2\pi f t + \varphi) + C_0, \quad (33a)$$

and

$$\varphi = \tan^{-1}\left(\frac{-B_0}{A_0}\right). \quad (33b)$$

where A_0 , B_0 , and C_0 are real numbers and they can be derived with the least-square method. And finally the phase of the all pixels on the CCD camera can be obtained. Based on their method, they demonstrated the two dimensional refractive index distribution of the GRIN lens and showed in figure 16.

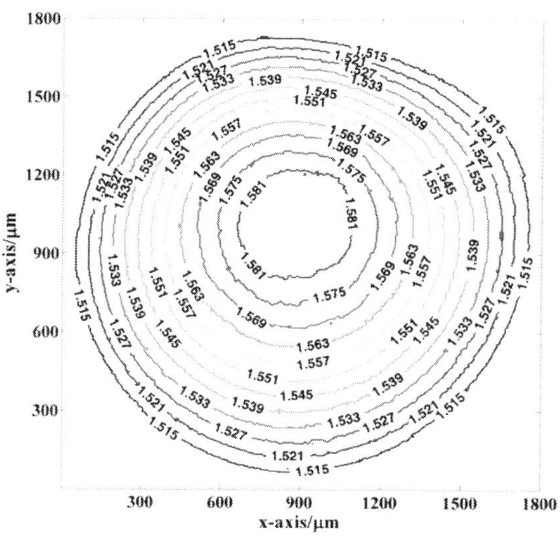

Figure 16. The refractive index contour of GRIN lens measured by full-field NICHI [38].

For the measurement of the optical constants of the thin film, K. H. Chen and C. C. Hsu [33] proposed a circular heterodyne refractometer. The optical configuration was shown in figure 17. The circular heterodyne light source was incident onto the sample at θ_0 and the light will be partially transmitted and reflected at the interface between the thin film and substrate. If the transmission axis of *AN* is located at α with respect to the x-axis, then the E-field of the light arriving at *D* is given

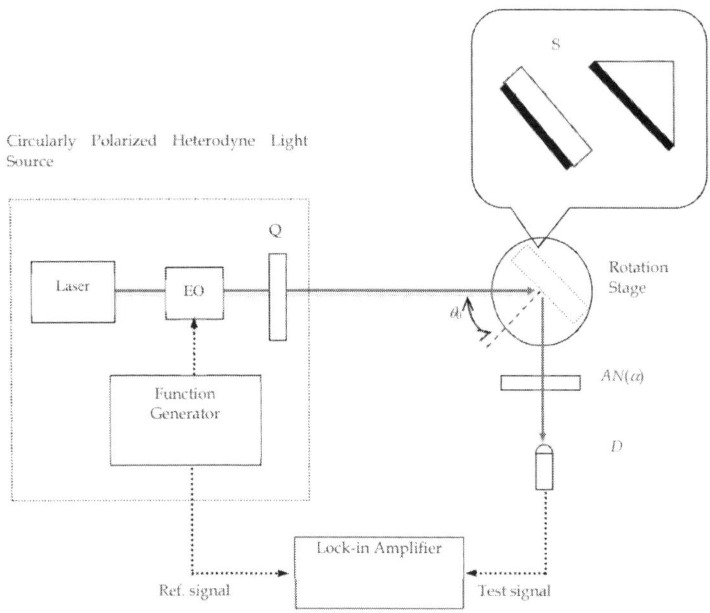

Figure 17. The circular heterodyne refractor.

$$E_t = \begin{pmatrix} \cos^2 \propto & \sin\alpha\cos\alpha \\ \sin\alpha\cos\alpha & \sin^2\alpha \end{pmatrix} \begin{pmatrix} r_p & 0 \\ 0 & r_s \end{pmatrix} \begin{pmatrix} \cos\frac{\omega t}{2} \\ -\sin\frac{\omega t}{2} \end{pmatrix} = \begin{pmatrix} r_p \cos^2\alpha \cos\frac{\omega t}{2} - r_s \sin\alpha\cos\alpha\sin\frac{\omega t}{2} \\ r_p \sin\alpha\cos\alpha\cos\frac{\omega t}{2} - r_s \sin^2\alpha\sin\frac{\omega t}{2} \end{pmatrix}. \quad (34)$$

Therefore, the testing signal detected by the detector D can be written as

$$I_t = |E_t|^2 = I_0[1 + \frac{\sqrt{A^2+B^2}}{I_0}\cos(\omega t + \phi)], \quad (35)$$

where I_0 and γ are the bias intensity and the visibility of the signal, and φ is the phase difference between the p- and s- polarizations coming from the reflection of the sample. They can be expressed as

$$I_0 = \frac{1}{2}\left(\left|r_p\right|^2 \cos^2\alpha + |r_s|^2 \sin^2\alpha\right), \quad (36a)$$

$$A = \frac{1}{2}(|r_p|^2 \cos^2 \alpha - |r_s|^2 \sin^2 \alpha), \tag{36b}$$

$$B = \frac{1}{2}(r_p r_s^* + r_s r_p^*) \sin \alpha \cos \alpha, \tag{36c}$$

and

$$\phi = \tan^{-1}(\frac{B}{A}) = \tan^{-1}[\frac{(r_p r_s^* + r_s r_p^*) \sin \alpha \cos \alpha}{(|r_p|^2 \cos^2 \alpha - |r_s|^2 \sin^2 \alpha)}]. \tag{36d}$$

The symbols r_p and r_s are the Fresnel equation (as equation 13); r_p^* and r_s^* are the conjugates of r_p and $r_{s,}$ respectively. It is obvious that the phase difference coming from the samples are function of the incident angle and the transmission axis of the analyzer *AN*. In practice, one can adjust three transmission axis of the analyzer at fixed incident angle or change three different incident angles with fixed transmission axis of the analyzer to get the corresponding phase difference φ. Therefore, substitute the phase difference into equation (36d) the optical constants of the sample can be obtained. According to Chen's results, they can successfully measure the thin metal film deposited on the glass substrate with lower measurement errors, which are 10^{-3} for the complex refractive index and 10^{-1} nm for the thickness.

Birefrigent crystals (BC) have been used to fabricate polarization optical components for a long time. To enhance their qualities and performances, it is necessary to determine the optical axis (OA) and measure the extraordinary index n_e and the ordinary index n_o accurately. There are many methods proposed to measure these parameters of the birefrigent crystal. Huang *et al.* [39] measured (n_e, n_o) of the wedge-shaped birefrigent crystal with transmission-type method. Therefore, the accuracy of thickness, flatness and parallelism of the two opposite sides of the birefrigent crystals are strongly required. D. C. Su and C. C. Hsu [37] proposed a novel method for measuring the extraordinary index n_e, the ordinary index n_o, and the azimuth angle of the birefrigent crystal with single apparatus which described in figure 18. Based on the circular heterodyne interferometer (CHI) and replaced the sample by the birefrigent crystal, the Jones vector of the E-field detected by *D* can be written as

$$E_t = AN(\beta)\cdot\begin{bmatrix} r_{pp} & r_{ps} \\ r_{sp} & r_{ss}\end{bmatrix}\cdot E_i$$

$$= \left(\left(r_{pp}\cos\beta + r_{sp}\sin\beta\right)\cos\frac{\omega t}{2} - \left(r_{ps}\cos\beta + r_{ss}\sin\beta\right)\sin\frac{\omega t}{2}\right)\begin{pmatrix}\cos\beta \\ \sin\beta\end{pmatrix} \quad (37)$$

where S is the Jones matrix for BC, r_{pp} and r_{ss} are the direct-reflection coefficients, and r_{ps} and r_{sp} are the cross-reflection coefficients [14], respectively. Based on Fresnel equations, r_{pp}, r_{ss}, r_{ps} and r_{sp} are the function of the n_e, n_o, and azimuth angle α of the birefrigent crystal. Therefore, the intensity of the testing signal can be expressed

$$I_t = |E_t|^2 = \frac{\left(r_{pp}\cos\beta + r_{sp}\sin\beta\right)^2 + \left(r_{ps}\cos\beta + r_{ss}\sin\beta\right)^2}{2}\left[1+\cos(\omega t+\phi)\right], \quad (38a)$$

and

$$\phi = \tan^{-1}\left(\frac{2\left(r_{pp}\cos\beta + r_{sp}\sin\beta\right)\left(r_{ps}\cos\beta + r_{ss}\sin\beta\right)}{\left(r_{pp}\cos\beta + r_{sp}\sin\beta\right)^2 + \left(r_{ps}\cos\beta + r_{ss}\sin\beta\right)^2}\right). \quad (38b)$$

Theoretically, it is difficult to obtain the n_e, n_o, and azimuth angle α of the birefrigent crystal by substituting the phase difference, which is arbitrary choose the measurement conditions

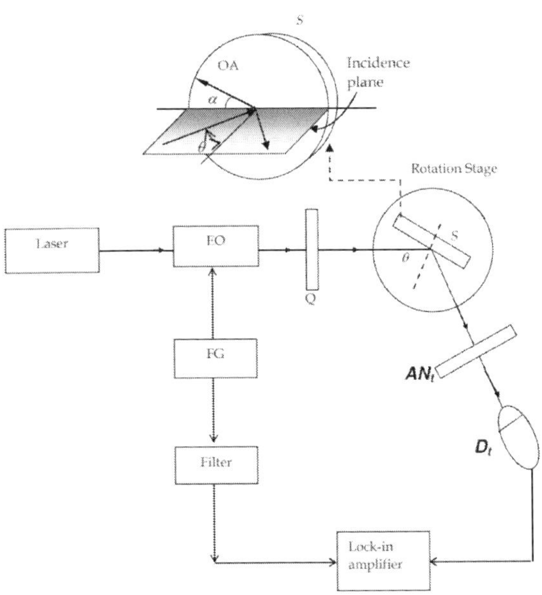

Figure 18. Optical configuration of the determination of the optical properties of birefrigent crystal with CHI.

of the incident angle and azimuth angle of the analyzer, into equation (38b). Therefore, Su developed a sequence for determining these parameters. First, let azimuth angle β of analyzer equal to 0° and phase difference can be written as

$$\phi = \tan^{-1}\left(\frac{2r_{pp}r_{ps}}{r_{pp}{}^2 - r_{ps}{}^2}\right). \tag{39}$$

As azimuth angle α of the birefrigent crystal at 0° or 90°, r_{ps} and r_{sp} will be equal to 0, and phase difference φ is equal to 0. But in this period, one cannot determine the azimuth angle α of brifrigent crystal exactly at 0° or 90°.

Second, fixed the azimuth angle of the birefrigent crystal and rotated azimuth angle β of analyzer to nonzero position. The phase difference φ can be expressed as

$$\phi = \tan^{-1}\left(\frac{\sin 2\beta \cdot r_{pp}r_{ps}}{r_{pp}{}^2 \cos^2\beta - r_{ss}{}^2 \sin^2\beta}\right). \tag{40}$$

At this period, the r_{pp} and r_{ss} will be under one of the conditions (i) α=90°or (ii) α=0°. Hence, we solved the n_e and n_o under conditions (i) and (ii) with two different incident angles.

Third, determine the correct solution with two justifications. (1) Rationality of the solution: In general, both n_e and n_o are within the range 1 and 5. If any estimated data of n_e and n_o is not within this range, it is obvious that the estimated data may be incorrect. (2) Comparison between n_e and n_o: Either a positive or negative crystal is tested, all two pairs of solutions of either group should meet with only either $n_e > n_o$ or $n_e < n_o$. If not, then that group is incorrect.

Based on Su's procedure, they have successfully determined the n_e, n_o, and azimuth angle α of the birefrigent crystal, which were positive crystal (quartz) and negative crystal (calcite), with lower error of the refractive index (~ 10^{-3}) and azimuth angle (~0.1) of BC.

J. F. Lin *et al* [40] proposed a transmission type circular heterodyne interferometer to determine the rotation angle of chiral medium (glucose solution). The optical setup was shown in figure 19 and the E-field of the testing signal derived from Jones calculation is

$$\mathrm{E} = \mathrm{A}(0) \cdot \mathrm{S}(\theta) \cdot Q_1(45) \cdot EO(90) \cdot P(45) \cdot E_{in}$$

$$= \begin{bmatrix} 1 & 0 \\ 0 & 0 \end{bmatrix} \begin{bmatrix} \cos\theta & \sin\theta \\ -\sin\theta & \cos\theta \end{bmatrix} \begin{pmatrix} \frac{1-i}{2} & \frac{1+i}{2} \\ \frac{1+i}{2} & \frac{1-i}{2} \end{pmatrix} \begin{bmatrix} e^{-i\omega t/2} & 0 \\ 0 & e^{-\omega t/2} \end{bmatrix} \begin{bmatrix} \frac{1}{2} & \frac{1}{2} \\ \frac{1}{2} & \frac{1}{2} \end{bmatrix} \begin{bmatrix} 0 \\ E_0 e^{iw_0 t} \end{bmatrix}, \tag{41}$$

where θ is the optical rotation angle of the chiral medium. And the intensity of the testing signal detected by the photodetector can be derived and written as

$$I_t = I_{dc}[1 - \sin(\omega t - 2\theta)], \tag{42}$$

Compare with the reference signal by lock-in amplifier, the phase difference between the reference and testing signals can be obtained. Theoretically, the optical rotation angle is strongly related to the concentration, temperature, and propagation length of the chiral medium (glucose solution). And that can be expressed as

$$C = \frac{100\theta}{L[\theta]}, \tag{43}$$

where the glucose concentration C (g/dl) in a liquid solution, θ (degree) is optical rotation angle, L (decimeter) is the propagation length in chiral medium.

Figure 20 showed that the optical rotation angle of the glucose solution which the concentration was varied from 0 to 1.2 g/dl. Their results showed the good linearity and high sensitivity which can achieve 0.273° /g/dl.

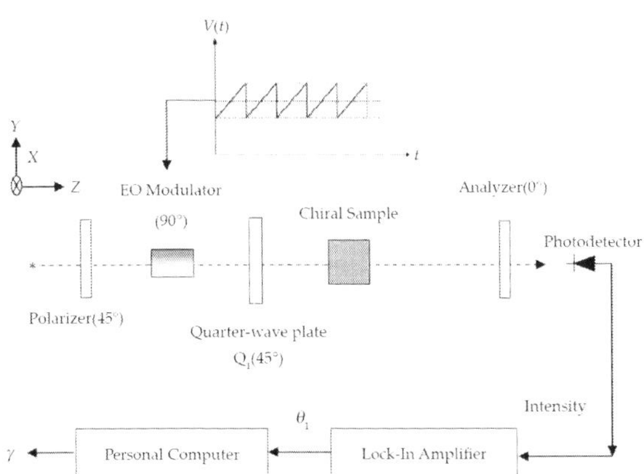

Figure 19. Schematic diagram of circular heterodyne interferometer for measuring the optical rotation angle in a chiral medium [40].

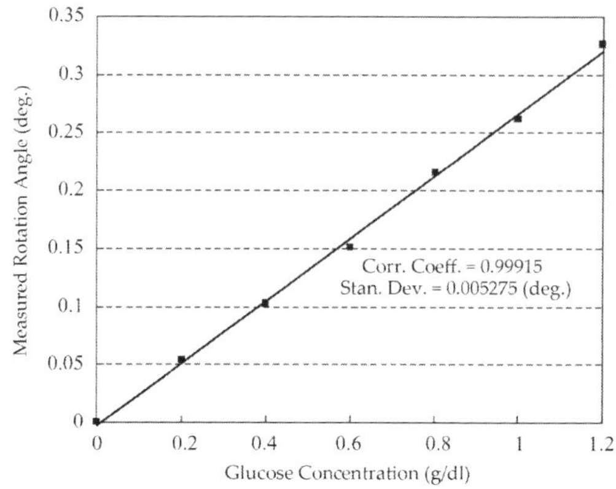

Figure 20. Optical rotation angle of glucose solution with concentration variation [40].

CONCENTRATION MEASUREMENT WITH HETERODYNE INTERFEROMETER

The concentration of solution is an important factor in food, chemical and biochemical industrial, especially in health care and disease prevention. For example, the blood glucose concentration is related to the diabetes. To control the blood glucose concentration within the normal level is critical issue to the diabetic daily care. Therefore, many researchers developed novel methods for measuring the solution concentration. Because of the advantages of the optical method such as high sensitivity, high resolution, non-contact, and quick response, optical measurement method become more popular in past few decades. And these methods can roughly divide into fiber type sensor [41, 44-45, 49] and non-fiber type sensor [42-43, 46-48, 50]. In this section, we will review some recent development in both type sensors for measuring concentration of the specific chemical compound [41-50].

M. H. Chiu [41] developed a D-shape fiber sensor with SPR property and integrated with heterodyne interferometer which could detect variation in the alcohol concentration of 2%. The optical setup was shown in figure 21. The heterodyne light source was guided into the sensor and suffered the attenuate total reflection (ATR) at the sensing region. Because of the refractive index of the sample will be varied as the concentration changed. And induce the phase difference between the p- and s- polarizations. To measure the phase difference can be obtained the concentration variation of the sample.

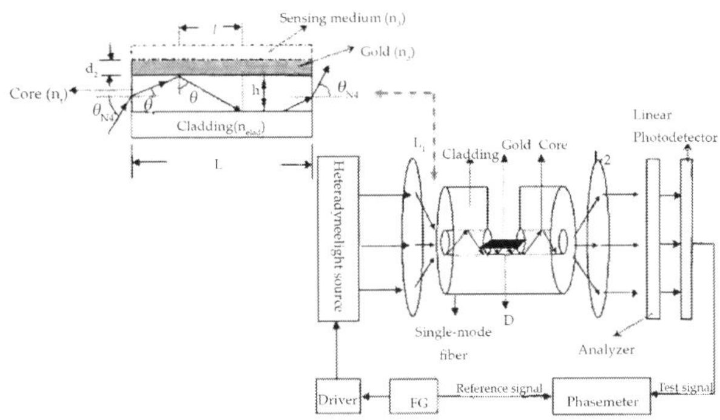

Figure 21. The scheme of the D-shape fiber sensor [41].

Figure 21 shows that the testing signal detected by photodetector and sent into the phasemeter. Therefore, the interference signal can be written as

$$I(t) = I_0\left\{1 + V\cos\left[\omega t + \left(\frac{L(\phi_p-\phi_s)}{2h\tan\theta_i}\right)\right]\right\}, \tag{44}$$

where I_0 and V are the average intensity and visibility; L and h are the sensing length and core diameter; (φ_p-φ_s) is the phase difference between p- and s-polarizations; θ_i is the incident angle at the interface between fiber core and metal film. Based on the Fresnel equation, one can derive the (φ_p-φ_s) from the amplitude reflection coefficient under ATR condition and it is obvious that (φ_p-φ_s) is the function of the refractive index of the sensing medium. Figure 22 shows that the results measured by the D-shape fiber sensor for different concentration of the alcohol.

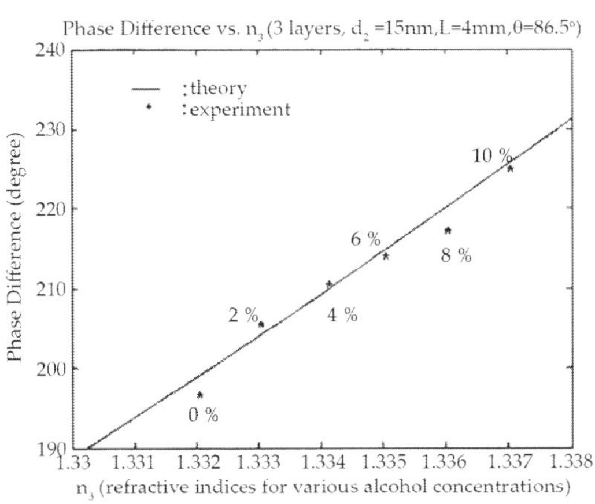

Figure 22. The experiment result of different concentration of alcohol measured by D-shape fiber sensor [41].

In Chiu's results, they can observe the concentration variation 2 %, in which the corresponding refractive, index variation is about 0.0009. Based on error analysis, their method can be reached 2 10^{-6} refractive index unit.

Recently, T. Q. Lin [44] and C. C. Hsu [45] developed a fiber sensor which immobilized glucose oxidase (GOx) on the fiber core for measuring the glucose concentration in serum and phosphate buffer solution (PBS). Their measurement method integrated the fiber sensor and heterodyne interferometer which showed in figure 23.

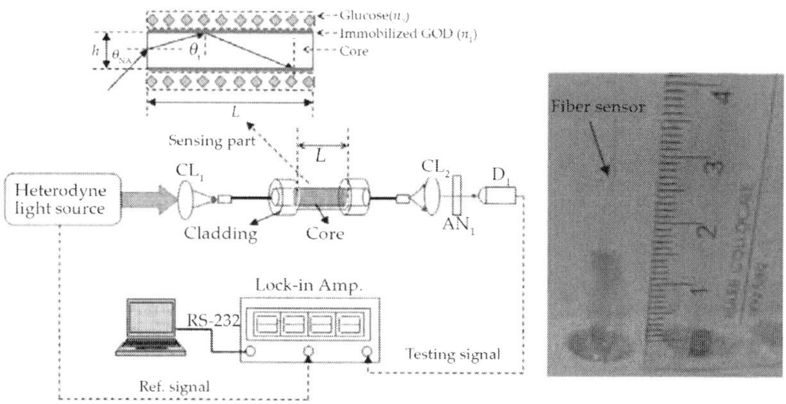

Figure 23. Schematic diagram of the measurement system and preliminary test of the glucose fiber sensor [44].

As the heterodyne light source enters the sensing part, the light beam undergoes total internal reflection (TIR) and the phase difference between the p- and s- polarization states can be written as

$$\phi_t = m\phi_{TIR} = \frac{L}{h\tan\theta_t}\cdot\tan^{-1}\left(\frac{\sqrt{\sin^2\theta_t-(\frac{n_2}{n_1})^2}}{\tan\theta_t\cdot\sin\theta_t}\right), \qquad (45)$$

where n_1 and n_2 are the refractive indices of the immobilized GOx and the testing solution. θ_t and m are the incident angle and the number of TIRs that occur at the interface between the GOx and the testing solution. After dripping the testing sample onto the sensor, the phase will vary as the glucose reacts with the GOx to be converted into gluconic acid and hydrogen peroxide. The chemical reaction can be formulated as follows:

$$\text{Glucose} + O_2 \xrightarrow{\text{GOx}} \text{gluconic acid} + H_2O_2. \quad (46)$$

It means that the refractive index (n_2) will change and consequently the phase will change. Besides, the refractive index n_2 is a function of the concentration of the testing sample. Different concentration of the solution has different refractive index. Therefore, one can determine the concentration variation by measuring the phase variation. In their methods, the phase difference can be carried in the heterodyne interference signal and written as

$$I_t = I_0[1 + \cos(\omega t) + \phi_t].$$

To deserve to be mentioned, they found that the pH property between the testing sample and sensor is critical issue for rapid measurement. Figure 24 shows that the response time and response efficiency of the fiber sensor. It is clear that the response time for measuring glucose solution was shorter than those for serum measurement. And the response efficiency for measuring glucose solution was faster than those for serum measurement at different GOx concentration.

Based on their results, this fiber sensor has good linearity of the calibration curve for glucose solution and serum sample. And they showed the best resolutions were 0.1 and 0.136 mg/dl for glucose solution and serum based sample, respectively.

One of the non-fiber type sensors is SPR (surface Plasmon resonance) sensor which has been applied in field such as pharmaceutical development and life sciences. And SPR provides ultra-high sensitivity for detecting tiny refractive index (RI) changes or other quantities which can be converted into an equivalent RI. The heterodyne interferometer detects the SPR phase by using a Zeeman laser or optical modulator, such as an acousto-optic modulator or electro-optic modulator and has been reported in the literature. Heterodyne phase detection techniques offer the high measurement performance high sensitivity and high resolution in real-time. J. Y. Lee [43] proposed wavelength-modulation circular heterodyne interferometer (WMCHI) with SPR sensor for measuring the different concentration of alcohol. The diagram of the WMCHI is shown in figure 25.

The SPR sensor had the Kretschmann configuration consists of a BK7 prism coated with a 50 nm gold film and integrated with micro-fluid channel which used to inject the testing

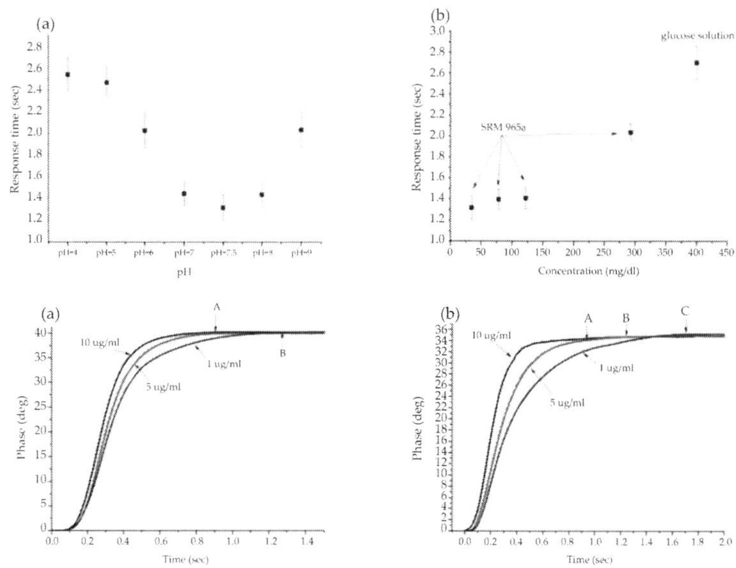

Figure 24. The response time and response efficiency of the fiber sensor for measuring both glucose solution and human serum [44].

sample. The E-field detected by D1 and D_2 can be written as

$$E_1 = P(45^\circ) \cdot J_{SPR} \cdot J_Q(45^\circ) \cdot E_h$$

$$= \frac{1}{2}\begin{bmatrix} 1 & 1 \\ 1 & 1 \end{bmatrix} \begin{bmatrix} |r_p| e^{i\phi p} & 0 \\ 0 & |r_s| e^{i\phi s} \end{bmatrix} \frac{1}{\sqrt{2}} \begin{bmatrix} 1 & i \\ i & 1 \end{bmatrix} \begin{bmatrix} e^{i(\phi_0-\omega t)/2} \\ e^{-i(\phi_0-\omega t)/2} \end{bmatrix}$$

$$= \frac{1}{2\sqrt{2}} \left[\left(|r_p| e^{i\phi p} + i|r_s| e^{i\phi s}\right) e^{\frac{i(\phi_0-\omega t)}{2}} + \left(i|r_p| e^{i\phi p} + |r_s| e^{i\phi s}\right) e^{-\frac{i(\phi_0-\omega t)}{2}} \right] \begin{bmatrix} 1 \\ 1 \end{bmatrix}, \tag{47a}$$

and

$$E_2 = P(-45^\circ) \cdot J_{SPR} \cdot J_Q(45^\circ) \cdot E_h$$

$$= \frac{1}{2\sqrt{2}} \left[\left(|r_p| e^{i\phi p} + i|r_s| e^{i\phi s}\right) e^{\frac{i(\phi_0-\omega t)}{2}} + \left(i|r_p| e^{i\phi p} + |r_s| e^{i\phi s}\right) e^{-\frac{i(\phi_0-\omega t)}{2}} \right] \begin{bmatrix} 1 \\ -1 \end{bmatrix}, \tag{47b}$$

where J_{SPR} is the Jones matrix of SPR sensor. They became two testing signals and sent into lock-in amplifier which can obtained the phase difference between them. The phase difference Φ of these two signals is obtained as

$$\Phi = \left(\Phi_0 + \tan^{-1}\frac{B}{A}\right) - \left(\Phi_0 - \tan^{-1}\frac{B}{A}\right) = 2\tan^{-1}\left(\frac{|r_p|^2 - |r_s|^2}{2|r_p||r_s|\cos\phi}\right). \tag{48}$$

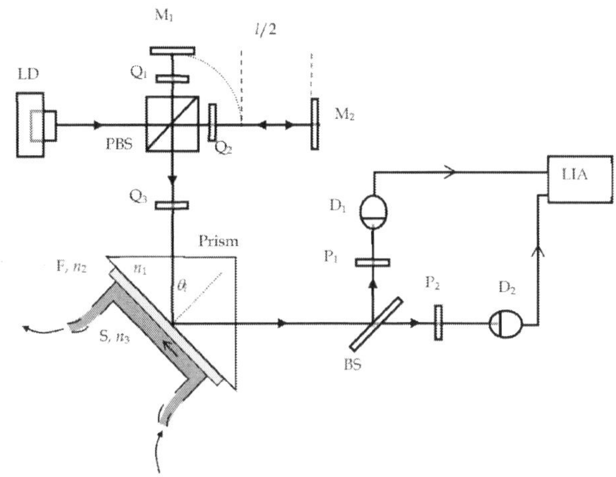

Figure 25 Schematic diagram of WMCHI for different concentration measurement [43].

Based on equation (44), it is clear that the resonant angles for φ and Φ are different. Obviously, Φ is the function of r_p, r_s and φ which varies with the refractive index of sample n_3 and the incident angle. The relationship between φ and Φ and the incident angle was shown in figure 26 It is obvious that the maximum of Φ is larger than that of φ. On the other hand, Φ can be larger than 10,000 in the incident angle interval between 66.25° and 66.75°. This means that Lee's method has a high angle toleration and larger dynamic range.

C. Chou [42] proposed a novel pair surface plasma wave biosensor which provided a larger dynamic measurement range for effective refractive index. In their system, it can avoid excess noise coming from laser intensity fluctuation and environment disturbance. It is important to retain the amplitude stability in this method for high detection sensitivity. Figure 27 showed the amplitude sensitivity PSPR method. In this figure, PBS separated the pair of p-polarization waves and the pair of s-polarization waves, which can be optical heterodyne interference signal at the photodetectors D_p and D_s. Then these two signals can be expressed as

$$I_{p1+p2}(\Delta\omega t) = A_{p1}A_{p2}\cos(\Delta\omega t + \phi_p), \tag{49a}$$

$$I_{s1+s2}(\Delta\omega t) = A_{s1}A_{s2}\cos(\Delta\omega t + \phi_s), \tag{49b}$$

where A_{P1} and A_{P2} are the attenuated amplitudes of the reflected P_1 and P_2 waves respectively; A_{S1} and A_{S2} are the attenuated amplitudes of the reflected S_1 and S_2 waves

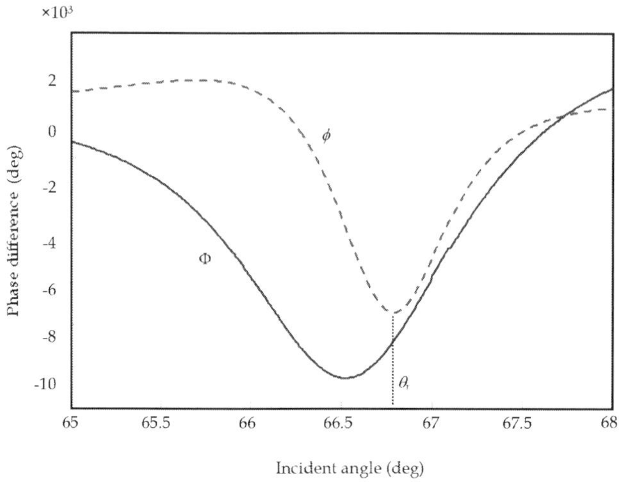

Figure 26. The relationship between φ and Φ and the incident angle [43].

respectively. φ_P and φ_S are the phase differences of the reflected P and P and the reflected S and S waves respectively. In equation (49), φ_P and φ_S are equal to 0 and these two interference signals will remain at maximum intensity under the SPR proceeded.

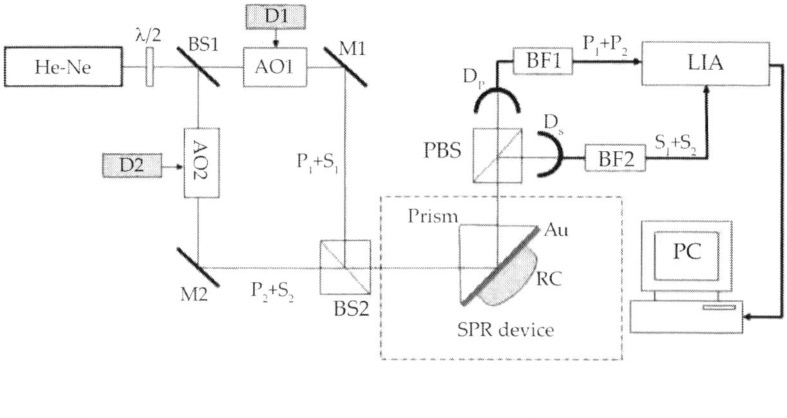

Figure 27. The schematic of the amplitude sensitivity PSPR [42].

Based on this method, Chou demonstrated three different testing samples with concentration variation which were sucrose, glycerin-water solution, and rabbit anti-mouse IgG. In figure 28, the best of these sample are 8×10-8, 7.6×10^{-7}, and 2×10^{-9}, respectively.

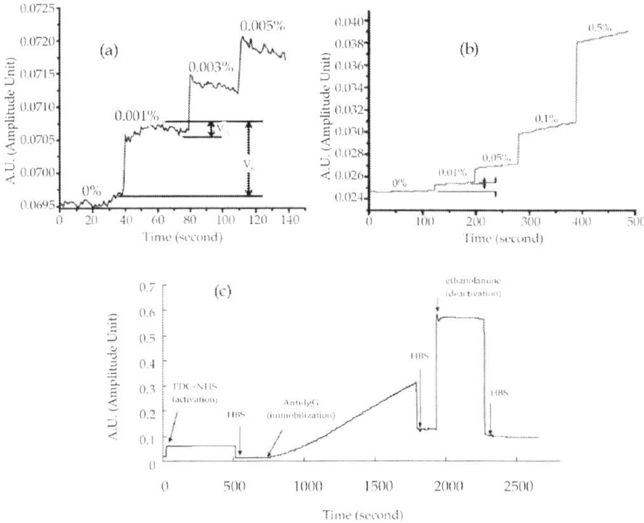

Figure 28. The measurement results of the PSPR method [42].

CONCLUSION

In this chapter, we reviewed some recent development or state of the art techniques. It shows that the heterodyne interferometry is a mature technique and can be applied to many different aspects. For example, the diffraction grating heterodyne interferometry (DGHI) provided nanometer resolution for precision positioning which can be integrated with motorized stage. Full-field circular heterodyne interferometry (FFCHI) can be used to investigate the two-dimensional optical properties, such as refractive index and birefrigence of testing sample. For this point of view, the heterodyne interferometry can be a refractometer with high accuracy. To integrate with optical sensor, the heterodyne interferometry can be used to diagnose the concentration of the body fluid such as blood glucose and glycerin or the protein interaction between body-antibody. Therefore, the heterodyne interferometry is a powerful, flexible, integrable, and reliable technique for precision metrology.

REFERENCES

1. L. C. De Backer, "In-plane displacement measurement by speckle interferometry," Non-destructive Testing 8, pp. 177-180 (1975).
2. B. Pan, K. Qian, H. Xie, and Anand Asundi, "Two-dimensional digital image correlation for in-plane displacement and strain measurement: a review," Measurement Science and Technology 20, pp. 062001 (2009).
3. S. Nakadate, "Vibration measurement using phase shifting time average holographic interferometry," Applied Optics 25, pp. 4155-4161 (1986).
4. G. Pedrini, W. Osten, and M. E. Gusev, "High-speed digital holographic interferometry for vibration measurement," Applied Optics 45, pp. 3456-3462 (2006).
5. W. B. Ribbens, "Surface roughness measurement by two wavelength holographic interferometry," Applied Optics 13, pp. 1085-1088 (1974).
6. C. J. Tay, M. Thakur, C. Quan, "Grating projection system for surface contour measurement," Applied Optics 44, pp. 1393-1400 (2005).
7. Haller, H. A. Huggins, and M. J. Freiser, "On the measurement of indices of refraction of nematic liquid," Molecular Crystals and Liquid Crystal 16, pp. 53-59 (1972).
8. W. Lukosz and P. Pliska, "Determination of thickness and refractive index of SiO2 films on silicon wafers using an Abbe refractometer," Optics Communications 85, pp. 381-384 (1991).
9. J. A. D. Feijter, J. Benjamins, and F. A. Veer, "Ellipsometry as a tool to study the adsorption behavior of synthetic and biopolymers at the air-water interface," Biopolymers 17, pp. 1759-1772 (1978).
10. F. L. McCrackin, E. Passaglia, R. R. Stromberg, and H. L. Steinberg, "Measurement of the thickness and refractive index of very thin films and the optical properties of surfaces by ellipsometry," Journal of Research of the National Bureau of Standards A. Physics and Chemistry 67A, pp. 363-377 (1963).
11. B. T. Liu, S. J. Tang, Y. Y. Yu, and S. H. Lin, "High refractive index polymer/inorganic hybrid films containing high TiO2 contents," Colloids and Surfaces A: Physicochemical and Engineering Aspects 377, pp. 138-143 (2011).
12. C. M. Jan, Y. H. Lee, K. C. Wu, and C. K. Lee, "Integrating fault tolerance algorithm and circularly polarized ellipsometer for point of care applications," Optics Express 19, pp. 5431-5441 (2011).

13. H. G. Tompkins and E. A. Irene, *Handbook of Ellipsometry*, Willian Andrew, New York, 2010
14. R. M. A. Azzam and N. M. Bashara, *Ellipsometry and Polarized Light*, Elsevier Science Publisher, Netherlands, 1992
15. J. A. Dahlquist, D. G. Peterson, and W. Culshaw, "Zeeman laser interferometer," Applied Physics Letters 9, pp. 181-183 (1966).
16. T. Suzuki and R. Hioki, "Translation of light frequency by a moving grating," Journal of the Optical Society of America 57, pp. 1551 (1967).
17. W. H. Stevenson, "Optical frequency shifting by means of a rotating diffraction grating," Applied Optics 9, pp. 649-652 (1970).
18. M. J. Ehrlich, L. C. Phillips, and J. W. Wagner, "Voltage-controlled acousto-optic phase shifter," Review of Scientific instrument 59, pp. 2390-2392 (1988).
19. T. Q. Banh, Y. Ohkubo, Y. Murai, and M. Aketagawa, "Active suppression of air refractive index fluctuation using a Fabry-Perot cavity and a piezoelectric volume actuator," Applied Optics 50, pp. 53-60 (2011).

CHAPTER 10

SYNTHETIC WAVELENGTH INTERFEROMETRY OF AN OPTICAL FREQUENCY COMB FOR ABSOLUTE DISTANCE MEASUREMENT

Guanhao Wu[1], Lei Liao[1], Shilin Xiong[1], Guoyuan Li[2], Zhijian Cai[3], and Zebin Zhu[1]

[1]State Key Laboratory of Precision Measurement Technology and Instruments, Department of Precision Instrument, Tsinghua University, Beijing, 100084, China.
[2]Satellite Surveying and Mapping Application Center, NASG, Beijing, 100048, China.
[3]College of Physics, Optoelectronics and Energy, Soochow University, Suzhou, 215006, China.

ABSTRACT

We present a synthetic-wavelength based heterodyne interferometer of optical frequency combs with wide consecutive measurement range for absolute distance measurement. The synthetic wavelength is derived from two wavelengths obtained by two band-pass filters. The interferometric phase of the synthetic wavelength is used as a marker for the pulse-to-pulse

Citation: (APA): Wu, G., Liao, L., Xiong, S., Li, G., Cai, Z., & Zhu, Z. (2018). Synthetic wavelength interferometry of an optical frequency comb for absolute distance measurement. Scientific reports, 8(1), 1-7. (7 pages)

alignment, which greatly improves the accuracy of traditional peak finding method. The consecutive measurement range is enlarged by using long fiber to increase the path length difference of the reference and measurement arms. The length of the long fiber is stabilized according to the interferometric phase of a CW laser. The experimental results show the present system can realize an accuracy of 75 nm in 350 mm consecutive measurement range.

INTRODUCTION

High accuracy distance measurement is essential for various applications such as industrial sensing of critical dimensions, and satellite formation in space missions[1]. Traditionally, single wavelength interferometry with fringe counting is employed for these applications. However, fringe counting is susceptible to cause loss of absolute position[2]. Multiple-wavelength interferometry can be used to realize absolute distance measurement avoiding fringe counting[3]. However, to realize large measurement range, a number of synthetic wavelengths are needed, which will result in a complicated system of multiple lasers or a low measurement speed due to wavelength scanning. Thanks to the advent of optical frequency comb which has led to revolutionary progress in absolute distance measurement[4].

An optical frequency comb emits evenly spaced ultra-short pulse train with a broad spectrum consisting of discrete, uniform mode-spacing narrow lines. When its mode frequencies are referenced to a frequency standard, it becomes an ultra-accurate ruler in space, time and frequency domains[5]. Such inherent advantages are attractive for absolute distance measurement. Since the first work undertaken by Minoshima *et al.*[6], various methods have been reported for such studies[7,8,9,10,11,12,13,14,15,16,17,18,19,20,21,22], e.g., using the comb as a wavelength or frequency standard for CW lasers[7,8,9,10,11], using a pair of frequency combs to realize time-of-flight measurement[12,13,14,15,16,17,18], using the comb as a light source of dispersive interferometry[19,20,21,22], and using the adjacent pulse repetition interval length (APRIL) as a scale[23,24,25,26,27,28,29,30,31,32]. Among the methods mentioned above, the APRIL is a simple and effective scheme based on an unbalanced Michelson interferometer[23,24] which has been widely used in absolute distance measurement[25,26,27,28,29,30,31,32]. By adjusting the repetition rate (f_{rep}) of the frequency comb to make the pulses from reference and measurement arms overlap, the target distance can be determined by accurate measurement of f_{rep}[23]. Generally, two factors are extremely important to the performance of this method. The first is how to judge the pulses overlap, because the accuracy of pulse-to-pulse alignment determines the accuracy of distance measurement[32]. The second one is that limited tunable range of f_{rep} limits the consecutive measurement range, which leads to large 'dead zone' of this method[33].

The accuracy of such distance measurement depends mainly on the knowledge of relative positions of the two overlapped pulses, i.e., pulse-to-pulse alignment. Conventionally, the peak position of interferogram envelope of the overlapped pulses is used for the pulse-to-pulse alignment[25,26,27]. Hereafter this method is referred to as peak finding method. It can realize micrometer or even sub-micrometer level alignment since the pulse width can be very narrow, but the accuracy is affected by the intensity noise of the interferogram envelope. Moreover, such accuracy is still not enough for ultra-precision applications. In order to improve the accuracy of pulse-to-pulse alignment, several attempts have been made on improving the peak finding method[30,31] or linking the peak finding with the interferometric phase[32]. In our previous study, we proposed a two-color heterodyne interferometry based on the fundamental and second harmonic (SH) of optical frequency combs[32]. The synthetic wavelength derived from the virtual second harmonic (half wavelength of the fundamental) and the real second harmonic (generated by nonlinear crystal) was used to bridge the gap between peak finding method and the single-wavelength heterodyne interferometry. However, the system that requires a fully stabilized comb, second harmonic generation and two acousto-optic modulators is complicated for practical applications.

When applying the APRIL method, the consecutive measurement range depends on the tunable range of the cavity length and the optical path length difference (OPLD) between the measurement and reference arms[33]. For example, we denote the optical length of the cavity (equal to the optical distance between adjacent pulses) as L_{pp}, and denote its tuable range as ΔL_{max}. Assuming that OPLD is k times of L_{pp}, it can realize a consecutive measurement range of $k\Delta L_{max}/2$ (consider the round trip of the beam) for distance measurement. Consequently, if $k>L_{pp}/\Delta L_{max}$, the consecutive measurement range is large enough that no 'dead zone' exist. When we measure a short distance, we can add additional length delay in the referenc/measurement arm to increase OPLD to realize 'multiplication effect' to remove the 'dead zone'. This concept has been applied in surface profilometry[34] and optical sampling by cavity tuning[35,36]. It is also promising to be combined with the APRIL method.

In the present study, we present a simplified synthetic-wavelength interferometry of optical frequency combs for absolute distance measurement.

The synthetic wavelength is derived from two wavelengths obtained by two band-pass filters. The interferometric phase of the synthetic wavelength is used as a marker for the pulse-to-pulse alignment. Thus, the carrier-envelope-offset frequency (f_{ceo}) is not sensitive to the measurement results in this system[37]. In order to extend the consecutive measurement range, we employ long fiber to increase the OPLD of the reference and measurement arms. Consequently, the present system can realize large consecutive range distance measurement with high accuracy.

THE MEASUREMENT PRINCIPLE

Experimental Setup

A home-made mode-locked Er:fiber ring laser is used as the frequency comb source (Fig. 1). Its optical power is 8 mW. The f_{rep} is stabilized referencing to a rubidium atomic clock (SIM940, Stanford Research Systems) and f_{ceo} is free running. The f_{rep} of the comb is 75.0 MHz, and it can be varied coarsely up to 1.9 MHz by a fiber delay line and finely up to 38 Hz by PZT-driven fiber stretching. The corresponding optical distance between adjacent pulses is $L_{pp} = 4$ m, and it can be varied up to $\Delta L_{max} = 100$ mm.

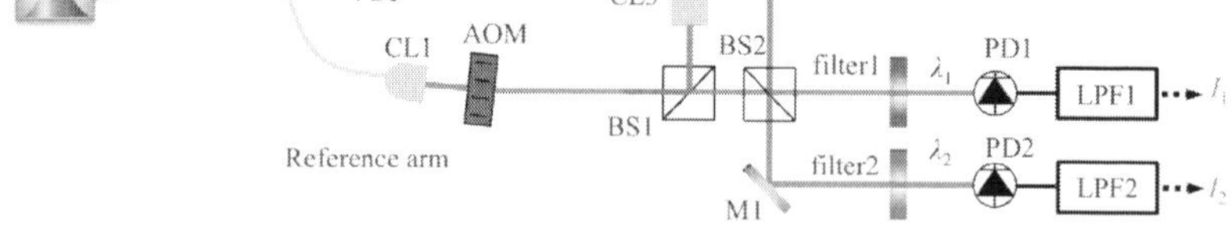

Figure 1. Schematic of experimental setup. Blue lines: fiber or fiber devices; red lines: optical paths in free space. C1–3: Fiber Coupler; CL1–3: Collimator lens. BS1–2: Beam splitter; M1–2: Mirror; CR: Corner reflector; PD1–3: Photodetector; LPF1–2: low-pass-filter; FBG: fiber Bragg grating filter. I and II indicate two positions of CR.

The pulse train from the comb is sent to a coupler (C1) and divided into the reference and measurement arms. In the reference arm, an acousto-optic modulator (AOM) shifts the optical frequency by 80 MHz. In the measurement arm, there is long fiber (51 m) composed of PM fiber and dispersion compensation fiber. A corner reflector (CR) is placed on a translation stage and used as the target mirror, whose displacement is monitored by a commercial heterodyne interferometer (PT-1105c, Pretios Inc., not shown in this figure). The OPLD between the reference and measurement arms equal to 17 L_{pp}. Thus, the consecutive measurement range of this system is $8.5\Delta L_{max} = 850$ mm. The pulses from the two arms overlap at a beam splitter (BS2). The interference signals are filtered by two band-pass filters (filter1, center wavelength $\lambda_1 = 1559$ nm, bandwidth 7 nm; filter2, center wavelength $\lambda_2 = 1567$ nm, bandwidth 7 nm) and detected by two Photodetectors (PD1 and PD2), respectively. In order to cancel out the drift and phase noise caused by the long fiber, a CW laser (center wavelength 1550 nm, Koheras Basik E15, NKT Photonics) which is locked to another fully stabilized comb is used to build another heterodyne interferometer. This interferometer shares the reference arm and the long fiber in measurement arm with the interferometer of the comb. The interference signal I_{cw} is detected by PD3. Before PD3, a 100 GHz (~0.8 nm) fiber Bragg grating (FBG) band-pass filter (center wavelength 1550 nm) is used for improving the signal-to-noise of I_{cw}. The phase of I_{cw} is used for the feedback control of the length of long fiber. We use temperature control (TEC) of the long fiber to compensate its drift and use a Piezoelectric Transducer (PZT) to stretch the fiber for fast control.

In the reference arm, the AOM shifts the optical frequency by $f_\Delta = 80$ MHz. The mth mode of the comb in reference arm can be expressed by:

$$f_{ref}(m) = mf_{rep} + f_{ceo} + f_\Delta. \tag{1}$$

The mth mode of the comb in the measurement arm can be expressed by:

$$f_{mea}(m) = mf_{rep} + f_{ceo}. \tag{2}$$

Therefore, when the pulses from the reference arm and the measurement arm overlap, the frequency of beat signal can be expressed by $f_\Delta - N.f_{rep} = (80 - N75)$ MHz (N is an integer). We use low-pass-filters (LPF1 and LPF2) after PD1 and PD2 to select the lowest-frequency beat signal ($f_\Delta - f_{rep} = 5$ MHz) as the heterodyne signals (I_1 and I_2). The phase of the heterodyne interference signals from PD1 and PD2 (denoted as ϕ_1 and ϕ_2) is measured by a lock-in amplifier (HF2, Zurich Instruments). The heterodyne interference signal of the CW laser equals to $f_\Delta = 80$ MHz. Its phase (ϕ_{cw}) is measured by another lock-in amplifier (SR844, Stanford Research Systems). ϕ_{cw} is stabilized by feedback control of length of long fiber.

Measurement Principle

During the measurement, the target mirror was moved from position I to position II by a distance D. At both positions, peak finding method was applied for I_1, i.e., the intensity of I_1 was maximized by adjusting f_{rep}. Here, it is not necessary to maximize the intensity precisely, because we will use the interferometric phase of the synthetic wavelength as a marker for the pulse-to-pulse alignment. The synthetic wavelength is in the magnitude of hundreds of micrometers, so it does not require very precise pre-alignment of the pulses.

After applying peak finding method for I_1 at position I, we assume pulse A of reference arm overlaps with pulse B of measurement arm. Normally, pulses A and B cannot overlap perfectly due to the accuracy limitation of peak finding method. The distance between the peaks of the two pulses is denoted as δ_1 (Fig. 2a). When the target mirror CR was moved by D to position II, resulting a relative displacement of D between the pulses A and B. In case of $D<L_{pp}/2n_g$ (n_g is the group refractive index of air), applying peak finding method at position II will compensate this relative displacement back to make the two pulses overlap again with a peak distance of δ_2 (Fig. 2b). From position I to position II, we denote the optical distance between adjacent pulses change from L_{pp1} to L_{pp2}, and the change value is $\Delta L_{pp} = L_{pp2} - L_{pp1} = c/f_{rep2} - c/f_{rep1}$ (c is the speed of light in vacuum, f_{rep1} and f_{rep2} are the repetition rate corresponding to positions I and II). We also record the interferometric phase change as $\Delta\phi_1$ and $\Delta\phi_2$ for the two wavelengths. The relation between target distance D and the parameters recorded can be given by:

$$\delta_2 - \delta_1 = 2D - 17\Delta L_{pp}/n_g = (N_1 + \Delta\phi_1/2\pi)\lambda_1, \tag{3}$$

$$\delta_2 - \delta_1 = 2D - 17\Delta L_{pp}/n_g = (N_2 + \Delta\phi_2/2\pi)\lambda_2, \tag{4}$$

where N_1 and N_2 are the fringe orders to be determined. Note that the difference of the refractive index of air between the two wavelengths are ignored. It is necessary to point out the phase of the heterodyne signal are the integration of phases from all comb modes within the pass band, it corresponds to an effective center wavelength of the heterodyne interference signal. Here, we use the center wavelength of the filters as the effective center wavelength.

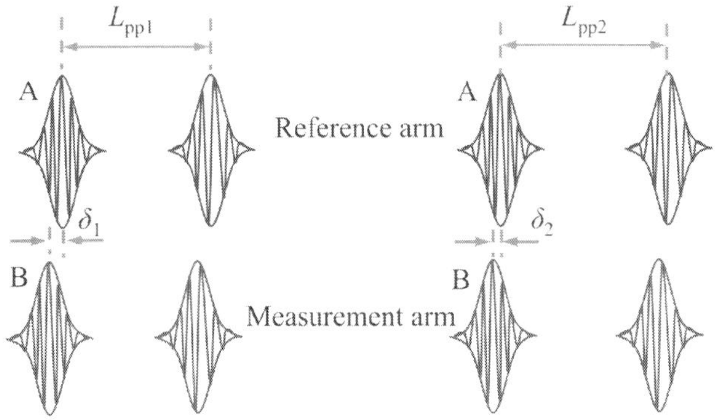

Figure 2. Demonstration of the pulses overlap from (**a**) position I to (**b**) position II.

From Eqs. (3) and (4), we can derive that

$$\delta_2 - \delta_1 = 2D - 17\Delta L_{pp}/n_g = (N_s + \Delta\phi_s/2\pi)\lambda_s, \tag{5}$$

where $\lambda_s = \lambda_2\lambda_1/(\lambda_2 - \lambda_1) = 305.4\ \mu m$ is the synthetic wavelength, the corresponding phase $\Delta\phi_s = \Delta\phi_2 - \Delta\phi_1$ and $N_s = N_2 - N_1$. Note that the peaking finding method can easily ensure δ_1 and δ_2 be smaller than several micrometers, which is much smaller than λ_s. Thus, $N_s = 0$. Therefore, we can obtain the target distance D according to Eq. (5) and the parameters recorded at positions I and II.

The optical spectrum of the comb is wide, and its full width at half maximum (FWHM) is more than 50 nm (Fig. 3a). Hence, there is wide space for selecting the center wavelengths (λ_1 and λ_2) of filters 1 and 2. To obtain an appropriate synthetic wavelength, we should pay attention to the following three points. (1) The synthetic wavelength should be long enough to avoid unambiguity of N_s in Eq. (5). (2) Based on point 1, shorter synthetic wavelength is favorable to obtain higher accuracy in distance measurement because the accuracy in phase measurement is fixed. (3) The transmission spectra of filters 1 and 2 should not overlap with that of the CW laser to avoid crosstalk. In the present system, the wavelength of the CW laser is 1550 nm, the transmission spectra of the filters are shown in Fig. 3. Please note that even under the three conditions mentioned above, there are a lot choices of the combination of λ_1 and λ_2. We did not make specific optimization in selecting λ_1 and λ_2 for the present system.

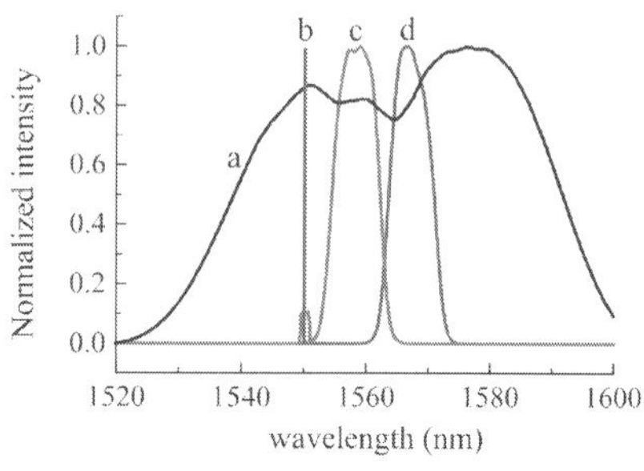

Figure 3. Optical spectra of the interference signals. (a) The optical spectrum of the comb; (b) indicates the wavelength of the CW laser; (c) the optical spectrum after filter1, there is a small bump at 1550 nm which indicates the spectrum of the CW laser is not suppressed perfectly; (d) the optical spectrum after filter2.

RESULTS

Evaluation of Long Fiber Stabilization

In the present system, we used long fiber to increase the OPLD between the measurement and reference arms to extend the consecutive measurement range. The drift or phase noise of the long fiber affects the interferometric phase measurement directly. Therefore, we stabilized the long fiber according to the interferometric phase (ϕ_{cw}) of the CW laser during the distance measurement. Figure 4 shows the stabilization results. In the first 400 s, the long fiber is not stabilized, ϕ_{cw} drifts nearly 5 thousands degrees (Fig. 4a) and interferometric phase (ϕ_s) of the synthetic wavelength drifts nearly 26 degrees (Fig. 4b), corresponding to nearly 20 μm in optical length change.

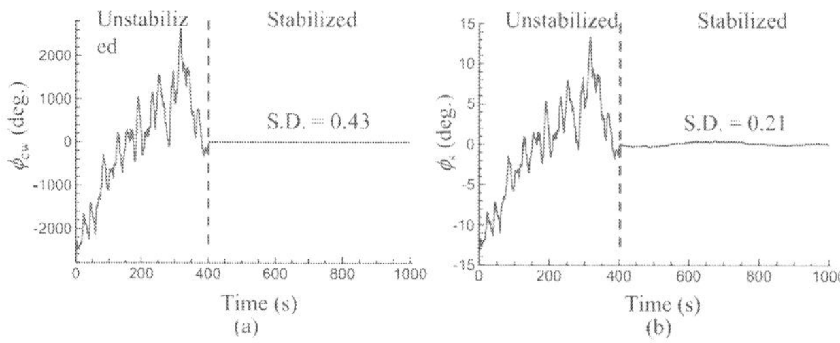

Figure 4. Performance of long fiber stabilization. (**a**) Interferometric phase (ϕ_{cw}) of the CW laser. (**b**) Interferometric phase (ϕ_s) of the synthetic wavelength.

When the long fiber is stabilized, ϕ_{cw} becomes very stable. It shows a standard deviation (S.D.) of 0.43 degree in 600 s (Fig. 4a), corresponding to 1.8 nm in optical length. In comparison, ϕ_s is not so flat when the long fiber is stabilized. It shows a S.D. of 0.21 degree in 600 s (Fig. 4b), corresponding to 180 nm in optical length. This is reasonable, because the interferometers of the CW laser and the comb are not in total common path. For example, the position of the target mirror CR may have some drift during the stabilization period of 600 s, and it will cause ϕ_s change. The residual instability of ϕ_s can be improved by decreasing the non-common path of the two interferometers and improving the environmental stabilities. Altogether, the long fiber stabilization decreases the interferometric phase drift greatly, which can ensure high-accuracy distance measurement of the present system.

Results of Absolute Distance Measurement

To evaluate the measurement accuracy of this system, we test the repeatability of the system firstly. The target mirror CR was moved by 100 mm, and the accurate displacement is given by a commercial heterodyne interferometer. The difference between the results obtained from the present method and the commercial interferometer indicates the measurement error (ε). We repeated the measurement 8 times with a time interval of 10 minutes. The measurement errors (ε) ranged from −83 nm to 92 nm with a standard deviation of 71 nm.

Secondly, we made a linear measurement comparison with a commercial heterodyne interferometer. The target mirror CR was moved by 7 steps at an increment of 50 mm. The whole experiment was accomplished in 450 s and ϕ_s was recorded during the experiment (Fig. 5). We measured ϕ_s at 8 positions in total (Fig. 5a). Once the target mirror moved, we should scan the repetition rate to make the pulses overlap again. Thus, ϕ_s changes randomly between two adjacent positions, which indicates the repetition rate scanning. Please note it is not necessary to record ϕ_s during repetition rate scanning, we only need to know ϕ_s at different positions and then we can obtain the absolute distance between the two positions. At each position, we should lock the repetition rate after the repetition rate scanning process (Fig. 5b). After that, ϕ_s became stable and it was recorded for nearly 10 s. The average value during the 10-s recording was used for distance calculation, and its standard deviation is about 0.1°, corresponding to 42 nm in distance. During the measurement, the temperature, humidity and air pressure were recorded for calculation of the refractive index of air.

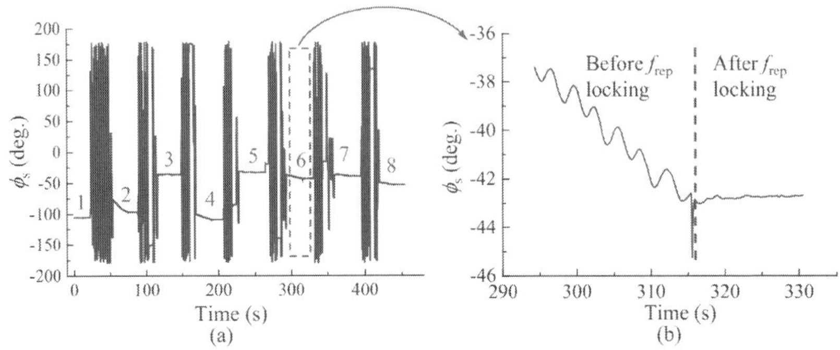

Figure 5. Interferometric phase (ϕ_s) of the synthetic wavelength recorded in the distance measurement. (**a**) ϕ_s recorded in the 7-step movement, corresponding to 8 positions. For each step measurement, it only costs tens of seconds. (**b**) Expanded view of ϕ_s at position 6. We only use the ϕ_s after repetition rate locking for distance calculation.

We made a comparison of the distances measured by using the present method and the results of the commercial interferometer (Fig. 6). The final distance measurement results (Fig. 6a) are in good accordance with the results of the reference interferometer. By applying linear fitting (Fig. 6b), the slope is 0.999999 and the correlation coefficient (R^2) is 1.000000. The residuals are ranged from −123 nm to 94 nm with a standard deviation of 75 nm (Fig. 6d). To demonstrate the advantages of present method, the residuals of the measurements by only applying the peak finding method are also shown (Fig. 6c) for comparison, which are ranged from −18 μm to 18 μm. Note that the accuracy of peak finding method here is lower than that in other reports[25,27]. In previous studies[25,27], piezo is used for fine scanning to get the envelope of the cross-correlation function and obtain the intensity peak. In contrast, as mentioned in the section of measurement principle, it is not necessary to maximize the intensity precisely for this system. Because the synthetic wavelength is hundreds of micrometers, so it does not require very precise pre-alignment of the pulses. We only adjust the motor-driven fiber delay-line in the cavity, and make the intensity of the interference signal close to the peak value. Consequently, we can accomplish the peak finding process quickly but with a lower accuracy. Fortunately, this accuracy is enough to link with the synthetic wavelength interferometry.

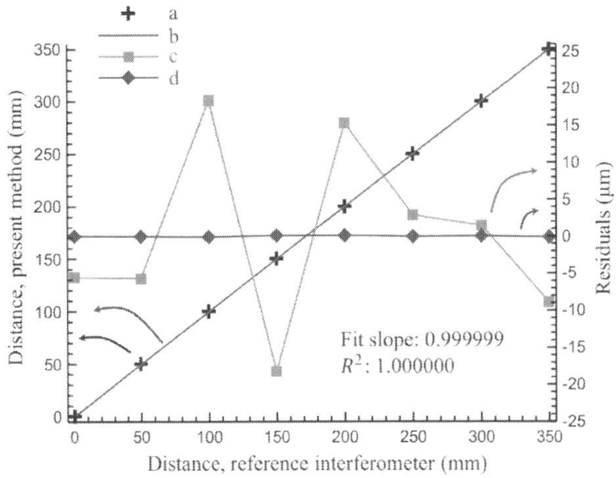

Figure 6. Experimental results versus the data obtained by the reference interferometer. (**a**) Distance data measured, (**b**) linear fitting of the data measured, (**c**) residuals of peak-finding method, (**d**) residuals of final results. a and b refer to the left longitudinal coordinates; c and d refer to the right one. They are indicated by the arrows.

DISCUSSION

In this experiment, we only demonstrated a consecutive measurement range of 350 mm. It is limited by the translation range of the stage. As analyzed in Section 2.1, the consecutive measurement range of the present system is 850 mm. It can be extended further by increasing the OPLD, e.g. increase the length of the long fiber. Once the consecutive measurement range reaches half L_{pp}, the system can realize absolute distance measurement without 'dead zone'.

It is necessary to point out f_{ceo} is free running during the experiment, which greatly decrease the complexity and cost of the laser source. Although f_{ceo} may drift during the measurement and cause some drift in ϕ_1 and ϕ_2, they are cancelled out in ϕ_s ($\phi_s = \phi_2 - \phi_1$). Therefore, using the synthetic wavelength as a marker for the pulse-to-pulse alignment and distance measurement is promising for practical applications. Although we cannot improve the accuracy furtherly by employing the interferometric phase of a single wavelength[32], the results that 350 mm consecutive measurement range with 75 nm accuracy (1σ) demonstrated here is enough for various applications. Note that we used a fully-stabilized comb as a reference for the CW laser. Actually, it is not necessary if we use a frequency-stabilized CW laser instead of the present CW laser.

Another advantage of this method is that the interferometric phase of the synthetic wavelength is used quantify a very small distance (e.g. δ_1 and δ_2 shown in Fig. 2) between the peaks of two overlapped pulses. No matter how long distance we measure, the fringe order (N_s) of the synthetic wavelength is always zero. Thus, the accuracy of the synthetic wavelength itself is not a severe factor in the measurement since its error will not be accumulated by the fringe order. This feature is different from traditional fringe counting interferometry, and it is very attractive for long distance measurement.

CONCLUSIONS

We proposed a synthetic-wavelength based heterodyne interferometer of optical frequency combs with wide consecutive measurement range for absolute distance measurement. The synthetic wavelength was derived from two wavelengths obtained by two band-pass filters, which simplified the complexity of synthetic wavelength generation. The interferometric phase of the synthetic wavelength was used as a marker for the pulse-to-pulse alignment, which greatly improved the accuracy of traditional peak finding method. Furthermore, the interferometric phase of the synthetic wavelength is immune to the drift of the carrier-envelope-offset frequency, which is attractive for practical applications. In order to extend the consecutive measurement range, we employed long fiber to increase the path length difference of the reference and measurement arms. The length of the long fiber was stabilized according to the interferometric phase of a CW laser. The repeatability measurement of 100 mm distance shows an accuracy better than 71 nm. By making a linear measurement comparison with a commercial interferometer, the present system showed an accuracy of 75 nm in 350 mm consecutive measurement range. This method can be widely used in various applications demanding high-accuracy distance measurement.

DATA AVAILABILITY

The datasets generated during and/or analyzed during the current study are available from the corresponding author on reasonable request.

ACKNOWLEDGEMENTS

This work was supported by the National Natural Science Foundations of China (61575105, 61611140125, 51405317); Natural Science Foundation of Jiangsu Province (BK20140358); the Special Fund for High Resolution Images Surveying and Mapping Application System (AH1601-8); Tsinghua University Initiative Scientific Research Program.

AUTHOR CONTRIBUTIONS

G.W. proposed the principle of the method and supervised the study; G.W., L.L. and Z.Z. constructed the setup; L.L., S.X., G.L. and Z.C. performed experiments, collected and analyzed data; G.W. interpreted the data and wrote the paper; all authors discussed the results and commented on the manuscript.

REFERENCES

1. Bender, P. L., Hall, J. L., Ye, J. & Klipstein, W. M. Satellite-satellite laser links for future gravity missions. *Space Science Reviews* **108**, 377–384 (2003).
2. Falaggis, K., Towers, D. P. & Towers, C. E. Multiwavelength interferometry: extended range metrology. *Optics Letters* **34**, 950–952 (2009).
3. Dandliker, R., Salvade, Y. & Zimmermann, E. Distance measurement by multiple-wavelength interferometry. *Journal of Optics-Nouvelle Revue D Optique* **29**, 105–114 (1998).
4. Kim, S.-W. Combs rule. *Nature Photonics* **3**, 313–314 (2009).
5. Cundiff, S. T. & Ye, J. Colloquium: Femtosecond optical frequency combs. *Reviews of Modern Physics* **75**, 325–342 (2003).
6. Minoshima, K. & Matsumoto, H. High-accuracy measurement of 240-m distance in an optical tunnel by use of a compact femtosecond laser. *Applied Optics* **39**, 5512–5517 (2000).
7. Schuhler, N., Salvade, Y., Leveque, S., Daendliker, R. & Holzwarth, R. Frequency-comb-referenced two-wavelength source for absolute distance measurement. *Optics Letters* **31**, 3101–3103 (2006).
8. Bitou, Y., Schibli, T. R. & Minoshima, K. Accurate wide-range displacement measurement using tunable diode laser and optical frequency comb generator. *Optics Express* **14** (2006).
9. Salvade, Y., Schuhler, N., Leveque, S. & Le Floch, S. High-accuracy absolute distance measurement using frequency comb referenced multiwavelength source. *Applied Optics* **47**, 2715–2720 (2008).
10. Baumann, E. *et al.* Comb-calibrated frequency-modulated continuous-wave ladar for absolute distance measurements. *Optics Letters* **38**, 2026–2028 (2013).
11. Wang, G. *et al.* Absolute positioning by multi-wavelength interferometry referenced to the frequency comb of a femtosecond laser. *Optics Express* **23**, 9121–9129 (2015).
12. Coddington, I., Swann, W. C., Nenadovic, L. & Newbury, N. R. Rapid and precise absolute distance measurements at long range. *Nature Photonics* **3**, 351–356 (2009).
13. Liu, T.-A., Newbury, N. R. & Coddington, I. Sub-micron absolute distance measurements in sub-millisecond times with dual free-running femtosecond Er fiber-lasers. *Optics Express* **19**, 18501–18509 (2011).

14. Lee, J. *et al.* Absolute distance measurement by dual-comb interferometry with adjustable synthetic wavelength. *Measurement Science and Technology* **24** (2013).
15. Zhang, H., Wei, H., Wu, X., Yang, H. & Li, Y. Absolute distance measurement by dual-comb nonlinear asynchronous optical sampling. *Optics Express* **22**, 6597–6604 (2014).
16. Hebert, N. B., Boudreau, S., Genest, J. & Deschenes, J.-D. Coherent dual-comb interferometry with quasi-integer-ratio repetition rates. *Optics Express* **22**, 29152–29160 (2014).
17. Wu, G. *et al.* Experimental optimization of the repetition rate difference in dual-comb ranging system. *Applied Physics Express* **7** (2014).
18. Wu, G., Xiong, S., Ni, K., Zhu, Z. & Zhou, Q. Parameter optimization of a dual-comb ranging system by using a numerical simulation method. *Optics Express* **23**, 32044–32053 (2015).
19. Joo, K. N. & Kim, S. W. Absolute distance measurement by dispersive interferometry using a femtosecond pulse laser. *Optics Express* **14**, 5954–5960 (2006).
20. Joo, K.-N., Kim, Y. & Kim, S.-W. Distance measurements by combined method based on a femtosecond pulse laser. *Optics Express* **16**, 19799–19806 (2008).
21. Cui, M., Zeitouny, M. G., Bhattacharya, N., van den Berg, S. A. & Urbach, H. P. Long distance measurement with femtosecond pulses using a dispersive interferometer. *Optics express* **19**, 6549–6562 (2011).
22. van den Berg, S. A., Persijn, S. T., Kok, G. J. P., Zeitouny, M. G. & Bhattacharya, N. Many-Wavelength Interferometry with Thousands of Lasers for Absolute Distance Measurement. *Physical Review Letters* **108** (2012).
23. Yamaoka, Y., Minoshima, K. & Matsumoto, H. Direct measurement of the group refractive index of air with interferometry between adjacent femtosecond pulses. *Applied Optics* **41**, 4318–4324 (2002).
24. Ye, J. Absolute measurement of a long, arbitrary distance to less than an optical fringe. *Optics Letters* **29**, 1153–1155 (2004).
25. Cui, M., Schouten, R. N., Bhattacharya, N. & van den Berg, S.A. Experimental demonstration of distance measurement with a femtosecond frequency comb laser. *Journal of the European Optical Society-Rapid Publications* **3** (2008).

26. Balling, P., Kren, P., Masika, P. & van den Berg, S. A. Femtosecond frequency comb based distance measurement in air. *Optics Express* **17**, 9300–9313 (2009).
27. Cui, M. *et al.* High-accuracy long-distance measurements in air with a frequency comb laser. *Optics Letters* **34**, 1982–1984 (2009).
28. Lee, J., Kim, Y.-J., Lee, K., Lee, S. & Kim, S.-W. Time-of-flight measurement with femtosecond light pulses. *Nature Photonics* **4**, 716–720 (2010).
29. Wei, D., Takamasu, K. & Matsumoto, H. A study of the possibility of using an adjacent pulse repetition interval length as a scale using a Helium-Neon interferometer. *Precision Engineering-Journal of the International Societies for Precision Engineering and Nanotechnology* **37**, 694–698 (2013).
30. Wu, H., Zhang, F., Cao, S., Xing, S. & Qu, X. Absolute distance measurement by intensity detection using a mode-locked femtosecond pulse laser. *Optics Express* **22**, 10380–10397 (2014).
31. Zhu, J., Cui, P., Guo, Y., Yang, L. & Lin, J. Pulse-to-pulse alignment based on interference fringes and the second-order temporal coherence function of optical frequency combs for distance measurement. *Optics Express* **23**, 13069–13081 (2015).
32. Wu, G., Takahashi, M., Inaba, H. & Minoshima, K. Pulse-to-pulse alignment technique based on synthetic-wavelength interferometry of optical frequency combs for distance measurement. *Optics Letters* **38**, 2140–2143 (2013).
33. Nakajima, Y. & Minoshima, K. Highly stabilized optical frequency comb interferometer with a long fiber-based reference path towards arbitrary distance measurement. *Optics Express* **23**, 25979–25987 (2015).
34. Joo, W.-D. *et al.* Femtosecond laser pulses for fast 3-D surface profilometry of microelectronic step-structures. *Optics Express* **21**, 15323–15334 (2013).
35. Hochrein, T. *et al.* Optical sampling by laser cavity tuning. *Optics Express* **18**, 1613–1617 (2010).
36. Wu, H. *et al.* Long distance measurement using optical sampling by cavity tuning. *Optics Letters* **41**, 2366–2369 (2016).

37. Hou, D., Lee, C. C., Yang, Z. & Schibli, T. R. Timing jitter characterization of mode-locked lasers with <1 zs/√Hz resolution using a simple optical heterodyne technique. *Optics Letters* **40**, 2985–2988 (2015).

CHAPTER 11

ANALYSIS OF THE NOISE REDUCTION CHARACTERISTICS OF A DIFFUSER WITH FIZEAU INTERFEROMETRY

Hongjun Wang, Xueliang Zhu, Bingcai Liu, and Ailing Tian

Xi'an Technological University, International Center for Optical Research and Education (iCORE), No. 2 Xuefu Middle Road, Xi'an, Shaanxi 710021, China

ABSTRACT

To solve the coherent noise problem of an interference image, the method of a rotating diffuser was adopted to change the coherence of a beam to reduce the noise of the interference system. The relationship between the speed of the diffuser and the signal-to-noise ratio (SNR) of the fringe contrast system was simulated to obtain the diffuser control parameters needed for the best interference fringe state. The fringe contrast of each image and the SNR of the system were analyzed. The results showed that the increased speed of the diffuser reduced the contrast of the interference image to a certain

Citation: (APA): Wang, H., Zhu, X., Liu, B., & Tian, A. (2021). Analysis of the Noise Reduction Characteristics of a Diffuser with Fizeau Interferometry. International Journal of Optics, 2021. (13 pages)

extent, but the increased speed also effectively improved the SNR and facilitated the subsequent interference image processing. Due to the coherent noise in the interferometric system, the method of the rotated diffuser reduced the coherence of the light beam to suppress the noise of the interference image. By analyzing the coherent noise reduction characteristics of the rotated diffuser with different surface roughnesses, the relationship between the surface roughness and the noise contrast for different rotation speeds was simulated, and the effective roughness range with the noise reduction effect was selected. A noise reduction system was built based on Fizeau interference, and the noise contrast of the interference image was collected and calculated. The effective range of σ_h/λ was 0.2–0.5 when the rotation speed was 10 r/s, while the effective range of σ_h/λ was 0.4–0.6 when the rotation speed was 100 r/s. The experimental results showed that the surface roughness and wavelength ratio σ_h/λ of the rotated diffuser increased when the noise contrast tended toward 1, but the effective range of the surface roughness decreased with the increase of the rotational speed of the diffuser.

INTRODUCTION

In an interference system, the defects on the surface of the optical element (such as depression, bubbles, dust, and scars) are a new light source. As a laser has a high degree of coherence, a large amount of coherent noise is generated by the light source formed by defects, which creates Newton rings in the interferogram, thus affecting the quality of the interferogram and further affecting the interference image processing, forming a measurement error [1, 2]. The coherent noise that affects the interference result is the intrinsic noise of the system. The suppression of intrinsic noise is a difficult problem in interferometer design. Therefore, to suppress intrinsic noise, many efforts have been made by experts and scholars over the years. For example, multiple fiber coupling is used to produce phase changes, optical wedges are used to change optical paths, and moving scatter or phase shifters are used to reduce laser coherence [3, 4]. In 1970, Asakura [5] studied the spatial coherence of a laser passing through a rotating ground glass. In 1980, Kwon et al. [6] studied the relationship between the contrast of infrared interference fringes and surface roughness. In 1983, Harwalkar [7] et al. carried out theoretical and experimental research on the scattered light passing through rotating ground glass. Chen [8] used the phase-shifting algorithm to calculate infrared interference fringe contrast and verified the theoretical relationship between the fringe contrast and the surface roughness. In 2000, Ziraki et al. [9] applied a rotating diffuser to a holographic system to suppress speckle noise. Pitter et al. [10] used a rotating diffuser for heterodyne interference microscopy to suppress coherent noise in coherent whole-field microscope images in 2004. Morris et al. [11] added a rotating diffuser in an interference system and then effectively reduced the coherence of the light source in 2010. In 2011, Xu et al. [12] and others calculated the statistical characteristics of the rotating ground glass in an interference system. Chen et al. [13] studied the use of a rotating diffuser in projection systems to reduce speckle distortion in 2012. In 2014, Qiao et al. [14] studied the denoising effect of rotating ground glass in a holographic system. In 2015, Wang Sen and others studied the effect of the rotational speed of ground glass on the imaging quality of a thermal ghost. These results showed that the intrinsic noise of the system could be reduced and the SNR of the system could be improved by using the rotating ground glass.

In this research, the noise reduction performance of the diffuser in a large-diameter Fizeau interferometer was mainly analyzed to improve the SNR of the interference image and improve the quality of the interference image. The optical modulation theory of diffuser was studied, included the theory of modulation of rough surface and rotation speed, and simulated the effect of surface roughness and rotation speed on the noise suppression effect of the interference image. The system of noise suppression based on Fizeau interference was built. According to the surface roughness and rotation speed of the diffuser, the noise suppression results of the interference image were verified. The noise contrast and image SNR were compared with the simulation results.

THEORETICAL ANALYSIS

Fizeau Interferometer

To eliminate coherent noises, the rotating diffuser was placed near the focal plane of the imaging system to modulate the interference beam that entered the imaging system, as shown in Figure 1. The terms RF and TF shown represent the reference mirror and the test mirror, respectively. The light emitted by the laser passed through the spatial filter to form a point light source and then through the beam expander to form a parallel light. After the parallel light passed through the reference mirror and the test mirror, an interference pattern was formed on the CCD.

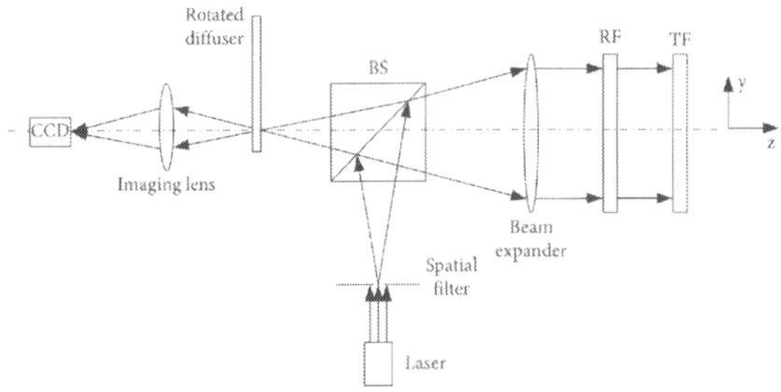

Figure 1: Fizeau interferometer with rotated diffuser.

The Influence of the Diffuser Rotation Speed on the Interference Pattern

It was assumed that the incident wavefront of diffuser was $u_0(x, y)$, the phase introduced by the diffuser was Φ_d, and the diffuser rotated around O at the angular velocity ω. Then, at time t, the wavefront u (x, y, t) could be expressed as

$$u(x, y, t) = u_0(x, y)\exp\left[i\Phi_d(r, \theta_0 - \omega t)\right]. \tag{1}$$

In the equation, r is the distance between the center of rotation and the point of incidence of light and θ_0 is the initial angle of rotation of the diffuser. The spot position was as shown in Figure 2. During the measuring process, r and w were constants. Then, $\exp[i\Phi_d(r, \theta_0 - \omega t)]$ was just a function of the time t in equation (1)

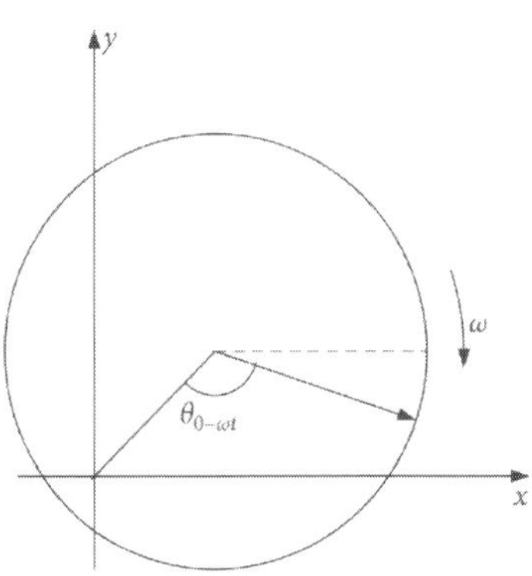

Figure 2: Spot position.

It was assumed that the imaging lens magnification was 1. Then, the light field $U(x_1, y_1, t)$ on the CCD was expressed with

$$U(x_1, y_1, t) = \iint_{-\infty}^{\infty} k(x + x_1, y + y_1) u(x, y, t)\, dxdy. \tag{2}$$

In the equation, k(x, y) is the amplitude point spread function of an imaging system. According to the speckle interference theory, the coherent time of the speckle, τ_c [15],

$$\tau_c = \frac{8\lambda z}{3\pi^2 \omega r D}. \tag{3}$$

In the equation, z is the distance between the diffuser and the pupil of the imaging system, D is the entrance pupil diameter, and λ is the wavelength. Within the exposure time T of the CCD, there are $N = T/\tau_c$ different speckle images which could be acquired. When N was large enough, the contrast of the speckle, C_n [15],

$$C_n \propto \frac{1}{\sqrt{N}} = \sqrt{\frac{8\lambda z}{3\pi^2 \omega r D T}}. \tag{4}$$

According to the interference theory, the contrast of the interference fringes, C_s [15],

$$C_s \propto \sin c\left(\frac{\tau_{RT}}{\tau_c}\right) = \sin c\left(\frac{3\pi^2 \omega r D \tau_{RT}}{8\lambda z}\right). \tag{5}$$

In the equation, τRT is the time delay of the test light and the reference light. The ratio of the interference fringe contrast to the speckle contrast was defined as the signal-to-noise ratio (SNR) of the interference system. Therefore, the SNR of the interference system with a rotating diffuser could be written as

$$\text{SNR} = \frac{C_s}{C_n} \propto \frac{\sin c\left(3\pi^2 \omega r D \tau_{RT}/8\lambda z\right)}{\sqrt{8\lambda z/3\pi^2 \omega r D T}}. \tag{6}$$

From equation (6), it could be seen that the SNR of the interference system was related to the rotational angular velocity of the diffuser, the exposure time of the detector, and the delay time of the different light path.

The Influence of the Diffuser Surface Roughness on the Interference Pattern

To study the corresponding relationship between the noise contrast and the surface roughness of diffuser, it was assumed that the scattering types were surface scattering, as shown in Figure 3.

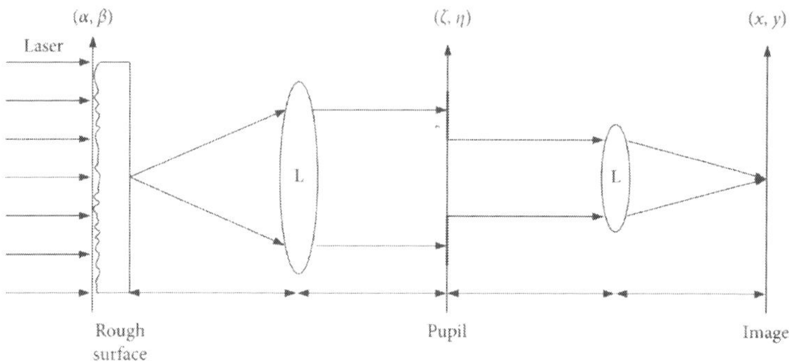

Figure 3: Imaging light path diagram for rough surface noise suppression.

Figure 4 shows a reflection light path diagram in a free space, for which $\hat{n}$ is a unit vector indicating the direction of the average normal surface, $\hat{i}$ is a unit vector representing the direction of the incident light, and $\hat{o}$ is the unit vector showing the direction of the observation point on the right plane.

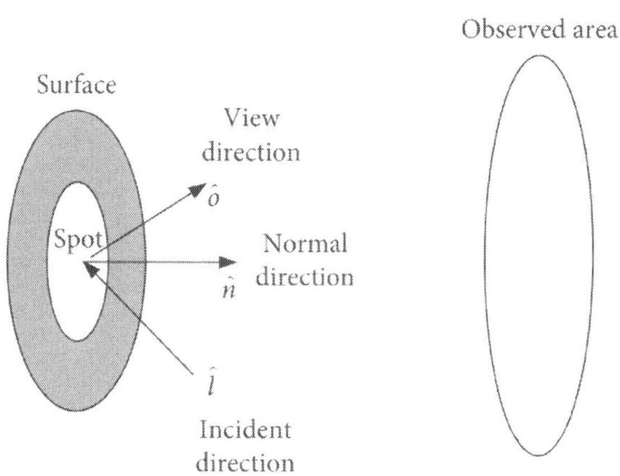

Figure 4: Free space reflection light path diagram [15].

When the light was perpendicularly incident on the surface of the diffuser, at any point on the surface, the phase delay Φ(α, β) associated with propagating to the surface and scattering from the surface was as follows [15]:

$$\Phi(\alpha, \beta) = \frac{2\pi}{\lambda}[1 + (\hat{o} \cdot \hat{n})]h(\alpha, \beta), \quad (7)$$

where $\hat{n}$ is the normal outward of the surface and $\hat{o}$ is the unit vector pointing to the observation point. For paraxial imaging, $\hat{o} \cdot \hat{n} \cong 1$, equation (7) could be simplified to

$$\Phi(\alpha, \beta) = \frac{4\pi}{\lambda} h(\alpha, \beta). \quad (8)$$

It was assumed that the incident intensity was 1 and the light transmittance was 1, so the outgoing light a (α, β) depended on the distribution of the rough surface and the expression was as follows:

$$a(\alpha, \beta) = \exp[j\Phi(\alpha, \beta)] = \exp\left[j\frac{4\pi}{\lambda} h(\alpha, \beta)\right]. \quad (9)$$

If the focal distances of the two lenses were the same, the magnification of the imaging system was 1, and the image A(x, y) in the (x, y) coordinate could be written as a convolution:

$$A(x, y) = \iint_{-\infty}^{\infty} h(x - \alpha, y - \beta)a(\alpha, \beta)\mathrm{d}\alpha\mathrm{d}\beta. \quad (10)$$

The relationship between the point spread function h(α, β) and the pupil function P(ξ, η) of the imaging system was

$$h(\alpha, \beta) = \frac{1}{\lambda \cdot f^2}\iint_{-\infty}^{\infty} P(\xi, \eta)\exp\left[j \cdot \frac{2\pi}{\lambda \cdot f}(\alpha\xi + \beta\eta)\right]\mathrm{d}\xi\mathrm{d}\eta. \quad (11)$$

It was assumed that the system had no aberrations, so $P(\xi, \eta) = p(\xi, \eta)$ was a real value. It was also assumed that p was 1 at the origin and the point spread function h was a real value.

The relationship between the surface roughness σ and the noise contrast C could be obtained as follows:

$$C = \sqrt{\frac{8(N-1) \cdot \{N - 1 + \cosh(\sigma_\phi^2)\}\sinh(\sigma_\phi^2/2)}{N[N - 1 + \exp(\sigma_\phi^2)]^2}}. \quad (12)$$

Based on equation (12), it could be seen that the different surface roughnesses of the diffuser had different influences on the noise contrast. We knew that the diffuser needed to be in the state of rotation to play a role in noise suppression, but the above theory did not consider the rotation speed, so the effects of different surface roughnesses on the noise suppression were analyzed for different rotation speeds, as described next.

It was assumed that the complex amplitude α (α, β) of the scattered wave above the surface was related to the surface height through a purely geometrically similar relationship. A phase was added to α that represented the phase delay that spread the surface and scattered from the surface:

$$a(\alpha, \beta) = rS(\alpha, \beta)e^{j\Phi(\alpha,\beta)}, \tag{13}$$

where r denotes the average amplitude reflectivity of the surface and $S(\alpha, \beta)$ indicates the complex amplitude of the illumination over the entire scattered spot and Φ represents the dot product in the expression, which was

$$\Phi(\alpha, \beta) = \frac{2\pi}{\lambda}(-\hat{i} \cdot \hat{n} + \hat{o} \cdot \hat{n}) \cdot h(\alpha, \beta). \tag{14}$$

When the illumination direction or observation direction deviated from the surface normal, the height fluctuation of the surface decreased due to the perspective relationship. The relationship between the variance σ^2_Φ of the phase shift and the variance σ^2_h of the surface height fluctuation was

$$\sigma_\phi^2 = \left[\frac{2\pi}{\lambda}(-\hat{i} \cdot \hat{n} + \hat{o} \cdot \hat{n})\right]^2 \sigma_h^2. \tag{15}$$

The relationship between the phase shift correlation function $\Gamma\Phi$ $(\Delta\alpha, \Delta\beta)$ and the normalized correlation function $\mu_h(\Delta\alpha, \Delta\beta)$ of the surface height fluctuation was

$$\Gamma_\Phi(\Delta\alpha, \Delta\beta) = \sigma_\phi^2 \mu_h(\Delta\alpha, \Delta\beta). \tag{16}$$

The normalized autocorrelation function of the light field was

$$\mu_\alpha(\Delta\alpha, \Delta\beta) = \overline{\exp[j(\varphi_1 - \varphi_2)]}. \tag{17}$$

Then, a mathematical hypothesis was made about the relative function μ_h of the surface height fluctuations:

$$\mu_h(\Delta\alpha, \Delta\beta) = \exp\left[-\left(\frac{r}{r_c}\right)^2\right], \tag{18}$$

where $r = \sqrt{\Delta\alpha^2 + \Delta\beta^2}$, and r_c is the radius of the normalized surface correlation down to $1/e$. Therefore,

$$\mu_\alpha(\Delta\alpha, \Delta\beta) = \exp\left\{-\sigma_\varphi^2\left\{1 - \exp\left[-\left(\frac{r}{r_c}\right)^2\right]\right\}\right\}. \tag{19}$$

The surface height of the diffuser was fluctuant, and r was the displacement of the coherent noise. When the diffuser rotated, the coherent noise displacement was equal to the product of the rotating speed of the diffuser and the exposure time; that is, r=vt, as shown in Figure 5. By substituting r=vt into equation (19), the following expression could be obtained:

$$\mu_\alpha(\Delta\alpha, \Delta\beta) = \exp\left\{-\sigma_\varphi^2\left\{1 - \exp\left[-\left(\frac{\mathrm{vt}}{r_c}\right)^2\right]\right\}\right\}. \tag{20}$$

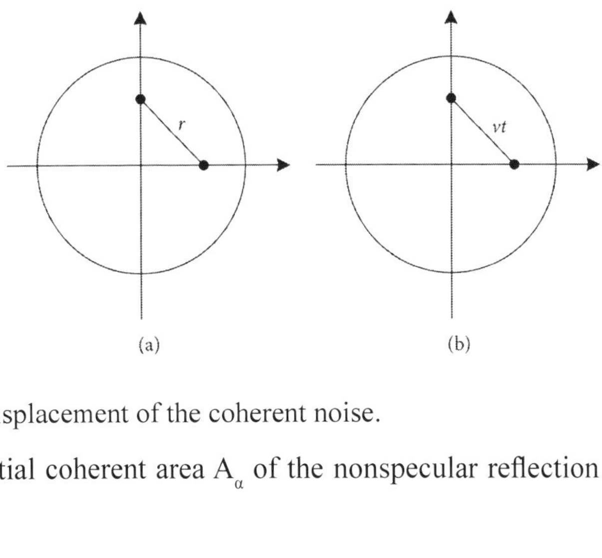

Figure 5: Displacement of the coherent noise.

The spatial coherent area A_α of the nonspecular reflection components was

$$A_\alpha = 2\pi \int_0^\infty v^2 t \mu_\alpha(t) \mathrm{d}t. \tag{21}$$

Hence,

$$A_\alpha = \frac{2\pi e^{-\sigma(2/\varphi)}}{1 - e^{-\sigma_\varphi^2}} \int_0^\infty v^2 t\left[\exp\left(\sigma_\varphi^2 \exp\left[\left(v_t/r_c\right)^2\right]1\right]\mathrm{d}t$$

$$= \frac{\pi e^{-\sigma_\varphi^2}}{1 - e^{-\sigma_\varphi^2}} \frac{\left[\mathrm{Ei}\left(\sigma_\varphi^2\right)\varepsilon \ln\left(\sigma_\varphi^2\right)\right]}{V^2}. \tag{22}$$

Here, Ei (x) is the exponential integral function and ε is the Euler constant. The number N′ of these completely incoherent noise fields is the ratio of the spatial coherence area A_k and the total illumination area A_α; it was found that

$$N' = \frac{\nu^2 N_0\left(e^{\sigma_\phi^2} - 1\right)}{Ei\left(e^{\sigma_\phi^2}\right) - \varepsilon - \ln\left(\sigma_\phi^2\right)}. \tag{23}$$

Bringing equation (23) into equation (12) produced a new numerical relationship between the surface roughness of the diffuser and the noise contrast:

$$C = \sqrt{\frac{8\left(N' - 1\right)\left\{N' - 1 + \cosh\left(\sigma_\varphi^2\right)\right\}\sinh\left(\sigma_\varphi^2/2\right)}{N'\left[N' - 1 + \exp\left(\sigma_\varphi^2\right)\right]^2}}. \tag{24}$$

SIMULATION

System Parameter

According to the measurement principle, the Fizeau interference experiment system was built. The time delay τ_{RT} between the testing light path and the reference light path was 4.76×10^{-9}s, the exposure time T of the CCD was 1/714s, the pupil diameter D was 7 mm, the wavelength λ was 650 nm, and the distance r between the center of rotation and the diameter of the incidence light was 20 m

Relationship between the Rotation Speed and the Image Quality

The relationship between the rotation speed of the diffuser and the coherent time could be calculated according to equation (3), as shown in Figure 6(a). As can be seen from Figure 6(a), the coherent time decreased gradually as the rotational speed of the diffuser increased, and the falling speed slowed down with the increase of the rotational speed of the diffuser. According to equation (6), the relationship between the rotation speed of the diffuser, the contrast of the fringe and speckle, and the signal-to-noise ratio of the system could be simulated, as shown in Figures 6(b)–6(d). Figure 6(b) shows that the image SNR was improved with the increase of the diffuser speed. When the diffuser speed was in the 40 r/s–90 r/s, fringe contrast was between 0.999–0.994, the relative value of decline was not obvious, and more fringe information was retained. At the same time, the relative value of the speckle contrast was between 0.5 and 0.3. Compared with the relative value of the fringe contrast, the decrease trend was obvious, indicating that properly increasing the rotational speed of diffuser could effectively suppress the coherent noise of the interference image.

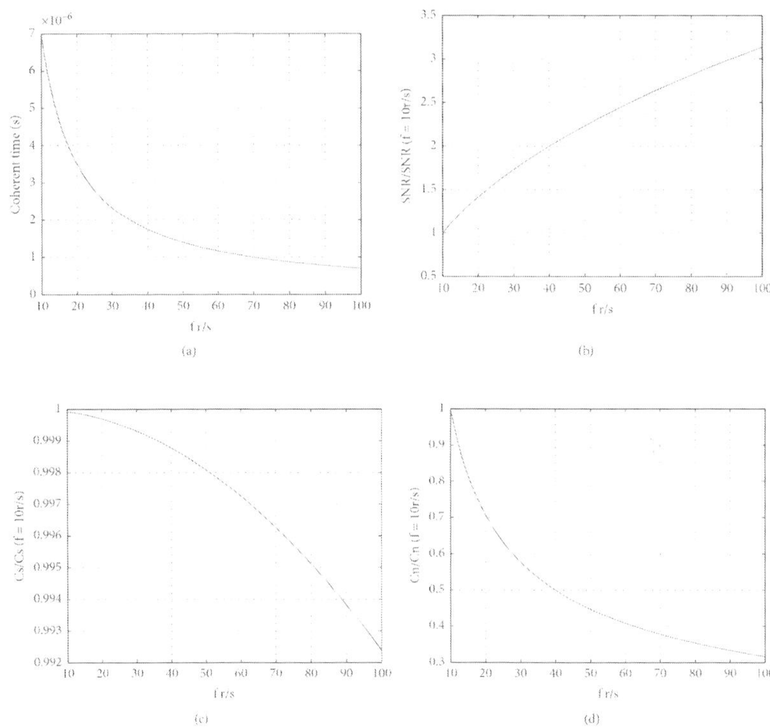

Figure 6: The influence of the rotation frequency on the image quality. (a) The influence on the coherent time. (b) The influence on the SNR. (c) The influence on the contrast of the fringes. (d) The influence on the speckle contrast.

Relationship between the Rotation Speed of the Diffuser and the Image Quality

Equation (24) describes the effect of the different surface roughnesses on the coherent noise contrast at different rotation speeds when N_0 was equal to 5, 10, 100, and 1000. The effect of different N_0 to noise contrasts at different rotation speeds was simulated in the range of σ_h/λ (surface height standard deviation of wavelength normalization) equal to 0.2–0.68. The simulated relationship curves are shown in Figures 7 and 8.

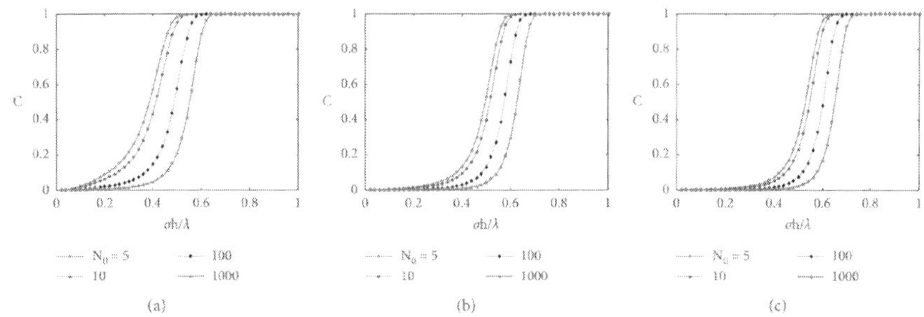

Figure 7: The relationship between the contrast C and σ_h/λ at different speeds v and N_0. (a) V=10 r/s. (b) V=50 r/s. (c) V=90 r/s.

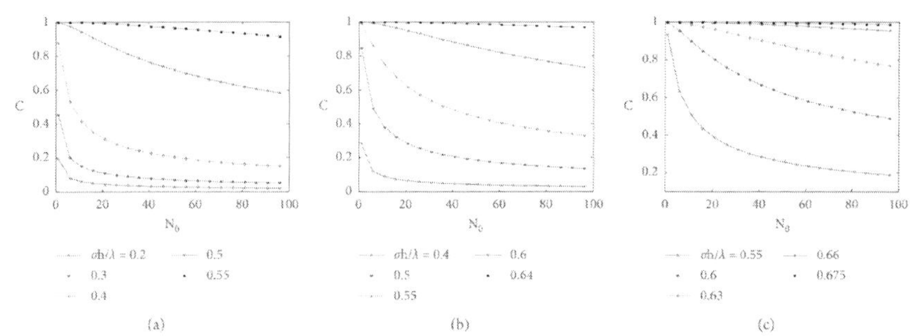

Figure 8: The relationship between the contrast C and N_0 at different speeds v and σ_h/λ. (a) V=10 r/s. (b) V=50 r/s. (c) V=90 r/s.

Figure 7 shows the relationship between the noise contrast C and the surface roughness at different rotational speed v and N_0. It can be seen from the figure that (1) for an ideally flat surface ($\sigma_h/\lambda=0$), the noise contrast was 0; (2) as σ_h/λ increased, the contrast was finally saturated at 1; and (3) as the rotational speed of the diffuser increased, the surface roughness corresponding to the saturated contrast also increased.

The relationship between the noise contrast C and N_0, different rotation frequencies v, and σ_h/λ for different diffusers is shown in Figure 8. It can be seen from the figure that for a small roughness value, the contrast C first increased with the parameter N_0 and then finally approached zero continuously. This was caused by the different dependence of the mirror reflection component and the diffuser component of light on the value of N_0. Especially in the light path, the mirror reflection component in the reflected light always appeared in the form of a small light spot on the optical axis in the focal plane of the first lens, while the diffuser light filled the entire pupil. It was necessary to change the size of the pupil to change N_0 and to decrease the pupil to increase N_0. As the mirror reflection component was always limited to the optical axis region, the mirror reflection component was not affected when the diameter of the pupil was reduced. However, the average intensity of the scattered light that was transmitted decreased in proportion to the pupil area as the pupil shrank. Therefore, when the value of σ_h/λ was small, N_0 increased from 1 or more so that the noise contrast increased from zero. Additionally, when N_0 finally increased further, the σ_I decreased and $\overline{I}$ remained constant, so the contrast decreased. Finally, when the pupil size was small enough, both the scattered light and the reflected light were affected, and the noise contrast C did not fall with N_0.

EXPERIMENTAL STUDY

As shown in Figure 9, in the constructed Fizeau interference system, the diffuser was placed near the focus of the imaging system, and its roughness was 0.3668 μm. Figure 10 shows the diffuser used in the experiment. The diameter of the diffuser was 50 mm, and the thickness was 3 mm.

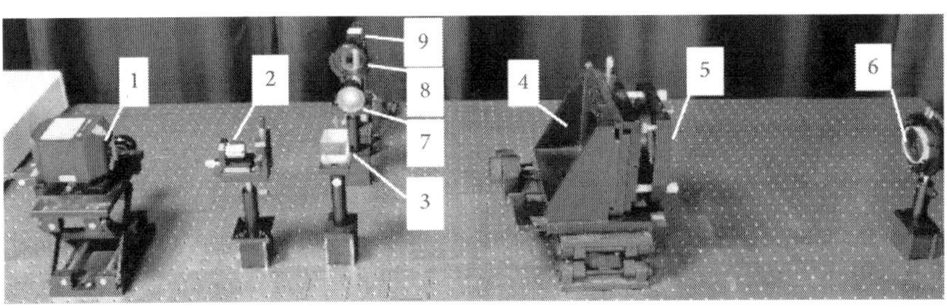

Figure 9: Experimental setup. 1, laser; 2, microscope objective; 3, beam splitter; 4, collimating lens; 5, reference flat; 6, testing flat; 7, diffuser; 8, imaging lens 9-CCD.

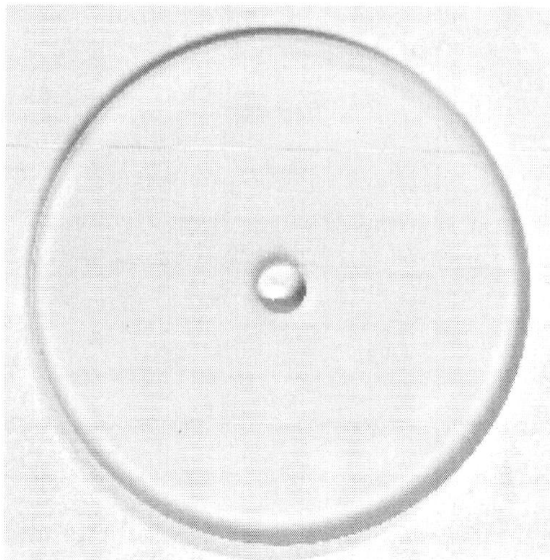

Figure 10: Diffuser.

To test different surface roughnesses of the diffuser for the noise suppression effect, a series of different roughnesses of the diffuser was used. The roughness of the diffusers was measured by Taylor Hobson PGI Optics Profilometer, as shown in Table 1.

Table 1: Diffuser with different surface roughnesses.

σ_h/λ	**Designed RMS roughness (nm)**	**Actual RMS roughness (nm)**	**Error (nm)**
0.2	126.56	135.9	9.34
0.3	189.84	199.8	9.96
0.35	221.48	216.7	−4.83
0.4	253.12	244.4	−8.7
0.47	297.4	300.3	2.9
0.5	316.4	305.1	−11.3
0.6	379.68	390.3	10.62
0.7	442.96	458.7	15.74

Based on the noise estimation processing, for the case of the known surface roughness, for different speeds of the diffuser, the fringe contrast, speckle contrast, image contrast, and image SNR were as shown in Figure 11. As can be seen from Figure 11, as the rotational speed of diffuser gradually increased, the fringe contrast, speckle contrast, and image contrast of the interference image all showed a downward trend, while the image signal-to-noise ratio increased. It was also shown that the rotating diffuser could reduce the contrast of the interference image while suppressing the coherent noise but improving the signal-to-noise ratio of the image. The interference images at different rotational speeds of the diffuser are shown in Figure 12. It can be seen from the figure that with the increase of the rotational speed of the diffuser, although the contrast of the image was reduced to a certain extent, the signal-to-noise ratio of the image was improved and the coherent noise was effectively suppressed.

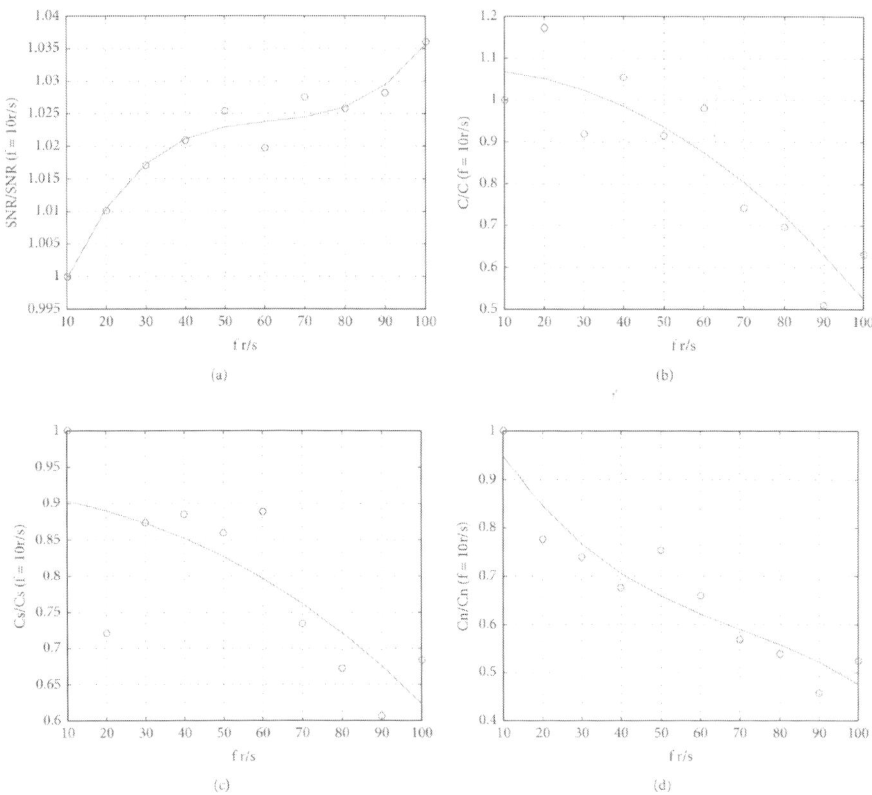

Figure 11: The relationship between the rotated diffuser and the parameters. (a) The influence on the SNR of the image. (b) The influence on the image contrast. (c) The influence on the contrast of the fringes. (d) The influence on the speckle contrast.

(a)

(b)

(c)

(d)

(e)

(f)

(g) (h)

Figure 12: The interference image for different speeds of the rotated diffuser. (a). (b). (c). (d). (e). (f). (g). (h).

For the condition of invariance in the diffuser surface roughness, the rotation speed of the motor was adjusted, and the interference images were acquired for different rotation speeds. Due to the need to analyze the influence of different surface roughnesses on the noise contrast, the intervention had to be separate. The Wiener filtering method was used to separate the image noise, and then the noise contrast for different rotation speeds was as shown in Figure 13.

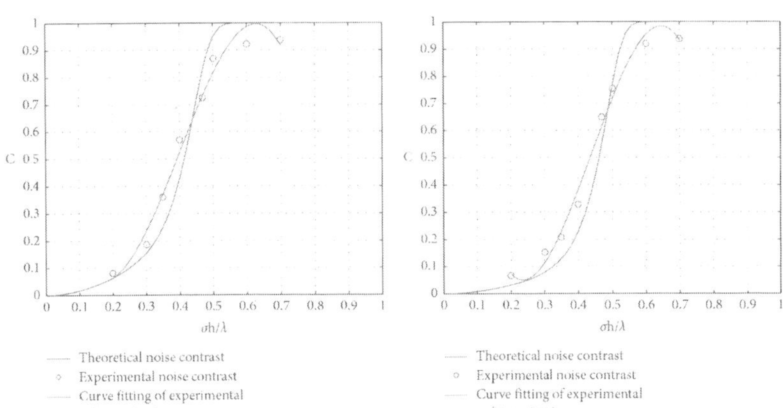

(a) (b)

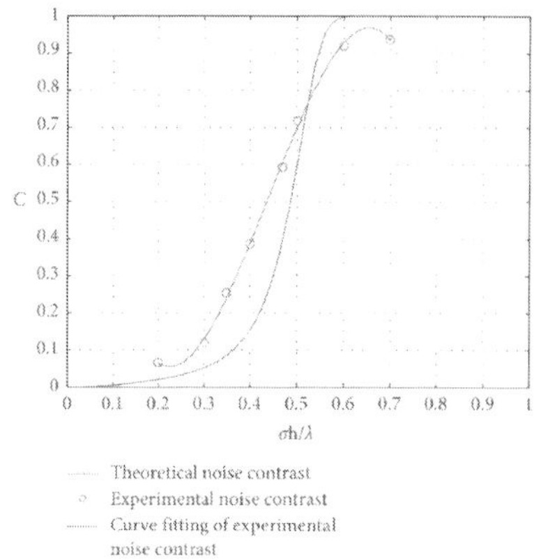
C
σh/λ
Theoretical noise contrast
Experimental noise contrast
Curve fitting of experimental noise contrast

(c)

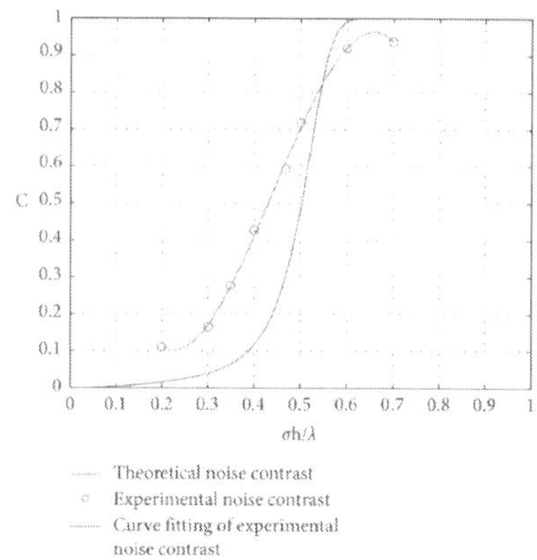
C
σh/λ
Theoretical noise contrast
Experimental noise contrast
Curve fitting of experimental noise contrast

(d)

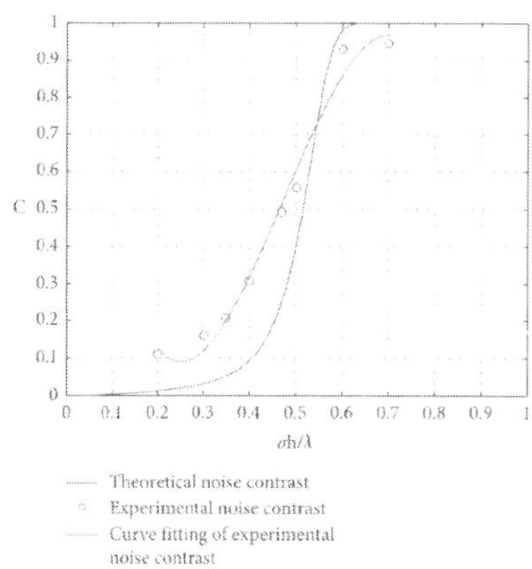
C
σh/λ
Theoretical noise contrast
Experimental noise contrast
Curve fitting of experimental noise contrast

(e)

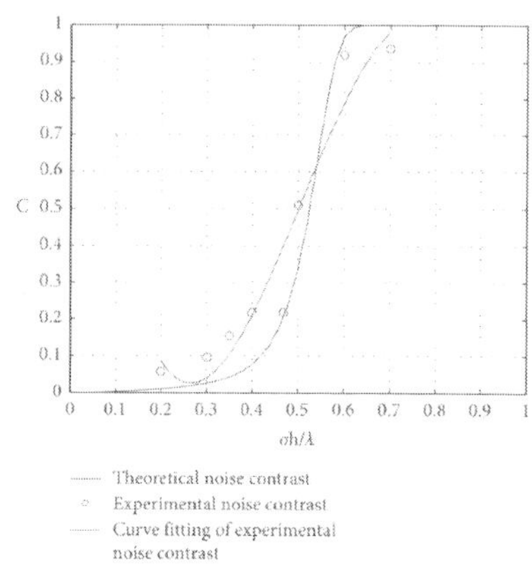
C
σh/λ
Theoretical noise contrast
Experimental noise contrast
Curve fitting of experimental noise contrast

(f)

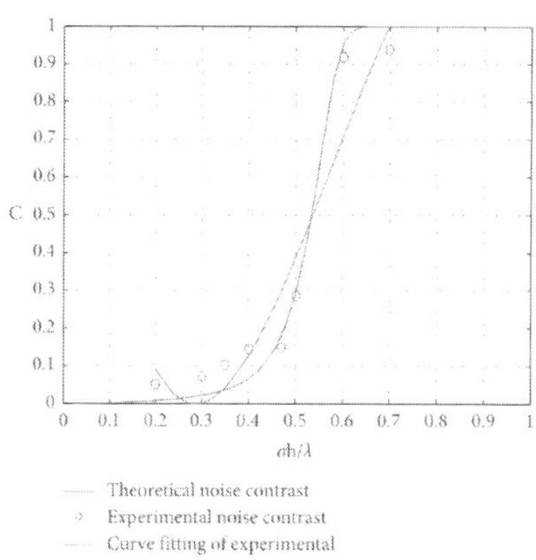
C
σh/λ
Theoretical noise contrast
Experimental noise contrast
Curve fitting of experimental noise contrast

(g)

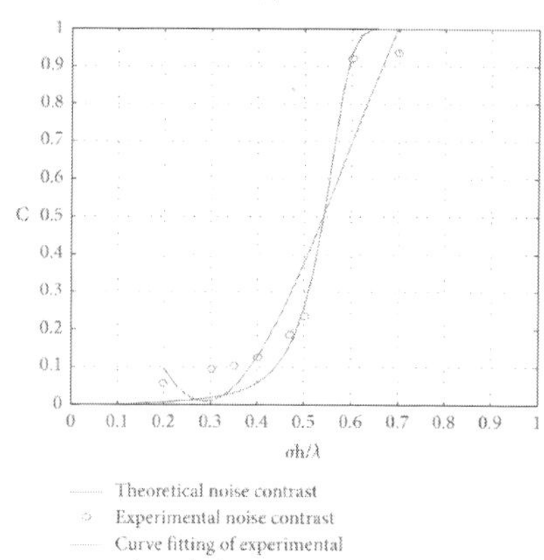
C
σh/λ
Theoretical noise contrast
Experimental noise contrast
Curve fitting of experimental noise contrast

(h)

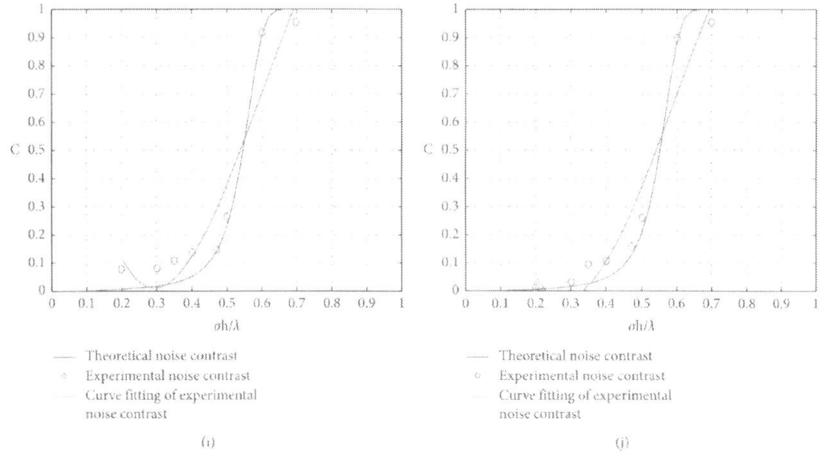

Figure 13: Effect of the surface roughness on the noise contrast at different rotational speeds. (a) V=10 r/s. (b) V=20 r/s. (c) V=30 r/s. (d) V=40 r/s. (e) V=50 r/s. (f) V=60 r/s. (g) V=70 r/s. (h) V=80 r/s. (i) V=90 r/s. (j) V=100 r/s.

As can be seen from Figure 13, the roughness of the diffuser could effectively suppress the noise increases with the increase of the rotation speed of the diffuser, but the effective range of the roughness of the surface that could effectively suppress noise gradually decreased, which was highly consistent with the simulation results. When the rotation speed was 10 r/s, the effective range of σ_h/λ was between 0.2 and 0.6. When the rotation speed was 100 r/s, the effective range of σ_h/λ was between 0.4 and 0.6. The smaller the value of σ_h/λ, the smaller the contrast of the noise and the less obvious the effect of the noise reduction. However, when σ_h/λ was bigger, the surface scattering characteristics of the diffuser were better. The entire light field was full of noise, which had no function of restraining noise but did seriously affect the quality of the interference image. There was a large error between the actual surface roughness and the designed surface roughness of the diffuser used in the experiment, leading to a large error in the numerical value between the experimental result and the simulation result. However, the experimental result was generally consistent with the simulation result, with a good consistency.

CONCLUSION

In this research, the relationship between the rotating speed of the diffuser in the Fizeau interferometer and the contrast of the interference fringes, speckle contrast, image contrast, SNR, and coherent time was analyzed and verified with experiments. The experimental results showed that the noise of the interference system was obviously suppressed by adding the rotating diffuser in the interference system. At the same time, the fringe contrast of the interference image decreased with the increase of rotational speed, and the image signal-to-noise ratio was improved, which showed the consistency of the experimental results and the simulation results. Therefore, when a rotating diffuser was used to reduce the noise of the interference system, we had to choose the appropriate surface roughness and the matching rotation speed. The noise contrast ratio had to be properly reduced so that the quality of the interference image could be optimized.

DATA AVAILABILITY

The simulation analysis data used to support the findings of this study are included within the article. The experimental measurement data used to support the findings of this study are included within the article.

CONFLICTS OF INTEREST

The authors declare that they have no conflicts of interest.

ACKNOWLEDGMENTS

This research was financially supported by the China Research Project (Grant no. JSZL2018411C001) and Shaanxi Science and Technology Department Research Project (Grant no. 2020GY-045). The authors thank LetPub (http://www.letpub.com) for its linguistic assistance during the preparation of this manuscript.

REFERENCES

1. G. Huang, *Research on Fizeau Phase-Shifting Laser interferometer*, Harbin Institute of Technology, Harbin, China, 2011.
2. M. Kuechel, "Apparatus and method for reducing the effects of coherent artifact in an interferometer," 2004, United Stated patents: 6804011.
3. Z. Chen, J. Li, and Y. Chen, "A practical method to suppress laser speckle noise in coherent imaging system," *Chinese Journal of Lasers*, vol. 12, no. 12, pp. 726–728, 1985.
4. W. Wu, *LRCS of Absorbing Coating Target under Coherent and Partially Coherent Light*, XIdian University, Xi'an, China, 2014.
5. T. Asakura, "Spatial coherence of laser light passed through rotating ground glass," *Opto-electronics*, vol. 2, no. 3, pp. 115–123, 1970.
6. O. Kwon, J. C. Wyant, and C. R. Hayslett, "Rough surface interferometry at 10.6 μm," *Applied Optics*, vol. 19, no. 11, pp. 1862–1869, 1968.
7. V. Harwalkar, H. Bohidar, and S. Chopra, "Theoretical and experimental investigations of light scattered from a rotating ground glass," *Applied Physics B Photophysics and Laser Chemistry*, vol. 31, no. 4, pp. 215–220, 1983.
8. L. Chen, *Research on Infrared Fizeau Interferometer and Automatic Infrared Interferogram Processing*, Nanjing University of Science & Technology, Nanjing, China, 1996.
9. M. Ziraki, R. Jones, P. M. W. French et al., "Whole-field coherent imaging through turbid media using Photorefractive holography]," *Applied Physics B*, vol. 70, no. 1, pp. 151–154, 2000.
10. M. C. Pitter, C. W. See, and M. G. Somekh, "Full-field heterodyne interference microscope with spatially incoherent illumination," *Optics Letters*, vol. 29, no. 11, pp. 1200–1202, 2004.
11. M. N. Morris, M. Naradikian, and J. Millerd, "Noise Reduction in dynamic interferometery measurements," *Interferometry XV: Techniques and Analysis*, vol. 7790, Article ID 77900, 2010.
12. J. Xu, Z. Liu, Y. Du et al., "Statistical analysis of interferometric imaging system with rotating diffuser," *High Power Laser and Particle Beams*, vol. 23, no. 3, pp. 702–706, 2011.
13. C.-Y. Chen, W.-C. Su, C.-H. Lin, M.-D. Ke, Q.-L. Deng, and K.-Y. Chiu, "Reduction of speckles and distortion in projection system by using a rotating diffuser," *Optical Review*, vol. 19, no. 6, pp. 440–443, 2012.

14. T. Qiao, Y. Yu, and W. Zhou, "A method to eliminate the noise of holographic system by using a rotated diffuser," *Optical Instruments*, vol. 36, no. 4, pp. 337–341, 2014.

15. J. W. Goodman, *Speckle Phenomena in Optics-Theory and Application*, Roberts & Company, Aurora, Canada, 2007.

CHAPTER 12

MICRO FABRY-PÉROT INTERFEROMETER AT RAYLEIGH RANGE

YusukeTsujiie & Yoshiyuki Kawamura

Department of Intelligent Mechanical Engineering, Faculty of Engineering, Fukuoka Institute of Technology, 3- 30-1 Wajirohigashi, Higashiku, Fukuoka, 811-0295, Japan.

ABSTRACT

The Fabry-Pérot interferometer is used in a variety of high-precision optical interferometry applications, such as gravitational wave detection. It is also used in various types of laser resonators to act as a narrow band filter. In addition, ultra-compact Fabry-Pérot interferometers are used in the optical resonators of semiconductor lasers and fiber-optic systems. In this work, we developed a micro-scale Fabry-Pérot interferometer that was constructed within the Rayleigh range of the optical focusing system. The high precision that is conventionally required for the optical parallelism and the surface accuracy of the mirrors was not so critical for this type of Fabry-Pérot interferometer. The interferometer was constructed using a gold-coated

Citation: (APA): Tsujiie, Y., & Kawamura, Y. (2018). Micro Fabry-Pérot Interferometer at Rayleigh Range. Scientific reports, 8(1), 1-6. (6 pages)

silicon microcantilever with reflectivity of 92% and a dielectric multilayer flat mirror with reflectivity of 85%. The focal spot size of the laser beam is 20 μm and the cavity length is approximately 20 μm. The finesse was measured to be approximately 25. The interferometric characteristics of the device were consistent with the theoretically calculated performance. The developed micro Fabry-Pérot interferometer has the potential to make a marked contribution to advances in optical measurements in various micro sensing system.

Keywords: Rayleigh Range, Microcantilever, Interferometer Characteristics, Focusing Optical System, Cavity Length

INTRODUCTION

In this work, we have developed a micro Fabry-Pérot (FP) interferometer with high sensitivity to realize high-performance feedback damping of the thermal vibration of a silicon microcantilever that is intended for use in an atomic force microscope (AFM). FP interferometers have been used in various high-precision optical interferometry applications, such as gravitational wave detection[1]. To date, there have been many studies of normal-sized FP interferometers[2,3], but only a few studies have addressed smaller types of FP interferometers[4,5].

There have been several studies of feedback cooling of the thermal vibration of micro cantilevers[6,7,8,9,10,11,12,13]. Recent studies found that the measured signal-to-noise ratio determines the limits of the feedback cooling performance[6,7,12]. We used an FP interferometer rather than a Michelson interferometer to improve the measurement sensitivity and thus increase the signal-to-noise ratio. In conventional FP interferometers, the polished end faces of optical fibers[10,11] and micro mirrors from the surface of a multilayer dielectric mirror[12,13] formed by focused ion beam microfabrication have been used as cavity mirrors. However, use of these methods for the mirror has led to issues such as low finesse due to optical diffraction from the fiber output aperture and problems with the parallelism of the optical alignment and the interferometer, along with difficulties in the microfabrication process. In addition, these methods do not use the merits of the Rayleigh range. In this work, we have developed a micro FP interferometer that uses the optical merits of the Rayleigh range of the focal system to simplify the optical system and improve the interferometric performance. The interferometric characteristics of this FP interferometer show good agreement with the theoretically calculated performance.

METHODS

Figure 1 shows the experimental system that was used to measure the interferometric characteristics of the micro FP interferometer. A He-Ne laser (wavelength of 632.8 nm; laser power of approximately 1 mW) was used as the light source for the interferometer. The micro FP interferometer is constructed using the gold-coated surface of a microcantilever and a dielectric multilayer flat mirror. We measured the vibration of a commercially available silicon microcantilever (OMCL-AC240TN, Olympus Corporation) that is intended for use in AFMs.

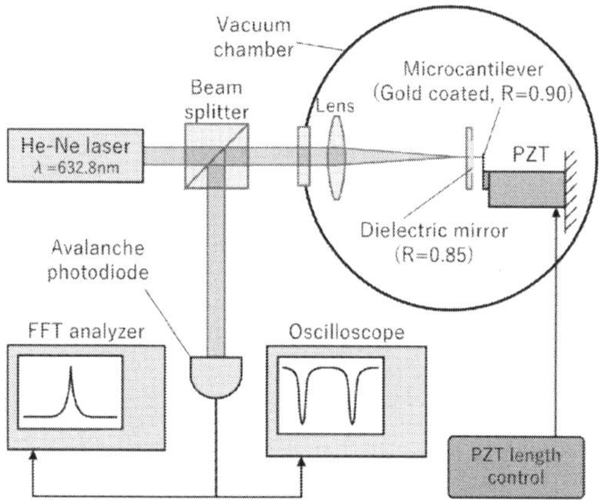

Figure 1: Micro Fabry-Pérot interferometer for vibration measurement system. The PZT is a piezoelectric actuator, and the FFT analyzer is a fast Fourier transform analyzer.

Figure 2 shows a scanning electron microscope image of this microcantilever. The micro cantilever's length, width, and thickness are 240 μm, 40 μm, and approximately 2.3 μm, respectively, and it is composed of single-crystal silicon. The natural oscillation frequency of the microcantilever is 77.6 kHz, and the catalog value of its spring constant is about 2 N/m. One single-side surface of the microcantilever was coated with gold to increase the laser reflectance using an ion-beam sputtering device that is commonly used for preprocessing before scanning electron microscope observation. The coating thickness was chosen to be as thin as possible while ensuring that sufficient reflectivity (92%) was obtained because reductions in both the natural oscillation frequency and the Q factor of the microcantilever were observed when a thick gold coating was used. The coating thickness was estimated to be approximately 25 nm based on the coating characteristic curve of the ion sputtering device. Ideally, the coating should be done only on the area, where the laser beam was irradiated. However the coating was done on the whole front surface of the cantilever, because it was easier than the partial coating. In case of the silicon micro cantilevers, we found dielectric multilayer coating was difficult on them. We tried several times, however the microlever were broken in all cases, probably due to the surface stress induced by the coating. It was one of the reason why we chose the gold coating.

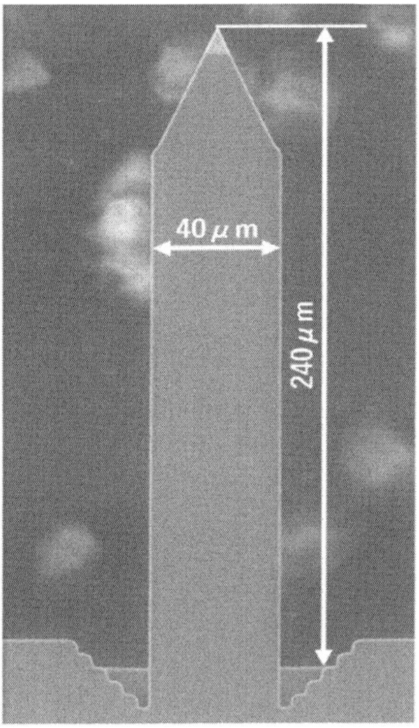

Figure 2: Scanning electron microscope image of the microcantilever.

The other side of the FP interferometer is formed by the dielectric multilayer flat mirror. The optical flatness and reflectance of this mirror were λ/10 and 85%, respectively. The diameter and thickness of the mirror were 30 mm and 1 mm, respectively.

A laser beam with a diameter of 4 mm was focused using a spherical lens, which has a focal length of 80 mm and an F number of 20. The focal spot size was estimated to be approximately 16 μm under the assumption of the diffraction limit. The Rayleigh range was estimated to be approximately 250 μm, and the cavity length was approximately 20 μm.

The optical system was set in a vacuum chamber at a pressure of approximately 4×10^{-3} Pa. The interferometric signal was separated using a beam splitter and measured using an avalanche photodiode. The microcantilever was driven using a lead zirconate titanate (PZT) piezoelectric actuator. The signal was measured using an oscilloscope and a fast Fourier transform (FFT) analyzer. The vacuum circumstance was not essential for this experiment. It was only for obtaining a clear thermal vibration signal of the micro cantilever. The same optical characteristics of it were also obtained in the atmospheric pressure.

RESULTS

The reflectance values of the microcantilever and the dielectric multilayer mirror were 92% and 85%, respectively. The reflectance of the microcantilever differs from that of the dielectric mirror. For the FP interferometer that was constructed using a pair of mirrors with different reflectances, the theoretical interferometric reflectance R was calculated to be

$$R = \frac{R_1 + R_2 - 2\sqrt{R_1R_2} + 4\sqrt{R_1R_2}\sin^2(\delta/2)}{1 + R_1R_2 - 2\sqrt{R_2R_2} + 4\sqrt{R_1R_2}\sin^2(\delta/2)}, \tag{1}$$

where R_1 and R_2 are the reflectances of the multilayer mirror and of the microcantilever, respectively14. δ is the phase shift of each transmitted light wave due to the change in the cavity length L_C and is given by $\delta=4\pi L_C/\lambda\delta=4\pi L_C/\lambda$.

Figure 3 shows the interferometric reflectance R as a function of the cavity length, as calculated using eq. (1) for various values of R_2. We can see that the minimum interferometric reflectance could not be 0% when R_1 and R_2 differ from each other. In the case where $R_1=R_2$, the minimum reflectance is 0%. In the case where $R_1>R_2$, the minimum reflectance increases as R_1 decreases.

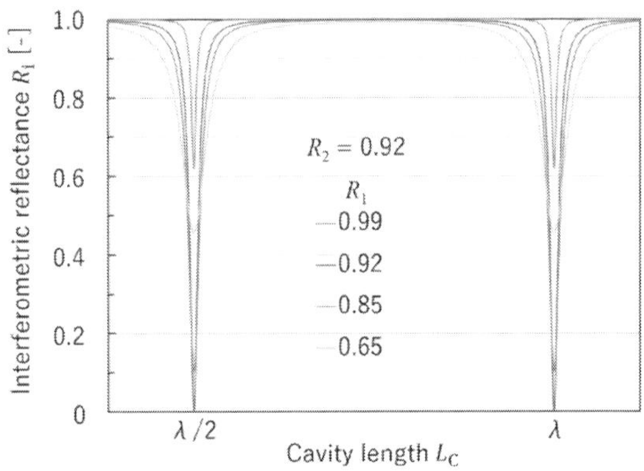

Figure 3: Interferometric reflectance ***R*** as a function of cavity length, calculated using eq. (2), for various values of $\boldsymbol{R_2}$.

Figure 4 shows the relative maximum sensitivity of the FP interferometer as a function of R_1 when R_2=0.92, where the slope of the interferometric curve is maximized. The sensitivity reached a maximum value at R_1=0.97 (i.e., not at R_1=0.92). For the various values of R_2, the maximum value of the sensitivity for R_2 differed from that for R_1 and was located between R_2 and 1. The open circle in Fig. 4 is related to the experimental conditions (where R_2=0.85).

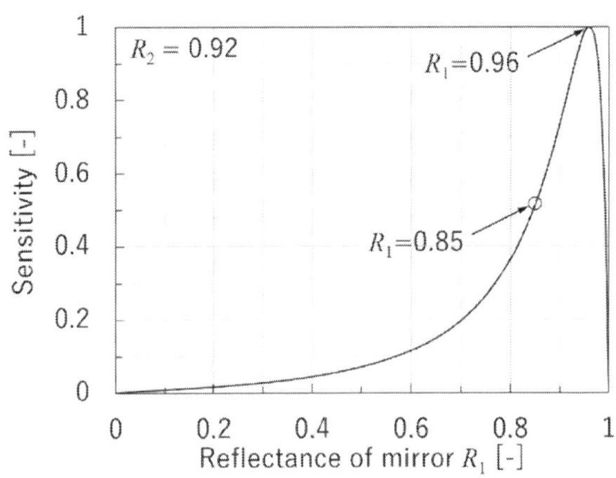

Figure 4: Calculated sensitivity for optimum FP design. The open circle represents the calculation for R_1=0.85.

Figure 5 shows the reflectance of the micro FP interferometer as a function of cavity length. R_1 and R_2 were 0.85 and 0.92, respectively. The FP interferometric characteristics were measured by varying the cavity length using the PZT actuator. The gray solid line indicates the theoretical calculation results obtained using eq. (1). The blue solid circles are the experimental results, which showed good agreement with the values on the theoretically calculated curve. Scale fitting was only performed for the horizontal scale. The finesse of the interferometer was measured to be 25.

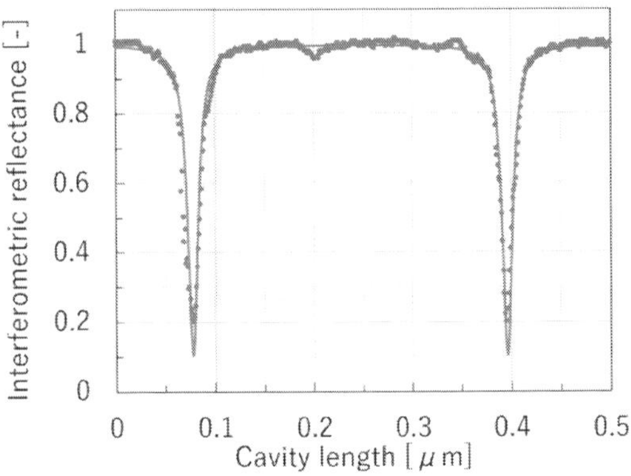

Figure 5: FP interferometric characteristics of the micro FP interferometer. The gray solid line shows the curve that was calculated theoretically using eq. (2). R_1 and R_2 are 0.85 and 0.92, respectively.

Figure 6 shows the FFT signal of the thermal vibration of the microcantilever, which is used as one of the mirrors of the micro FP interferometer, at maximum sensitivity. The frequency resolution of the FFT analyzer is 0.5 Hz. The data are averaged over 1000 measurements. The gray solid line is fitted to the experimental results using a Lorentzian curve. The quality factor Q was measured to be approximately 2000. The thermal vibration amplitude was approximately 5 pm.

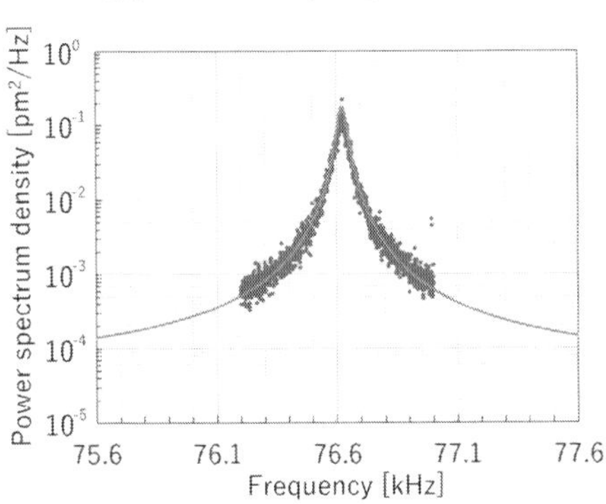

Figure 6: Power spectral density of the thermal vibration of the microcantilever. The solid line curve was fitted using a Lorentzian function.

DISCUSSIONS

In the vicinity of the focal point of the focusing optical system, the laser beam wave fronts are sufficiently flat to allow the FP interferometer to be constructed.

The Rayleigh length l_L is given by

$$l_L \cong \lambda \left(\frac{f}{D} \right)^2, \quad (2)$$

where λ is the wavelength of the laser, f is the focal length and D is laser beam spot diameter on the lens. In this experiment, it was estimated to be about 250 μm, which is long enough than the cavity length (20 μm). It is the reason why flat mirrors can be used as the reflectors of the FP interferometer placed in the focusing optical system.

Figure 7 shows a comparison of the retro reflectivity properties of the two types of optical reflecting systems, when the mirrors of the FP interferometer are not located in parallel with each other; this behavior is caused by the retroreflective effect. In case (b), the optical axis of the reflected beam is oriented parallel to the optical axis of the incident beam by the retroreflective effect, which makes it possible for the two beams to interfere. Consequently, in the micro FP interferometer, the requirement for parallel orientation of a pair of mirrors is greatly reduced when compared with the normal-type FP interferometer ((a)). We could observe the interference fringes, even when the reflected laser beam pattern from the interferometer was not completely overlapped with that of incident laser beam.

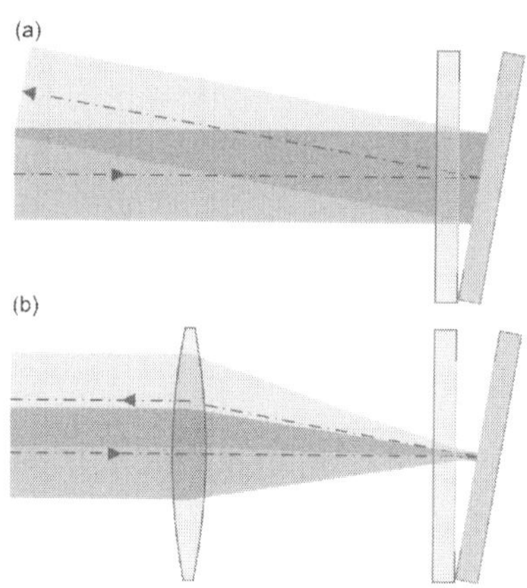

Figure 7: Retroreflectivity comparison of the two types of optical reflecting system. (**a**) Standard FP interferometer. (**b**) Micro FP interferometer.

The demand for optical flatness in the micro FP interferometer is also much weaker when compared with that for the normal-type FP interferometer because of the reduced cross-sectional area of the laser beam. The optical flatness of the reflection mirror were only λ/10, by which we cannot obtained the fines of 25 in case of the normal type FP interferometer.

Another characteristic of the micro FP interferometer is that it has a large free spectral range because of its short cavity length.

For the practical fabrication of this type of FP interferometer, we think that it is one of the method to contact a small dielectric multilayer mirror to the basement of the microcantilever with a thin spacer using the optical contact bonding.

CONCLUSIONS

We have developed a micro Fabry-Pérot interferometer that is constructed within the Rayleigh range of the optical focusing system and demonstrated that the interferometric characteristics of this interferometer were consistent with the theoretically calculated characteristics. The conventional high precision required for the optical parallelism and the surface flatness of the mirrors was not so essential for the micro FP interferometer. We believe that the proposed micro FP interferometer has the potential to make a marked contribution to advances in optical measurements in various micro sensing system.

DATA AVAILABILITY

The authors declare that all the relevant data supporting this finding of this study are available within the article or from the corresponding author upon request.

REFERENCES

1. Harry, G. M. Advanced LIGO: the next generation of gravitational wave detectors. Class. Quantum Grav. 27, 084006-1–12 (2010).
2. Sakane, T. An actively mode-locked TEA CO2 laser with multiline output. Opt. Commun. 12, 21–23 (1974).
3. Kawamura, Y. et al. Stable subnanosecond modelock multiline tea CO2 laser controlled by a NaCl etalon. Opt. Commun. 31, 178–180 (1979).
4. Murphy, K. A., Gunther, M. F., Vengsarkar, A. M. & Claus, R. O. Quadrature phase-shifed extrinsic Fabry–Perot optical fber sensors. Opt. Lett. 16, 273–275 (1991).
5. Wada, A., Tanaka, S. & Takahashi, N. Fast and high-resolution simultaneous measurement of temperature and strain using a FabryPerot interferometer in polarization-maintaining fber with laser diodes. J. Lightwave Technol. 36, 1011–1017 (2018).
6. Poggio, M., Degen, C. L., Mamin, H. J. & Rugar, D. Feedback Cooling of a Cantilever's Fundamental Mode below 5 mK. Phys. Rev. Lett. 99, 017201-1–4 (2007).
7. Kawamura, Y. & Kanegae, R. Feedback damping of a microcantilever at room temperature to the minimum vibration amplitude limited by the noise level. Sci. Rep. 6, 1–5 (2016).
8. Kawamura, Y. & Kanegae, R. Mode-selective control of thermal Brownian vibration of micro-resonator (Generation of a thermal no-equilibrium state by mechanical feedback control). Appl. Phys. Lett. 111, 133112-1–3 (2017).
9. Okamoto, H. et al. Cavity-less on-chip optomechanics using excitonic transitions in semiconductor heterostructures. Nat. Commun. 6, 1–6 (2015).
10. Metzger, C. H. & Karrai, K. Cavity cooling of a microlever. Nature 432, 1002–1005 (2004).
11. Vogel, M., Mooser, C. & Karrai, K. Optically tunable mechanics of microlevers. Appl. Phys. Lett. 83, 1337–1339 (2003).
12. Dustin, K. & Bouwmeester, D. Sub-kelvin optical cooling of a micromechanical resonator. Nature 444, 75–78 (2006).
13. Gigan, S. et al. Self-cooling of a micromirror by radiation pressure. Nature 444, 67–70 (2006).

14. Ismail, N., Kores, C. C., Geskus, D. & Pollnau, M. Fabry-Pérot resonator: spectral line shapes, generic and related Airy distributions, linewidths, fnesses, and performance at low or frequency-dependent refectivity. Opt. Express 24, 16366–16389 (2016).

CHAPTER 13

AN ATOMIC FABRY–PEROT INTERFEROMETER USING A PULSED INTERACTING BOSE–EINSTEIN CONDENSATE

P. Manju, K. S. Hardman, P. B. Wigley, J. D. Close, N. P. Robins & S. S. Sziget

Atomlaser and Quantum Sensors Group, Department of Quantum Science, Research School of Physics, The Australian National University, Canberra 2601, Australia

ABSTRACT

We numerically demonstrate atomic Fabry–Perot resonances for a pulsed interacting Bose–Einstein condensate (BEC) source transmitting through double Gaussian barriers. These resonances are observable for an experimentally-feasible parameter choice, which we determined using a previously-developed analytical model for a plane matter-wave incident on a double rectangular barrier system. Through numerical simulations using the non-polynomial Schödinger equation—an effective one-dimensional Gross–Pitaevskii equation—we investigate the effect of atom number, scattering length, and BEC momentum width on the resonant transmission peaks.

Citation: (APA): Manju, P., Hardman, K. S., Wigley, P. B., Close, J. D., Robins, N. P., & Szigeti, S. S. (2020). An atomic Fabry–Perot interferometer using a pulsed interacting Bose–Einstein condensate. Scientific reports, 10(1), 1-11. (11 pages)

For 8585Rb atomic sources with the current experimentally-achievable momentum width of $0.02\hbar k_0$ [$k_0=2\pi/(780\ \text{nm})$], we show that reasonably high contrast Fabry–Perot resonant transmission peaks can be observed using (a) non-interacting BECs, (b) interacting BECs of 5×10^4 atoms with *s*-wave scattering lengths $a_s=\pm0.1a_0$ (a_0 is the Bohr radius), and (c) interacting BECs of 10^3 atoms with as=$\pm1.0a_0$. Our theoretical investigation impacts any future experimental realization of an atomic Fabry–Perot interferometer with an ultracold atomic source.

INTRODUCTION

Understanding the different and complementary properties of atoms compared with photons has advanced both fundamental and applied physics. In direct analogy to optical systems, atomic matter-waves can be coherently focussed1, reflected2, diffracted3 and interfered4. These basic atom-optical elements have been combined to construct more sophisticated analogue systems such as atomic waveguides5, atom lasers6·7 and atom interferometers8·9. Atomic properties such as mass, tunable dispersion and differing degrees of freedom make these analogue systems versatile measurement tools. Atom interferometers, for example, have enabled state-of-the-art measurements of the fine structure constant10·11 and inertial fields such as gravity12·13·14 and rotations15·16.

In this paper, we consider the atomic analogue of a Fabry–Perot interferometer. Optical Fabry–Perot interferometry is used for many fundamental scientific and industrial applications, including linewidth measurements of continuous wave (CW) and pulsed lasers17, laser phase and frequency stabilisation18 and precision sensing19·20. An atomic Fabry–Perot interferometer could offer new sensing capabilities by exploiting the atomic mass and tunable dispersion. Furthermore, the analogous mirrors, formed using optical potentials, allow for real-time and versatile control of the system. Previous theoretical work has investigated the resonance properties of Fabry–Perot interferometry using matter-waves21 and their potential use in velocity selection22·23·24 and in the identification of bosonic and fermionic isotopes of an element25·26. In order to fully exploit the benefits of this atomic analogue, the transmission characteristics of an atomic Fabry–Perot interferometer in an experimentally-realisable regime must be understood.

Much like the optical Fabry–Perot interferometer, the atomic analogue requires a narrow linewidth source to fully exploit the interference effects of the system. The properties of an optical laser make it a superior source for a Fabry–Perot interferometer compared with a broad-band light source. Bose–Einstein condensates (BECs) display many properties analogous to a laser, including high coherence and narrow momentum width, which have proven advantageous for precision atom interferometry[27,28,29]. Although many theoretical proposals have shown that atomic Fabry–Perot interferometers can be developed using CW atom laser beams constructed from interacting and non-interacting BEC sources[23,30,31,32,33,34], a true CW atom laser has yet to be experimentally realised[35]. In contrast, a pulsed atom laser, formed by releasing and propagating a BEC, is readily achievable in ultracold-atom laboratories.

This paper investigates the properties of an atomic Fabry–Perot interferometer in an experimentally-realisable parameter regime[36,37]. We use a simple analytic model[38] for rectangular barrier potentials to study the dependence of cavity finesse and transmission coefficient on barrier width, barrier height, cavity length, initial momentum of the atoms, and the momentum width of the atomic cloud. This allows us to identify a parameter regime where high contrast, narrow peaks are observable in the transmission spectrum. Using these established parameters, we numerically simulate the propagation of a pulsed ^{85}Rb BEC through a cavity formed via double Gaussian barriers. ^{85}Rb is ideal for this study since its inter-atomic interactions are tunable via a Feshbach resonance[39,40,41]. We investigate the resonant transmission process for an interacting and non-interacting BEC and study the effect of the condensate's momentum width and inter-atomic interactions on resonant transport through the double barrier system. Using the experimentally feasible parameters we have determined, we demonstrate resonant transmission, requiring a momentum width achievable using delta-kick cooling[37,42,43]. We show that the transmission coefficient and finesse can be improved by reducing the cloud's initial momentum width, allowing for high contrast resonant transmission in the weakly-interacting regime.

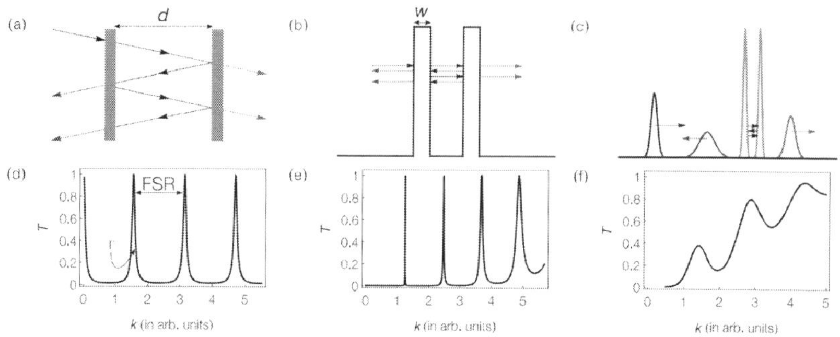

Figure 1: Schematic diagrams of Fabry–Perot interferometers and their respective transmission spectra. The top row depicts the transmission and reflection of **(a)** narrow band light through an optical Fabry–Perot interferometer, **(b)** a narrow-band matter-wave through a rectangular double barrier, and **(c)** a broad-band Gaussian wavepacket (i.e. a non-interacting BEC with finite momentum width) through double Gaussian barriers. Outside the cavity, the black arrow represents the incident wave, whilst blue and red denote the reflected and transmitted parts of the wave, respectively. **(d)**, **(e)** and **(f)** show corresponding transmission resonance peaks. Here d is the cavity length, w is the barrier width, T is the transmission coefficient, FSR is the free spectral range of the resonance spectrum, and Γ is the cavity linewidth.

FABRY–PEROT INTERFEROMETRY

In order to build an understanding of the parameter dependencies, we initially consider an idealised optical Fabry–Perot cavity, extending this to an analytic model describing the atomic analogue. Using this analytic model in tandem with known experimental limitations, we determine a feasible parameter regime for realising an atomic Fabry–Perot interferometer. These parameters are then used to simulate the more complex model involving barriers described by Gaussian potentials and a finite momentum width BEC, including inter-atomic interactions. This progression is illustrated in Fig. 1.

An ideal optical Fabry–Perot interferometer is made of two parallel mirrors separated by a distance, d (cavity length), as shown in Fig. 1a. The light entering the cavity undergoes multiple reflections from the mirrors and interferes with itself. Constructive interference enhances the light inside the cavity, leading to resonant transmission out of the cavity. Here we consider the ideal case where the mirror reflectivity is independent of the wavelength of light. Resonant transmission peaks, illustrated in Fig. 1d, are obtained by scanning the wave number (k) of the incident light. These peaks have a linewidth44

$$\Gamma = \frac{c}{2\pi d}\frac{1-\sqrt{R_1R_2}}{(R_1R_2)^{1/4}}, \qquad (1)$$

and are separated by the free spectral range

$$\mathrm{FSR} = \frac{c}{2d}, \qquad (2)$$

where R_1 and R_2 describe the reflectivity of each mirror forming the cavity, and c is the velocity of light in vacuum. These quantities, along with the finesse

$$\mathscr{F} = \frac{\mathrm{FSR}}{\Gamma} = \frac{\pi (R_1R_2)^{1/4}}{1-\sqrt{R_1R_2}}, \qquad (3)$$

are the figures of merit for an optical Fabry-Perot interferometer44.

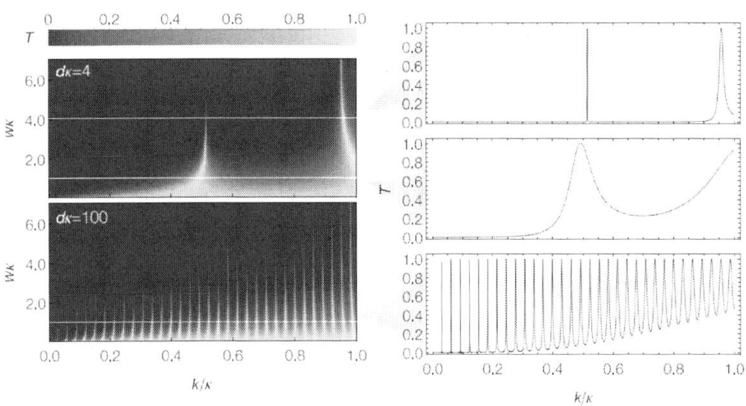

Figure 2: Transmission coefficient of a narrow-band beam of particles passing through a double rectangular barrier system [Eq. (4)—see also Fig. 1b, d] for (top left) dκ=4 and (bottom left) dκ=100. The three plots on the right-hand side show cross sections from the left plot at (top) wκ=4,dκ=4, (middle) wκ=1,dκ=4, and (bottom) wκ=1,dκ=100. For a fixed barrier height (fixed κ), the cavity linewidth decreases with increasing barrier width and/or cavity length, whereas the FSR decreases with increasing cavity length. Here w, d, k and κ are the barrier width, cavity length, and wave vectors of the particles and barriers, respectively.

In an atomic Fabry-Perot interferometer, the incoming light is replaced by atomic matter-waves and the mirrors are replaced by laser-induced potential barriers. The atomic system provides key differences including atomic mass, inter-atomic interactions, and mirrors where the key parameters can be tuned. This results in changes to the transmission characteristics of the atomic system compared with that of the optical system. This is qualitatively illustrated by Fig. 1e, f. In contrast to the ideal optical Fabry-Perot interferometer, the atomic system displays changes in the FSR and Γ as the wave number of the incoming particles changes. To analyse the characteristics of a matter-wave Fabry–Perot interferometer, we use a previously developed analytical model38. This model assumes a plane wave of non-interacting particles, with energy *E*, transmitting through two symmetric rectangular barriers of width *w*, height V_0, and separation distance *d* (see Fig. 1b).

The transmission coefficient, *T*, can be described using the dimensionless parameters dκ, wκ, k/κ, where $k = \sqrt{2mE/\hbar}$ and $\kappa = \sqrt{2mV_0/\hbar}$ are the wave vectors of particles and barriers, respectively. Specifically,

$$T(d\kappa, w\kappa, k/\kappa) = \frac{1}{1 + \frac{4}{\pi^2}[\mathscr{F}(w\kappa, k/\kappa)]^2 \sin^2[d\kappa \times k/\kappa + \phi(w\kappa, k/\kappa)]}, \tag{4}$$

where

$$\mathscr{F}(w\kappa, k/\kappa) = \frac{\pi\sqrt{R(w\kappa, k/\kappa)}}{1 - R(w\kappa, k/\kappa)}, \tag{5}$$

$$R(w\kappa, k/\kappa) = \frac{[M_+(k/\kappa)]^2}{[M_-(k/\kappa)]^2 + \coth^2\left(w\kappa\sqrt{1-(k/\kappa)^2}\right)}, \tag{6}$$

$$M_\pm(k/\kappa) = \frac{1}{2}\left(\frac{\sqrt{1-(k/\kappa)^2}}{(k/\kappa)} \pm \frac{(k/\kappa)}{\sqrt{1-(k/\kappa)^2}}\right), \tag{7}$$

$$\phi(w\kappa, k/\kappa) = \frac{\pi}{2} - \tan^{-1}\left[M_-(k/\kappa)\tanh\left(w\kappa\sqrt{1-(k/\kappa)^2}\right)\right]. \tag{8}$$

Here F and *R* are interpreted as the cavity's finesse and mirror (rectangular barrier) reflectivity, respectively. The relationship of the transmission coefficient to wκ and k/κ is illustrated for two different values of dκ in Fig. 2. The three plots on the right-hand side show cross sections from the left plot for dκ=4 with wκ=4 (top) and wκ=1 (middle) in addition to dκ=100 with wκ=1 (bottom). The first two cross sections show that the cavity linewidth decreases with increasing barrier width, for a fixed barrier height and cavity length. The second and third cross sections show that both FSR and linewidth decrease with increasing cavity length, for a fixed barrier height and width. In contrast to the ideal optical Fabry–Perot interferometer, the linewidth increases with increasing *k* for fixed barrier height and barrier width. This behaviour is due to the strong wavelength-dependence of the mirror reflectivity in the atomic Fabry–Perot interferometer. Similar effects would be seen when considering an optical cavity with strong wavelength-dependent reflectivities.

Now we consider the finesse of the cavity, which is given by Eq. (5). The finesse only depends upon the dimensionless parameters wκ and k/κ. As shown in Fig. 3, the finesse decreases with increasing k/κ and increases with increasing wκ. Therefore, for a fixed barrier height, an increase in particle momentum causes a decrease in finesse, whereas an increase in barrier width causes an increase in finesse. Again, this behaviour is different to the ideal optical Fabry–Perot interferometer, and is due to the mirror reflectivity's dependence on barrier height, barrier width, and the energy of the incoming atoms.

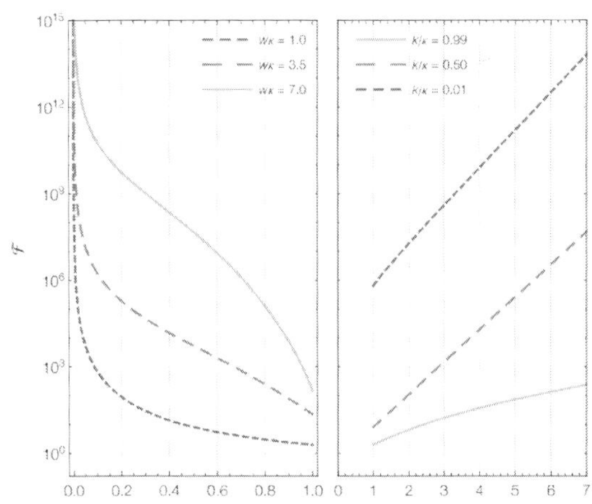

Figure 3: For a beam of particles transmitting through a double rectangular barrier system, finesse is plotted as a function of dimensionless particle wave vector (k/κ) and barrier width wκ, where $\kappa = \sqrt{2mV_0}/\hbar$ is the wave vector corresponding to the barrier. For a fixed barrier height, finesse decreases with increasing momentum of the particle, but increases with increasing barrier width.

The above analysis reveals the parameter regimes needed to make an atomic Fabry–Perot interferometer with desirable qualities such as high finesse and narrow linewidth. However, we do not have complete freedom in our parameter choice, as the limitations of current cold-atom technology impose additional constraints. We discuss these additional constraints below, and combine them with the results of the above analytic model to determine an experimentally-feasible parameter regime for realising an atomic Fabry–Perot interferometer. The operation of an atomic Fabry–Perot interferometer depends sensitively on the following parameters:

- *Momentum width of the atomic cloud.* The above analytical study assumes a plane matter-wave with infinitely narrow momentum width. In reality, all atomic clouds have finite momentum widths. In order to observe Fabry–Perot resonances, ideally the momentum spread of the atomic source needs to be much less than the FSR and linewidth of the cavity. To date, the smallest experimentally-achieved momentum width for an atomic cloud is 0.02ħk037, which was obtained by delta-kick cooling a rubidium BEC (here $k_0=2\pi/(780\ \text{nm})=8.06\times10^6\text{m}^{-1}$ is the wave vector of the light used to impart momentum through Bragg spectroscopy on the ^{85}Rb D_2 transition45·46·47). Hence, we aim to select parameters that give a cavity linewidth and FSR larger than this value.
- *Barrier width.* Although narrow resonances and a high finesse are desirable, the lower bound on experimentally-achievable BEC momentum width requires us to operate in a regime where the cavity linewidth is relatively broad. We find that selecting wκ=1 is a good compromise, since it gives resonance peaks that are wide enough to be observable yet still narrow enough to be potentially useful. Additionally, the minimum achievable barrier width is restricted by the diffraction limit of the laser, which for our system is on the order of 1 μm. We fix the barrier width to this minimum, i.e. w=1 μm; a significantly larger choice for *w* would require a smaller κ, resulting in a cavity spectrum that could only be observed by scanning unachievably small values of *k*.
- *Barrier height.* Fixing *w* and wκ completely determines the barrier height. Explicitly, $\kappa=1\mu\text{m}^{-1}=0.12414k_0$, corresponding to a barrier height of $V_0=\hbar^2\kappa^2/(2m)=3.944\times10^{-32}$ J.
- *Cavity length.* The finesse of the cavity does not depend on the cavity length. However, in order to reduce the overlap between the two laser-induced barriers experimentally, we need a cavity length that is larger than the barrier width. Additionally,

momentum width considerations require a regime where the cavity linewidth is relatively broad (see above). Figure 2 shows that as $d\kappa d\kappa$ decreases, both FSR and linewidth increase. Therefore, in order to observe at least two broad linewidth transmission peaks, we choose $d\kappa=4$. This gives a cavity length of $d=4\mu m$, which is larger than the chosen barrier width ($d=4w$).

- *Momentum imparted to the atoms.* Finally, in order to observe resonance peaks, we must be able to scan the incident energy of the atoms. This can be done by imparting momentum to the atoms through, for example, Bragg transitions[13,46]. For the above parameter choices, at least two peaks are observable by scanning $k/\kappa k/\kappa$ from 0.01 to 1.2 (see Fig. 2). This corresponds to a *k* range of $k=0.0013k_0$ to $k=0.15k_0$.

Based on the above experimentally-feasible parameter choice, our simple analytic model predicts that resonant peaks should be observable with $FSR=0.0687\hbar k_0$ (which is greater than the cloud momentum width, $0.02\hbar k_0$) and linewidth $\Gamma=0.0097\hbar k_0$. This linewidth is about a factor of two smaller than the momentum width of the cloud, which is not ideal. Nevertheless, as we show below with more detailed theoretical modelling, Fabry-Perot resonances are observable in this regime. Indeed, the experimental system includes much more complexity than the simple analytical model considered above. For example, the barriers are created experimentally using blue-detuned lasers, which are more accurately modelled as Gaussian barriers, in contrast to the rectangular barriers used in the analytical model. Furthermore, BECs are finite momentum width sources that typically have non-negligible inter-atomic interactions. These inter-atomic interactions couple different momentum components of the cloud and also have a non-trivial effect on the transmission dynamics[48,49,50,51]. Therefore, although the analytic study can guide our parameter choice, it cannot provide detailed modelling of an interacting BEC's transmission dynamics through double Gaussian barriers. This demands a numerical investigation.

THEORETICAL MODEL FOR NUMERICAL SIMULATION

The mean-field dynamics of a weakly-interacting BEC in a quasi-1D geometry (e.g. a waveguide potential with tight radial confinement) are well-described by the non-polynomial Schrödinger equation (NPSE)[36,52,53,54]. This is an effective 1D model of the Gross–Pitaevskii equation that incorporates spatial and temporal variations in the BEC's width in the tight radial direction. It assumes a cylindrically-symmetric tight harmonic radial confinement of frequency $\omega\perp$. Specifically,

$$i\hbar\frac{\partial\psi(z,t)}{\partial t}=\left[-\frac{\hbar^2}{2m}\frac{\partial^2}{\partial z^2}+V(z)+\frac{g}{2\pi a_\perp^2}\frac{|\psi(z,t)|^2}{\sigma(z,t)^2}-i\frac{\hbar K_3}{6\pi^2 a_\perp^4}\frac{|\psi(z,t)|^4}{\sigma(z,t)^4}+\frac{\hbar\omega_\perp}{2}\left(\frac{1}{\sigma(z,t)^2}+\sigma(z,t)^2\right)\right]\psi(z,t), \tag{9}$$

where $\sigma(z,t)^2=\sqrt{1+2a_s|\psi(z,t)|^2}$, $\psi(z,t)$ is the effective 1D macroscopic condensate wave function normalised to the total particle number, $N(t)=\int dz|\psi(z,t)|^2$, $V(z)$ is the external potential (a double Gaussian barrier potential during evolution), m is the atomic mass, $a_\perp=\sqrt{\hbar/(m\omega_\perp)}$, $g=4\pi\hbar^2 a_s/m$, is the two-body interaction strength, determined via the *s*-wave scattering length, a_s, and K_3 is the three-body recombination loss rate coefficient. In the case of an atomic Fabry-Perot interferometer, the three-body recombination loss rate can become significant due to the high density formed by the multiple reflections of the atoms between barriers. We set the three-body recombination loss rate coefficient to $K_3=4\times10^{-41}m^6/s$, which is the value determined from our previous experiments with ^{85}Rb BEC36·55. All our simulations of Eq. (9) were performed for condensates of ^{85}Rb atoms using the open-source software package XMDS256 with an adaptive 4th-5th order Runge-Kutta interaction picture algorithm.

The initial wave function for the simulations is a Gaussian,

$$\psi_0(z)=\frac{\sqrt{N}}{(\pi\sigma_c^2)^{1/4}}e^{-(z-z_0)^2/(2\sigma_c^2)}e^{ikz}, \tag{10}$$

where z0z0 is the initial position of the atomic cloud, $\sigma_c/\sqrt{2}$ is the standard deviation of the density profile, corresponding to a *k*-space density standard deviation of $1/(\sqrt{2}\sigma_c)$ (full width at half maximum (FWHM) of $\Delta k=2\sqrt{\ln 2}/\sigma_c$) , and ħk is the condensate's initial mean momentum. A Gaussian wave packet of this functional form could be engineered experimentally by delta-kick cooling37·42·43 the cloud after turning off the axial confinement, and then imparting a momentum kick ħk to the atoms via a shallow angle Bragg transition29·46·47.

The laser barriers that form the Fabry-Perot cavity are modelled as two Gaussian potentials:

$$V(z)=V_0\left[e^{-(z-z_{01})^2/(2\sigma_b^2)}+e^{-(z-z_{02})^2/(2\sigma_b^2)}\right], \tag{11}$$

where σb is the standard deviation of each barrier and z01 and z02 are the position of first and second barriers, respectively. We choose

$z_{01} = z_0 + 3(\sigma_c + \sigma_b) + 15\mu\text{m}$ and $z_{02} = z_{01} + 3\sigma_b + d + 3\sigma_b$ so that introducing the barrier potentials does not perturb the initial atomic cloud. Guided by our previous analytic analysis, we choose cavity length d=4μm, barrier height V_0=3.944×10^{-32} J, and barrier width $\sigma_b = 1\mu\text{m}/\sqrt{2\pi}$ (this gives a Gaussian barrier with equal area to a square barrier of height V_0 and width 1μm).

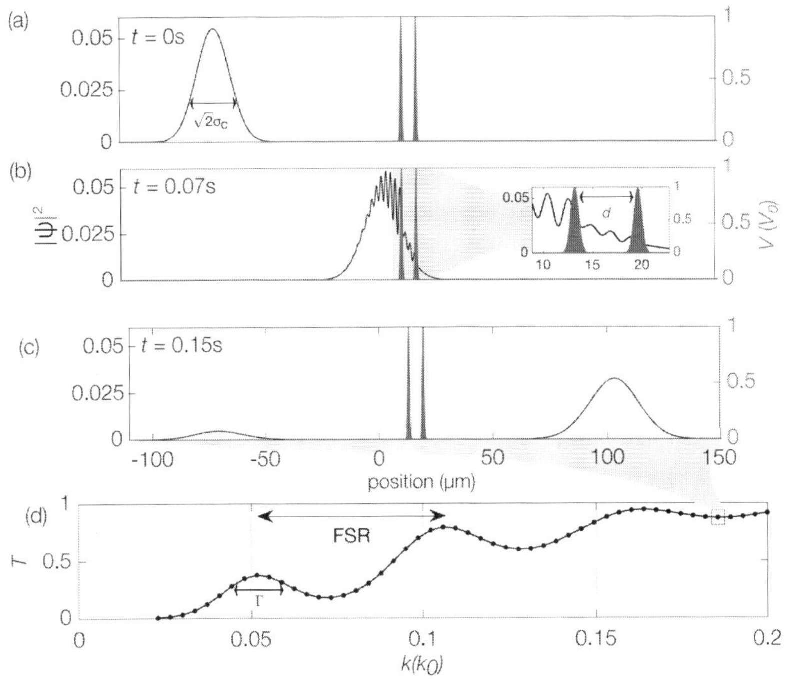

Figure 4: **(a)–(c)** show the propagation of a non-interacting BEC having mean momentum ħk=0.182ħk_0 and FWHM momentum width ħΔk=0.02ħk_0 through two Gaussian barriers, in three snapshots in time, simulated using the 1D Schrödinger equation (i.e. NPSE Eq. (9) with g=0 and K_3=0). The red curves indicate the Gaussian barriers and the black curves represent the density profile of the Gaussian cloud. **(a)** The BEC starts on the left side of the barriers, **(b)** propagates towards the barriers, and enters the cavity. The dynamics around the cavity region is expanded and illustrated in the inset plot. **(c)** Some parts of the cloud are transmitted through the barriers, whereas other parts are reflected, depending on the momentum of the cloud. **(d)** The transmission coefficient as a function of the mean momentum of the wave packet [here k_0=2π/(780nm)].

Our simulations allow us to determine the number of atoms transmitted (N_T) and reflected (N_R) through the double barrier system via

$$N_T = \int_{z_T}^{\infty} dz\, |\psi(z, t_{\text{end}})|^2, \quad (12)$$

$$N_R = \int_{-\infty}^{z_R} dz\, |\psi(z, t_{\text{end}})|^2, \quad (13)$$

where $z>z_T=(z_{02}+3\sigma_b)$ and $z<z_R=(z_{01}-3\sigma_b)$ are the transmitted and reflected regions, respectively. The stopping time, tendtend, for the simulation is chosen such that there are no atoms left in the cavity (i.e. $N-N_T-N_R<1$) and both N_T and N_R have reached a constant value (more precisely, do not change by more than 0.1 in a given time step). For the non-interacting case g=0 and $K_3=0$, N is simply a normalisation factor that no longer influences the dynamics. In this case, tendtend is chosen such that $(N_T+N_R)/N<10^{-5}$ and both N_T/N and N_R/N do not change by more than 10^{-6} in a given time step. We investigate the resonant transmission by computing the total transmission coefficient

$$T = \frac{N_T}{N_T + N_R}. \quad (14)$$

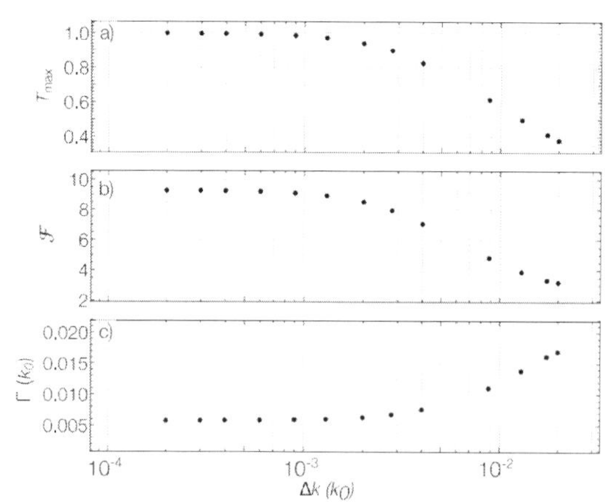

Figure 5: The influence of a non-interacting BEC's momentum width on the **(a)** height, **(b)** finesse, and **(c)** linewidth of the first resonance peak. The FWHM of the initial momentum distribution, $\hbar\Delta k = 2\sqrt{\ln 2}\hbar/\sigma_c$, and linewidth are given in units of $k_0=2\pi/(780\text{nm})$. Peak height and finesse increase and Γ decreases with decreasing Δk.

ANALYSIS OF RESONANT TRANSMISSION

Non-interacting case

Using the parameters determined above, we first simulate the resonant transmission of a non-interacting BEC ($g=0$ and $K_3=0$) passing through the double Gaussian barrier system described previously. We choose an initial cloud with a FWHM momentum width close to the smallest experimentally-realised value of $\Delta k=0.02k_0$ 37, which corresponds to a spatial width of $2\sigma c\approx21\,\mu m$. The propagation of this non-interacting BEC after momentum kick $\hbar k=0.1855\hbar k_0$ is schematically shown in Fig. 4a–c, where (a), (b) and (c) correspond to three snap shots in time: prior to interaction (t=0s), during interaction (t=0.07s), and after interaction (t=0.15t=0.15s) with the barrier, respectively. The interference caused from the overlapping incident and reflected cloud components is clearly seen in Fig. 4b. As expected, this behaviour is present both after the initial reflection, outside of the cavity, and through multiple reflections inside the cavity. Resonant transmission is observed for the *k* values which are resonant with the cavity. The transmission coefficient as a function of momentum kick given to the cloud is plotted in Fig. 4d. This confirms that Fabry–Perot resonances can indeed be observed for our parameter choice.

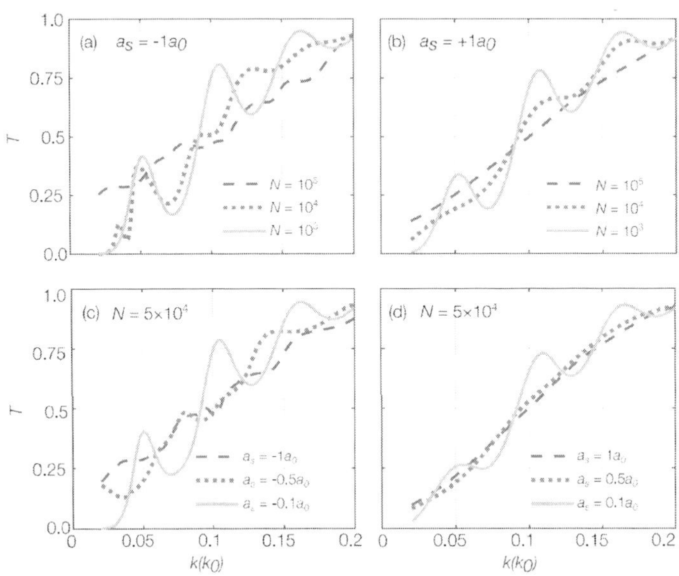

Figure 6: Plots showing that reducing inter-atomic interactions improves the resonant transmission peaks. Inter-atomic interactions are reduced by reducing

peak density (by reducing initial atom number) in plots **(a)** and **(b)** for $a_s=-1a_0$ and $a_s=+1a_0$, respectively. In **(c)** and **(d)**, the interaction strength is reduced by reducing the magnitude of the negative and positive scattering lengths, respectively, for BECs with an initial atom number of $N=5\times10^4$.

In contrast to the incident plane-wave source case, where the Fabry–Perot resonance peaks occur at a maximum of T_{max}=138·57, the transmission peaks observed here are suppressed, i.e, $T_{max}<1$, reducing the contrast of the resonance peaks. A previous theoretical investigation observed this behaviour for the resonance of a non-interacting CW atom laser beam23. The reduction in resonant transmission arises due to the finite momentum width of the source BEC. This also causes broadening of the peaks, which reduces finesse. As the peaks are suppressed due to the momentum spread of the cloud, we expect them to improve by reducing the BEC's momentum width. To investigate this further, we study the transmission profile for a range of initial cloud momentum widths. Fig. 5 shows the height, finesse and linewidth of the first resonant peak for a non-interacting BEC with momentum width ranging from $\Delta k=2\times10^{-4}k_0$ to $0.02\ k_0$. The resonance peaks are improved (peak height and finesse increase and linewidth decrease) by reducing the cloud momentum width, as one would expect.

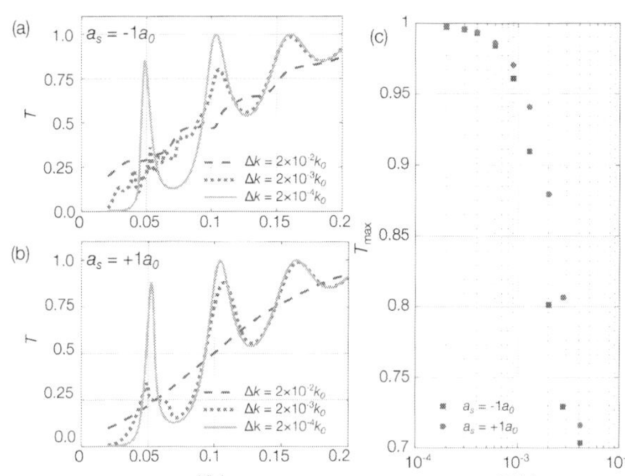

Figure 7: Plots illustrating that reducing the initial momentum width of the cloud improves the transmission resonance peaks. **(a)** and **(b)** show the resonant peaks for $a_s=-1a_0$ and $a_s=+1a_0$, respectively, for three values of cloud momentum width. **(c)** Plots the height of the second peak in the transmission spectrum (T_{max}) as a function of BEC momentum width, for scattering lengths as=$-1a_0$ and as=$+1a_0$. It shows that peak height increases with decreasing momentum width. In all plots the initial atom number is $N=5\times10^4$.

Interacting case

In the presence of inter-atomic interactions, the effect of three-body recombination losses become crucial. Due to the repeated reflections of the atoms between the two barriers, the density becomes very high inside the cavity, increasing the three-body recombination loss of atoms from the condensate. The overall atom loss due to three-body recombination depends non-trivially on the transmission dynamics, and so will vary with scattering length, the initial momentum kick, and spatial width of the BEC. Specifically, as the scattering length goes from positive to negative, the three-body recombination loss increases due to the difference in the propagation dynamics of the BEC as it approaches the barrier. Condensates with positive and negative scattering lengths undergo expanding or focussing, respectively, under free propagation50,58,59. These dynamics modify the cloud density and therefore the overall three-body loss as well. The momentum kick imparted to the BEC further modifies the atom loss by changing the total interaction time. For larger momentum kicks, the atomic cloud propagates faster and spends less time in the high density region inside the cavity. This reduces the possibility of three-body recombination loss. Hence, as the average momentum of the cloud increases the loss rate decreases. Finally, for a fixed initial atom number, increasing the spatial width of the cloud (i.e. decreasing the momentum width) decreases the overall three-body loss due to the reduction in initial density.

The transmission resonances corresponding to clouds having $a_s=-1a_0$ and $a_s=+1a_0$ and a momentum width of $\Delta k=0.02k_0$ are illustrated by the blue dashed curves in Fig. 6a, b. In the presence of inter-atomic interactions the peaks are either further suppressed or not well-defined, as compared to the non-interacting cloud. This reduction in contrast is caused by the additional interaction-induced expansion of the BEC's momentum distribution, scattering-length-dependent distortions in the momentum distribution that occur during interaction with barriers50, and non-trivial intra-cavity dynamics due to the presence of inter-atomic interactions.

In order to mitigate the loss of contrast caused by inter-atomic interactions, we can either decrease the interactions or reduce the initial momentum width. Both approaches effectively reduce the initial interaction energy of the cloud. Figure 6 shows the effects of two methods to reduce interactions, firstly by reducing initial atom number and secondly by reducing the magnitude of the scattering length. Figure 6a, b show the resonance peaks for $a_s=-1a_0$ and $+1a_0$, respectively, for initial atom numbers $N=10^5, 10^4$ and

10^3. Resonance peak contrast increases substantially by reducing initial atom number for both attractive and repulsive clouds. For comparison, Fig. 6c, d illustrates the effect of reducing the magnitude of the scattering length on the resonance peaks for attractive and repulsive clouds, respectively. Similar to the effect seen from the reduction in initial atom number, reducing the magnitude of the scattering length also improves the resonant peaks for both attractive and repulsive clouds.

Although reducing the initial momentum width pushes the cloud outside of the current experimentally-realisable regime, it is important to understand the system dynamics in this narrow momentum width regime. The effect of reducing momentum width on the Fabry-Perot transmission spectrum for clouds having scattering length $a_s = -1a_0$ and $+1a_0$ are shown in Fig. 7a, b, respectively. Figure 7c is produced by selecting the maximum transmission of the middle resonance. Here, the general trend towards $T_{max} = 1$ is evident as the momentum width of the cloud approaches zero.

This numerical analysis shows that an atomic Fabry–Perot interferometer with a pulsed BEC source can be experimentally achieved using current technology. Since the inter-atomic interactions reduce the resonant transmission, it is ideal to use a non-interacting cloud or a very weakly-interacting cloud with small to moderate atom numbers. Additionally, as current cooling techniques improve, allowing narrower momentum width sources, it may be possible to operate a 'good' quality atomic Fabry–Perot interferometer with more strongly-interacting cloud.

CONCLUSION

We have compared the properties of optical and atomic Fabry–Perot interferometers. By analysing the dependence of finesse and transmission coefficient on barrier height, barrier width, cavity length, incident atomic energy and cloud momentum width, we have determined an experimentally-feasible parameter regime for observing atomic Fabry–Perot resonances. Using these parameters, we numerically simulated the transmission dynamics of a ^{85}Rb BEC through two Gaussian barriers. The simulations showed that the Fabry-Perot resonances can be achieved for a non-interacting BEC with a momentum width around $0.02\hbar k_0$. Due to the finite momentum width of the BEC, the transmission peaks are suppressed ($T_{max} < 1$) leading to wider peaks and reduced finesse. Consequently, reducing momentum width can increase finesse and improve resonance peaks in the atomic Fabry–Perot spectrum. The introduction of inter-atomic interactions further modify and suppress the resonance peaks. We have investigated different possibilities for improving the quality of the resonant peaks of an interacting BEC, which includes reducing the interactions (by reducing initial atom number and/or the magnitude of the scattering length) and reducing the initial momentum width of the BEC. We have shown that both methods can improve the quality of resonances and we have illustrated that almost complete transmission ($T_{max} \approx 1$) is achievable for BECs having weak attractive and repulsive interactions. Our investigation shows that Fabry-Perot resonances can be observed only for atomic species with very low interactions or with tunable interactions, such as ^{85}Rb. This study paves the way to experimentally realise an atomic Fabry–Perot interferometer using a weakly-interacting and non-interacting pulsed BEC, that could potentially be used for many applications including velocity filtering, accelerometry and for identifying bosonic and fermionic isotopes of an element.

ACKNOWLEDGEMENTS

We acknowledge useful discussions with S.A. Haine and A.C. White. This research was supported by funding from the Australian Research Council (ARC) Discovery Project no. DP160104965 and was undertaken with the assistance of resources and services from the National Computational Infrastructure (NCI), which is supported by the Australian Government. S.S.S. was supported by an ARC Discovery Early Career Researcher Award (DECRA), Project no. DE200100495.

CONTRIBUTIONS

The theoretical analysis and numerical simulations were carried out by P.M. under the supervision of K.S.H., P.B.W., N.P.R., and S.S.S. J.D.C. and all other authors contributed to the interpretation of results and the writing of the manuscript.

REFERENCES

1. Gallatin, G. M. & Gould, P. L. Laser focusing of atomic beams. *JOSA B* **8**, 502–508. https://doi.org/10.1364/JOSAB.8.000502 (1991).
2. Balykin, V. I., Letokhov, V. S., Ovchinnikov, Y. B. & Sidorov, A. I. Reflection of an atomic beam from a gradient of an optical field. *JETP Lett.* **45**, 353–356 (1987).
3. Moskowitz, P. E., Gould, P. L., Atlas, S. R. & Pritchard, D. E. Diffraction of an atomic beam by standing-wave radiation. *Phys. Rev. Lett.* **51**, 370–373. https://doi.org/10.1103/PhysRevLett.51.370 (1983).
4. Becker, U. Molecular physics: Matter-wave interference made clear. *Nature* **474**, 586–587. https://doi.org/10.1038/474586a (2011).
5. Bongs, K. *et al.* Waveguide for Bose-Einstein condensates. *Phys. Rev. A* **63**, 031602. https://doi.org/10.1103/PhysRevA.63.031602 (2001).
6. Mewes, M.-O. *et al.* Output coupler for Bose-Einstein condensed atoms. *Phys. Rev. Lett.* **78**, 582–585. https://doi.org/10.1103/PhysRevLett.78.582 (1997).
7. Robins, N. P., Figl, C., Jeppesen, M., Dennis, G. R. & Close, J. D. A pumped atom laser. *Nat. Phys.* **4**, 731–736. https://doi.org/10.1038/nphys1027 (2008).
8. Contributors. in (Berman, P. R. ed.) *Atom Interferometry*, ix–xi, https://doi.org/10.1016/B978-012092460-8/50000-X (Academic Press, San Diego, 1997).
9. Keith, D. W., Ekstrom, C. R., Turchette, Q. A. & Pritchard, D. E. An interferometer for atoms. *Phys. Rev. Lett.* **66**, 2693–2696. https://doi.org/10.1103/PhysRevLett.66.2693 (1991).
10. Bouchendira, R., Cladé, P., Guellati-Khélifa, S., Nez, F. m. c. & Biraben, F. m. c. New determination of the fine structure constant and test of the quantum electrodynamics. *Phys. Rev. Lett.* **106**, 080801, https://doi.org/10.1103/PhysRevLett.106.080801 (2011).
11. Parker, R. H., Yu, C., Zhong, W., Estey, B. & Müller, H. Measurement of the fine-structure constant as a test of the Standard Model. *Science* **360**, 191–195. https://doi.org/10.1126/science.aap7706 (2018).
12. Peters, A., Chung, K. Y. & Chu, S. High-precision gravity measurements using atom interferometry. *Metrologia* **38**, 25 (2001).
13. Altin, P. A. *et al.* Precision atomic gravimeter based on Bragg diffraction. *New J. Phys.* **15**, 023009. https://doi.org/10.1088/1367-2630/15/2/023009 (2013).

14. Freier, C. *et al.* Mobile quantum gravity sensor with unprecedented stability. *J. Phys. Conf. Ser.* **723**, 012050. https://doi.org/10.1088/1742-6596/723/1/012050 (2016).
15. Stockton, J. K., Takase, K. & Kasevich, M. A. Absolute geodetic rotation measurement using atom interferometry. *Phys. Rev. Lett.* **107**, 133001. https://doi.org/10.1103/PhysRevLett.107.133001 (2011).
16. Savoie, D. *et al.* Interleaved atom interferometry for high-sensitivity inertial measurements. *Sci. Adv.* **4** (2018).
17. Xue, J. *et al.* Pulsed laser linewidth measurement using Fabry-Pérot scanning interferometer. *Results Phys.* **6**, 698–703. https://doi.org/10.1016/j.rinp.2016.10.004 (2016).
18. Drever, R. W. P. *et al.* Laser phase and frequency stabilization using an optical resonator. *Appl. Phys. B* **31**, 97–105. https://doi.org/10.1007/BF00702605 (1983).
19. Taylor, H. F. Principles and applications of fiber-optic Fabry-Perot sensors. in *Technical Digest. Summaries of Papers Presented at the Conference on Lasers and Electro-Optics. Conference Edition. 1998 Technical Digest Series, Vol.6 (IEEE Cat. No.98CH36178)*, 312, https://doi.org/10.1109/CLEO.1998.676211 (1998).
20. Islam, M. R., Ali, M. M., Lai, M.-H., Lim, K.-S. & Ahmad, H. Chronology of Fabry-Perot interferometer fiber-optic sensors and their applications: A review. *Sensors* **14**, 7451–7488. https://doi.org/10.3390/s140407451 (2014).
21. Dutt, A. & Kar, S. Smooth double barriers in quantum mechanics. *Am. J. Phys.* **78**, 1352–1360. https://doi.org/10.1119/1.3481701 (2010).
22. Ruschhaupt, A., Delgado, F. & Muga, J. G. Velocity selection of ultra-cold atoms with Fabry-Perot laser devices: Improvements and limits. *J. Phys. B At. Mol. Opt. Phys.* **38**, 2665–2674. https://doi.org/10.1088/0953-4075/38/15/007 (2005).
23. Damon, F., Vermersch, F., Muga, J. G. & Guéry-Odelin, D. Reduction of local velocity spreads by linear potentials. *Phys. Rev. A* **89**, 053626. https://doi.org/10.1103/PhysRevA.89.053626 (2014).
24. Valagiannopoulos, C. Quantum fabry-perot resonator: Extreme angular selectivity in matter-wave tunneling. *Phys. Rev. Appl.* **12**, 054042 (2019).
25. Lee, S. K., Lee, H. S., Kim, J. M. & Cho, D. Optical dipole trap using a Fabry-Perot interferometer as a power buildup cavity. *J. Phys. B*

At. Mol. Opt. Phys. **38**, 1381–1389. https://doi.org/10.1088/0953-4075/38/8/023 (2005).

26. Wilkens, M., Goldstein, E., Taylor, B. & Meystre, P. Fabry-Pérot interferometer for atoms. *Phys. Rev. A* **47**, 2366–2369. https://doi.org/10.1103/PhysRevA.47.2366 (1993).
27. Debs, J. E. *et al.* Cold-atom gravimetry with a Bose-Einstein condensate. *Phys. Rev. A* **84**, 033610 (2011).
28. Szigeti, S. S., Debs, J. E., Hope, J. J., Robins, N. P. & Close, J. D. Why momentum width matters for atom interferometry with Bragg pulses. *New J. Phys.* **14**, 023009 (2012).
29. Hardman, K. *et al.* Simultaneous precision gravimetry and magnetic gradiometry with a Bose-Einstein condensate: A high precision. *Quantum Sens. Phys. Rev. Lett.* **117**, 138501. https://doi.org/10.1103/PhysRevLett.117.138501 (2016).
30. Rapedius, K. & Korsch, H. J. Barrier transmission for the one-dimensional nonlinear Schrödinger equation: Resonances and transmission profiles. *Phys. Rev. A* **77**, 063610. https://doi.org/10.1103/PhysRevA.77.063610 (2008).
31. Carusotto, I. Nonlinear atomic Fabry-Perot interferometer: From the mean-field theory to the atom blockade effect. *Phys. Rev. A* **63**, 023610. https://doi.org/10.1103/PhysRevA.63.023610 (2001).
32. Paul, T., Richter, K. & Schlagheck, P. Nonlinear resonant transport of Bose-Einstein condensates. *Phys. Rev. Lett.* **94**, 020404. https://doi.org/10.1103/PhysRevLett.94.020404 (2005).
33. Paul, T., Hartung, M., Richter, K. & Schlagheck, P. Nonlinear transport of Bose-Einstein condensates through mesoscopic waveguides. *Phys. Rev. A* **76**, 063605. https://doi.org/10.1103/PhysRevA.76.063605 (2007).
34. Ernst, T., Paul, T. & Schlagheck, P. Transport of ultracold Bose gases beyond the Gross-Pitaevskii description. *Phys. Rev. A* **81**, 013631. https://doi.org/10.1103/PhysRevA.81.013631 (2010).
35. Robins, N. P., Altin, P. A., Debs, J. E. & Close, J. D. Atom lasers: Production, properties and prospects for precision inertial measurement. *Phys. Rep.* **529**, 265–296 (2013).
36. Everitt, P. J. *et al.* Observation of a modulational instability in Bose-Einstein condensates. *Phys. Rev. A* **96**, 041601. https://doi.org/10.1103/PhysRevA.96.041601 (2017).

37. Kovachy, T. *et al.* Matter wave lensing to Picokelvin temperatures. *Phys. Rev. Lett.* **114**, 143004. https://doi.org/10.1103/PhysRevLett.114.143004 (2015).
38. Xiao, Z., Huang, H. & Lu, X.-X. Resonant tunneling dynamics and the related tunneling time. *Int. J. Mod. Phys. B* **29**, 1550052. https://doi.org/10.1142/S0217979215500526 (2015).
39. Roberts, J. L. *et al.* Resonant magnetic field control of elastic scattering in cold R85b. *Phys. Rev. Lett.* **81**, 5109–5112. https://doi.org/10.1103/PhysRevLett.81.5109 (1998).
40. Kuhn, C. C. N. *et al.* A Bose-condensed, simultaneous dual-species Mach-Zehnder atom interferometer. *New J. Phys.* **16**, 073035. https://doi.org/10.1088/1367-2630/16/7/073035 (2014).
41. McDonald, G. D. *et al.* Bright solitonic matter-wave interferometer. *Phys. Rev. Lett.* **113**, 013002. https://doi.org/10.1103/PhysRevLett.113.013002 (2014).
42. Ammann, H. & Christensen, N. Delta kick cooling: A new method for cooling atoms. *Phys. Rev. Lett.* **78**, 2088–2091. https://doi.org/10.1103/PhysRevLett.78.2088 (1997).
43. McDonald, G. D. *et al.*80ħk80ħk momentum separation with Bloch oscillations in an optically guided atom interferometer. *Phys. Rev. A* **88**, 053620. https://doi.org/10.1103/PhysRevA.88.053620 (2013).
44. Vaughan, J. M. *The Fabry-Perot Interferometer History, Theory, Practice and Applications*. Series on Optics and Optoelectronics (Taylor & Francis, 1989).
45. Adam Steck, D. *Rubidium 85 D Line Data* (2013).
46. Stenger, J. *et al.* Bragg spectroscopy of a Bose-Einstein condensate. *Phys. Rev. Lett.* **82**, 4569–4573. https://doi.org/10.1103/PhysRevLett.82.4569 (1999).
47. Ernst, P. T. *et al.* Probing superfluids in optical lattices by momentum-resolved Bragg spectroscopy. *Nat. Phys.* **6**, 56–61. https://doi.org/10.1038/nphys1476 (2010).
48. Salasnich, L., Parola, A. & Reatto, L. Pulsed macroscopic quantum tunneling of falling Bose-Einstein condensates. *Phys. Rev. A* **64**, 023601. https://doi.org/10.1103/PhysRevA.64.023601 (2001).
49. Carr, L. D., Holland, M. J. & Malomed, B. A. Macroscopic quantum tunnelling of Bose–Einstein condensates in a finite potential

well. *J. Phys. B At. Mol. Opt. Phys.* **38**, 3217–3231. https://doi.org/10.1088/0953-4075/38/17/012 (2005).

50. Manju, P. *et al.* Quantum tunneling dynamics of an interacting Bose-Einstein condensate through a Gaussian barrier. *Phys. Rev. A* **98**, 053629. https://doi.org/10.1103/PhysRevA.98.053629 (2018).
51. Wales, O. J. *et al.* Splitting and recombination of bright-solitary-matter waves. *Commun. Phys.* **3**, 51 (2020).
52. Salasnich, L., Parola, A. & Reatto, L. Effective wave equations for the dynamics of cigar-shaped and disk-shaped Bose condensates. *Phys. Rev. A* **65**, 043614 (2002).
53. Wigley, P. B. *et al.* Precise wave-function engineering with magnetic resonance. *Phys. Rev. A* **96**, 013612 (2017).
54. Di Carli, A. *et al.* Excitation modes of bright matter-wave solitons. *Phys. Rev. Lett.* **123**, 123602 (2019).
55. Altin, P. A. *et al.* Collapse and three-body loss in a 8585 Rb Bose-Einstein condensate. *Phys. Rev. A* **84**, 033632. https://doi.org/10.1103/PhysRevA.84.033632 (2011).
56. Dennis, G. R., Hope, J. J. & Johnsson, M. T. Xmds2: Fast, scalable simulation of coupled stochastic partial differential equations. *Comput. Phys. Commun.* **184**, 201–208 (2013).
57. Uma Maheswari, A., Prema, P., Mahadevan, S. & Shastry, C. S. Quasi-bound states, resonance tunnelling, and tunnelling times generated by twin symmetric barriers. *Pramana* **73**, 969. https://doi.org/10.1007/s12043-009-0173-x (2009).
58. Dekel, G., Farberovich, O. V., Soffer, A. & Fleurov, V. Nonlinear dynamic phenomena in macroscopic tunneling. *Phys. D: Nonlinear Phenomena* **238**, 1475–1481. https://doi.org/10.1016/j.physd.2008.06.013 (2009).
59. Carr, L. D., Holland, M. J. & Malomed, B. A. Macroscopic quantum tunnelling of Bose-Einstein condensates in a finite potential well. *J. Phys. B At. Mol. Opt. Phys.* **38**, 3217–3231. https://doi.org/10.1088/0953-4075/38/17/012 (2005).

CHAPTER 14

TWO-PHOTON INTERFERENCE OF TEMPORALLY SEPARATED PHOTONS

Heonoh Kim[1], Sang Min Lee[1,2] & Han Seb Moon[1]

[1]Department of Physics, Pusan National University, Geumjeong-Gu, Busan 609-735, Korea.

[2]Present address: Korea Research Institute of Standards and Science Daejeon 305-340 Korea

ABSTRACT

We present experimental demonstrations of two-photon interference involving temporally separated photons within two types of interferometers: a Mach-Zehnder interferometer and a polarization-based Michelson interferometer. The two-photon states are probabilistically prepared in a symmetrically superposed state within the two interferometer arms by introducing a large time delay between two input photons; this state is composed of two temporally separated photons, which are in two different or the same spatial modes. We then observe two-photon interference

Citation: (APA): Kim, Heonoh, Sang Min Lee, and Han Seb Moon. "Two-photon interference of temporally separated photons." Scientific reports 6, no. 1 (2016): 1-9. (9 pages)

fringes involving both the Hong-Ou-Mandel interference effect and the interference of path-entangled two-photon states simultaneously in a single interferometric setup. The observed two-photon interference fringes provide simultaneous observation of the interferometric properties of the single-photon and two-photon wavepackets. The observations can also facilitate a more comprehensive understanding of the origins of the interference phenomena arising from spatially bunched/anti-bunched two-photon states comprised of two temporally separated photons within the interferometer arms.

INTRODUCTION

The Hong-Ou-Mandel (HOM) interference effect1 and the interference of path-entangled two-photon states2, i.e., the so-called N00N state3, have played an important role in fundamental investigations of quantum mechanics and the exploration of quantum information technology4·5. Since the late 1980s, various kinds of two-photon interference experiments have been performed in order to distinguish quantum mechanical treatment of optical interference phenomena from conventional classical optics6·7. These experiments have successfully shown that the interferences of correlated photons cannot be explained by any classical wave theory; instead, they should be viewed as interference between superposed probability amplitudes. The coherent superposition of states and the interference between probability amplitudes for indistinguishable processes in the total detection process have a crucial role in quantum mechanics and experimental quantum optics to observe interference phenomena. Thus, a number of experiments have been performed to elucidate two-photon quantum interference effects, such as the HOM effect1·8·9·10·11·12·13·14·15 and the N00N-state interference2·3·11·16·17·18·19·20·21. Recently, we have reported that the two kinds of two-photon interference effects can be observed in the most generalized two-photon interferometric scheme, including a fully unfolded HOM scheme as well as a N00N-state interferometer22.

Since the early 1990s, various apparatus for two-photon interference experiments have been utilized to investigate two-photon wavepacket interference phenomena, e.g., the Mach-Zehnder interferometer (MZI)2·17·19·21·23·24 and the Michelson interferometer (MI) 25·26·27. The majority of the experiments involving these devices were performed using two identical photons as the input state, where the two photons simultaneously entered the input port of an interferometer. However, the most interesting behavior occurs when two correlated photons are incident on the interferometer with a large time delay that is considerably longer than their coherence time[10,28,29]. Although, even in that case, if the two-photon states are still in symmetrically superposed states so as to exhibit two-photon interference effects, we can simultaneously observe the coherence properties of the single-photon and two-photon wavepackets by examining the full measured interferogram of the two-photon interference fringes. Although a number of studies have examined two-photon quantum interference experiments in a MZI or MI, further studies are required to fully elucidate the two-photon states within the interferometer arms. In addition, these studies should aim to reveal the origin of the rather complex interference fringe patterns, which contain the shapes of both the single- and two-photon wavepackets. Therefore, we aim to conduct a more comprehensive analysis on the origins of the two-photon wavepacket interference phenomena that occur when two temporally separated photons within the interferometer probabilistically generate spatially bunched/anti-bunched two-photon states.

In this paper, we report on an experimental demonstration of quantum interference effects using two kinds of two-photon state in a conventional MZI and a polarization-based MI (PMI). The two distinct two-photon states are prepared by introducing a time delay between two incident photons at the input ports of the interferometer. The two photons are well separated by a time-like interval that is longer than the coherence time of both the individual single photons and the two-photon states. Here, we consider two kinds of two-photon state within the interferometer arms in order to distinguish from the conventional HOM and N00N states, which are the *temporally separated and spatially anti-bunched* (TSSA) state and the *temporally separated and spatially bunched* (TSSB) state. The TSSA state is defined as a superposed input state with a large time delay between two single photons in two different spatial modes[10]. However, the overall state of the two photons is a symmetrically superposed state in the two spatial modes. On the other hand, the TSSB state involves two single photons with a large time delay and in the same spatial mode[21]. Experimental demonstrations employing the TSSA state have previously been performed using the polarization-entangled state[10] and, also, the frequency-entangled state[30,31]. Recently, the two-photon quantum interference of the TSSB state was successfully demonstrated revealing that the temporal separation between two sequential photons in the same spatial mode does not degrade the phase super-resolution, as in the case of the conventional N00N state[21].

RESULTS

Generation of two-photon states with temporally separated photons

The conceptual scheme for the generation of the TSSA and TSSB two-photon states is depicted in Fig. 1. As is well known, the conventional two-photon N00N state can be easily generated via the HOM interference effect, when two identical single photons enter a balanced beamsplitter (BS) simultaneously, as shown in Fig. 1 2·17. In this case, the two output photons are always probabilistically bunched at one of the two spatial modes as described by

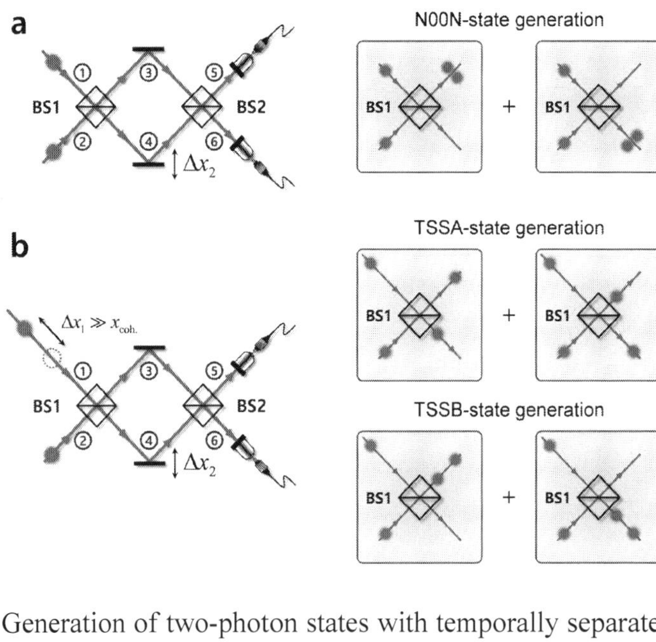

Figure 1: Generation of two-photon states with temporally separated photons. Conceptual scheme for generation of (**a**) conventional N00N state from HOM interference effect at BS1 output port and (**b**) superposed state of TSSA and TSSB states with large time delay between two photons at BS1 input port.

$$|1, 1\rangle_{1,2} \xrightarrow{\text{BS1}} \frac{1}{\sqrt{2}}(|2, 0\rangle_{3,4} + |0, 2\rangle_{3,4}), \quad (1)$$

where the subscripts denote the two spatial modes of the first BS (BS1) input and output. On the other hand, the TSSA and TSSB two-photon states can be prepared by introducing a time delay, $\Delta\tau_1 = \Delta x_1/c$ where c denotes the speed of light and Δx_1 is optical path-length difference between the two photons at the BS1 input stage, as shown in Fig. 1b. When the two photons are sufficiently separated from each other when relative to their coherence length, the HOM bunching effect at the BS1 output ports is no longer active. Then, the two output photons are in a state with the form

$$|1(\Delta x_1), 1\rangle_{1,2} \xrightarrow{\text{BS1}} \frac{1}{\sqrt{2}}(|\Psi\rangle_{\text{TSSA}} + |\Psi\rangle_{\text{TSSB}}), \tag{2}$$

where

$$|\Psi\rangle_{\text{TSSA}} = \frac{1}{\sqrt{2}}\left[|1\rangle_3|1(\Delta x_1)\rangle_4 - |1(\Delta x_1)\rangle_3|1\rangle_4\right],$$
$$|\Psi\rangle_{\text{TSSB}} = \frac{i}{\sqrt{2}}\left[|1(\Delta x_1), 1\rangle_3|0\rangle_4 + \exp(i2\phi)|0\rangle_3|1(\Delta x_1), 1\rangle_4\right]. \tag{3}$$

Here, ϕ is the relative single-photon phase difference between the two arms of the interferometer, which can be introduced by adjusting the path-length difference Δx_2.The two kinds of two-photon state represented by Eq. (3) are probabilistically coexistent within the interferometer arms, which construct TSSA and TSSB states, respectively. Then, the final state at the second BS (BS2) output port is composed of three states, such that

$$|\Psi\rangle_{\text{out}} = \frac{1}{2}\left\{\begin{array}{l}\sin\phi\,[|0\rangle_5|1(\Delta x_1), 1\rangle_6 - |1(\Delta x_1), 1\rangle_5|0\rangle_6] \\ -\cos\phi\,[|1(\Delta x_1)\rangle_5|1\rangle_6 + |1\rangle_5|1(\Delta x_1)\rangle_6] + [|1(\Delta x_1)\rangle_5|1\rangle_6 - |1\rangle_5|1(\Delta x_1)\rangle_6]\end{array}\right\}, \tag{4}$$

where the subscripts 5 and 6 denote the two spatial modes of the BS2 output. Here, the first term on the right-hand side represents a phase-sensitive TSSB state, while the last two terms correspond to phase-insensitive TSSA states. In the case of $\varphi = 0$ or $\varphi = \pi$, the output state has the form $|\Psi\rangle_{\text{out}} = |1\rangle_5|1(\Delta x_1)\rangle_6$ or $|\Psi\rangle_{\text{out}} = |1(\Delta x_1)\rangle_5|1\rangle_6$, respectively, which is an identical form to the BS1 input state. On the other hand, when $\phi = \pi/2$, the output state has a similar form to the BS2 input state. Although the MZI only was considered here, so as to show the output state for the input states given in Eq. (3), this result also applies to the PMI.

In this study, the observation of the two-photon interference effects obtained for the TSSA and TSSB states was performed at the MZI and PMI output ports, with Δx_2 being varied. It has been already shown that the TSSA-state interference has a phase-insensitive effect10·30, whereas the TSSB state can generate a resolution-enhanced phase-sensitive fringe pattern21. From Eq. (4), when the two input photons are injected into BS1 with a large time delay (Fig. 1b), the coincidence detection probability $P(\Delta x_2)$ at the two MZI output ports can be expressed as a superposition of the TSSA and TSSB-states interference fringes32, such that

$$P(\Delta x_2) = N_0\left\{2 + V\left[f(\Delta x_2) + g(\Delta x_2)\cos\left(\frac{2\pi}{\lambda_p}\Delta x_2\right)\right]\right\}, \tag{5}$$

where N_0 is a constant, V is the two-photon fringe visibility, λ_p is the centre wavelength of the pump laser, and $f(\Delta x_2)$ and $g(\Delta x_2)$ are envelope functions corresponding to the spectral properties of the detected single- and two-photon wavepackets, respectively (see the Methods section).

Experimental Setup

Mach-Zehnder interferometer

Figure 2 shows the experimental setup used to demonstrate the two-photon interference effects in an MZI. Correlated photon pairs at a telecommunication wavelength of 1.5 μm were generated through a quasi-phase-matched spontaneous parametric down-conversion (QPM-SPDC) process in a type-0 periodically-poled lithium niobate (PPLN) crystal. We used a mode-locked picosecond fiber laser (PriTel, FFL-20-HP-PRR and SHG-AF-200) as the QPM-SPDC pumping source, which had a 3.5-ps pulse duration at a 775-nm centre wavelength with a 20-MHz repetition rate. In our experiments, the average pump power was set to 20 mW. Using this setup, degenerate photon pairs were emitted with a full-opening angle of 4.6° in the noncollinear regime.

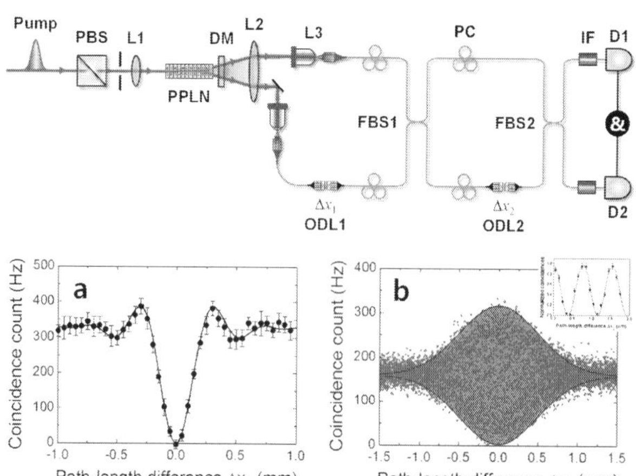

Figure 2: Experimental setup.

Mach-Zehnder interferometer: Pump: picosecond mode-locked fibre laser (3.5 ps, 20 MHz, 775 nm, 20 mW); PBS: polarizing beamsplitter; L1, L2: spherical lenses with 200-mm focal length; PPLN: periodically-poled lithium niobate crystal (length 10 mm, grating period 19.2 μm, temperature 40 °C); DM: dichroic mirror (T1550 nm/R775 nm); L3: aspherical lens with 8-mm focal length; PC: polarization controller; ODL: optical delay line; FBS: fibre beamsplitter 50/50; IF: interference filter with 6.25 nm bandwidth, D1, D2: gated-mode single-photon detection modules (Id Quantique id-210). (**a**) HOM-dip fringe measured at FBS1 output port as a function of delay length Δx_1. (**b**) N00N-state fringe measured at FBS2 output port as a function of path-length difference Δx_2. The inset shows the measured interference fringe at $\Delta x_2 \approx 0$.

The experimental setup was composed of two fibre interferometers, and the fibre length of each MZI arm was approximately 4 m. The first fibre BS (FBS1) acted as a "state preparator" to produce the state shown in Eq. (2), while the second fibre BS (FBS2) acted as a "two-photon interferometer". The FBS2 output photons were detected after they passed through interference filters (with 6.25-nm bandwidth) via two InGaAs/InP single-photon detection modules (Id Quantique id-210), which were operated in the gated mode. Electronic trigger signals were sent from the pump to the detector gates via electric delay lines. The detector quantum efficiency and dead time were set to 15% and 10 μs, respectively. The coincidence resolving time window was set to 10 ns, which was shorter than the pulse period of 50 ns. Under these experimental conditions, the coincidence to accidental coincidence ratio was approximately 4.13. The pair production probability per pulse was obtained by dividing the accidental coincidence by the measured coincidence, which was estimated to be approximately 0.24 per pulse.

Figure 2a shows the HOM interference fringe measured at the FBS1 output port. The net visibility was found to be 99.74±2.03% from the fit of a sinc function. The fringe width was determined to be approximately 0.38 mm, which was estimated from the rectangular-shaped 6.25-nm-bandwidth interference filter. Figure 2b shows the measured two-photon interference fringe arising from the conventional N00N-state input, representative of Eq. (1) when $\Delta x_1 = 0$. The red squared symbols indicate the measured coincidence counts and are plotted as functions of Δx_2, while the gray area corresponds to the phase-sensitive oscillatory fringe pattern with 98% visibility. To observe the phase-sensitive oscillatory fringe, we measured the coincidence counts at $\Delta x_2 \approx 0$ (see inset in Fig. 2b). The measured coincidence counts are normalized and the error bars represent the Poisson statistics of the coincidence counting rates. From the sinusoidal fit to the measured data points, the net visibility is found to be $99.62 \pm 0.02\%$. The blue solid lines denote the envelope curves obtained from the Gaussian function having a full-width at half-maximum (FWHM) of 1.17 mm, which was determined based on the two-photon coherence length. Note that this length is dependent on the pump pulse duration and the group velocity dispersion (GVD) in the SPDC pair generation process22.

Experimental results

Two-photon interference in Mach-Zehnder interferometer with temporally separated photons

Figure 3 shows the experimental results for the two-photon interference experiments with the two different kinds of input states shown in Fig. 1b. The conditions for the two input states before FBS2 were controlled by adjusting the first optical delay line (ODL1; Δx_1) before FBS1, and the two-photon coincidence fringes were measured for varying Δx_2 (with a 1-μm step size). When the input state was a superposition of the TSSA and TSSB states with the introduction of a large delay, $\Delta x_1 \gg x_{\mathrm{coh}}$, as shown in Fig. 1b, TSSA and TSSB two-photon interference fringes were observed simultaneously (Fig. 3b–e). In particular, it is worth noting that the two kinds of interference effects did not influence each other. Consequently, the two detectors (D1 and D2) probabilistically recorded the total coincidences resulting from the two kinds of interference phenomena, as expected from Eq. (5). The squared symbols represent the measured coincidence counts, which are plotted as functions of Δx_2. The gray areas correspond to the phase-sensitive oscillatory fringe patterns estimated from the cosine term in Eq. (5), while the solid lines denote the envelope curves in Eq. (5), which were determined by both the single- and two-photon spectral properties.

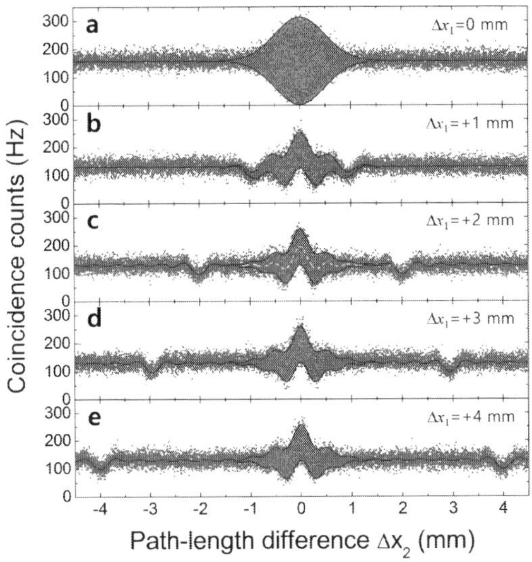

Figure 3: Two-photon interference fringes measured in Mach-Zehnder interferometer. Interference fringes obtained for **(a)** conventional N00N-state input $(\Delta x_1 = 0)$ and **(b–e)** simultaneous inputs of TSSA and TSSB states for various Δx_1 delay positions. The side dips represent ordinary HOM interference fringes obtained for path-length difference $\Delta x_2 = \pm\Delta x_1$.

In Fig. 3c–e, the widths of the central fringes for simultaneous inputs of TSSA and TSSB states are equal to that of the N00N state in Fig. 3a; therefore, this width is determined by the TSSB-state interference because the coherence length of the two-photon state is much larger than those of the single-photon wavepacket in our experiment. On the other hand, the complex sinuous fringe shape is caused by the single-photon spectral property $f(\Delta x_2)$, therefore, this shape is additionally influenced by the TSSA-state interference. The dashed line in Fig. 3c represents the interference peak of the TSSA state with 49% visibility. This peak has the same width as the single-photon wavepacket, as shown in Fig. 2a, which can be obtained by randomizing the relative phase between the two interferometer arms32·33. Here, $f(\Delta x_2) = \text{sinc}(\Delta x_2/\sigma_s)$, where σ_s is related to the single-photon bandwidth, provides a measure of the TSSA-state interference fringe. Thus, σ_s is determined by the interference filter used in the experiment only. Further, $g(\Delta x_2) = \exp[-\Delta x_2^2/(2\sigma_T^2)]$, where σ_T is the two-photon bandwidth, determines the size of the TSSB-state interference fringe. Therefore, the shapes and sizes of the central fringes shown in Fig. 3c–e are simultaneously determined by both the single- and two-photon coherence properties, and do not vary, even when Δx_1 is significantly longer than two-photon coherence length. When $\Delta x_2 = \pm\Delta x_1$, ordinary HOM-dip fringes are observed with 24.5% visibility, because only one-quarter of the total two-photon amplitudes contributes to the conventional HOM interference (see the side dips shown in Fig. 3b–e and the Methods section for details).

Experimental setup

Polarization-based Michelson interferometer

For the PMI, the conditions for the generation of correlated photon pairs are identical to those for the MZI. Figure 4 shows the experimental setup used to demonstrate the two-photon interference effects in a PMI. Two orthogonally polarized photons, H (horizontal polarization) and V (vertical polarization) from the SPDC source were combined using a fibre polarizing beamsplitter (FPBS) with a large delay length Δx_1; thereafter, the polarization directions of the two photons were rotated by ±45° using a half-wave plate (HWP). The PMI was composed of a polarizing beamsplitter (PBS) and two quarter-wave plates (QWPs), which had axes oriented at 45°. One of the mirrors (M2) was attached to a piezoelectric transducer (PZT) mounted on a linear translation stage, which could be used to scan the phase-sensitive interference fringe by varying Δx_2, and to extract the TSSA state from the TSSB state by randomizing the relative phase difference between the two PMI arms. To observe the two-photon interference fringe in the PMI, another PBS and a HWP with its axis oriented at 22.5° were placed at the PMI output port. Output photons from the PMI were detected by two single-photon detectors after they passed through coarse wavelength-division multiplexing (CWDM, 18-nm bandwidth) filters. In our experiment, two filter combinations (1550 and 1550 nm; 1530 and 1570 nm) were used to measure the two-photon coincidence fringes for both the degenerate and nondegenerate photon pairs. In particular, when the two-photons had different centre wavelengths, the two-photon state is in a frequency-entangled state30·31 and it has the form of $|\Psi\rangle = 1/\sqrt{2}\,(|\omega_1\rangle_H|\omega_2\rangle_V + |\omega_2\rangle_H|\omega_1\rangle_V)$. Thus the TSSA state with the frequency-entangled photons and the TSSB states involving two sequentially distributed photons with different frequencies could be obtained within the PMI arms.

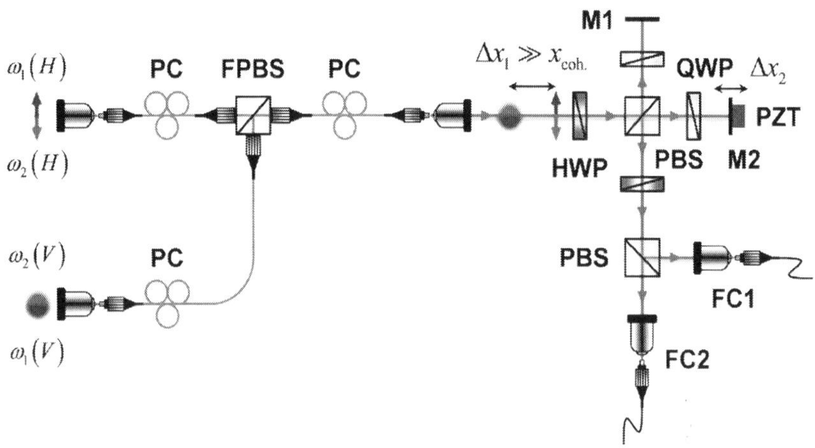

Figure 4: Experimental setup.

Polarization-based Michelson interferometer: Two orthogonally polarized photons were combined using a fibre polarizing beamsplitter (FPBS) with a delay length Δx_1. HWP: half-wave plate; PBS: polarizing beamsplitter; QWP: quarter-wave plate; M: mirror; FC: single-mode fibre coupler. A piezoelectric transducer (PZT) was used to scan the phase-sensitive interference fringe and to randomize the relative phase between two interferometer arms.

Experimental results

Two-photon interference in polarization-based Michelson interferometer with temporally separated photons

Figure 5 shows the two-photon interference fringes measured for two detected photons having the same wavelength of 1550 nm (a–c), and for those having different central wavelengths of 1530 and 1570 nm (d–f). When two orthogonally polarized photons with a delay of Δx_1 are injected into the PBS, as shown in Fig. 6, the two-photon states within the two PMI arms can be expressed in the same form as Eq. (3), although the polarization directions are also incorporated. For degenerate photon pairs

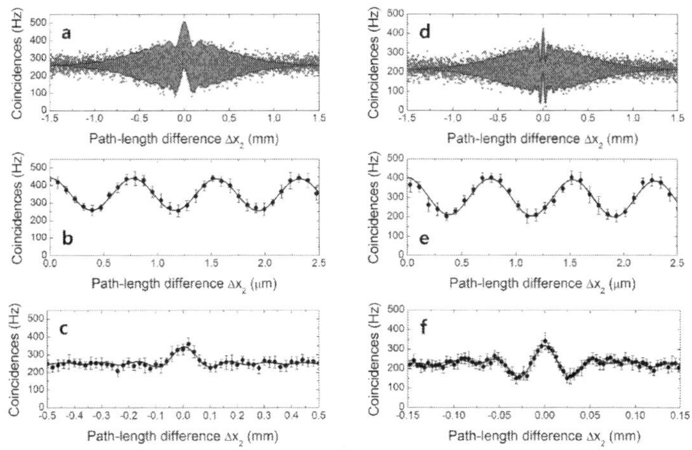

Figure 5: Two-photon interference fringes measured in polarization-based Michelson interferometer. Interference fringes obtained for simultaneous inputs of TSSA and TSSB states with (**a–c**) degenerate photon pairs and (**d–f**) nondegenerate photon pairs.

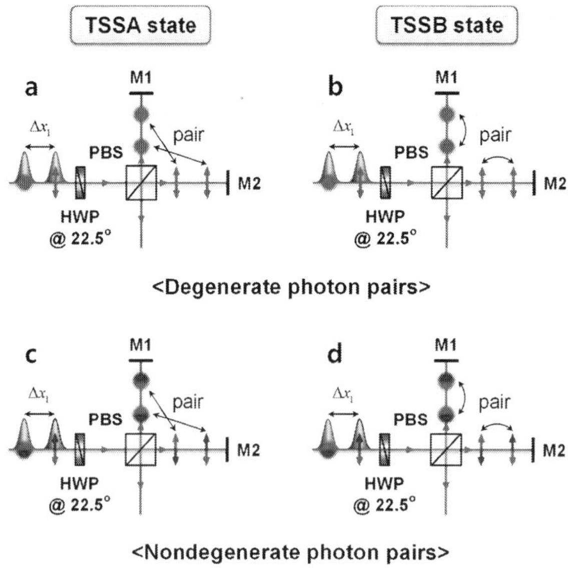

Figure 6: Generation of two-photon states with temporally separated photons in polarization-based Michelson interferometer. For (**a,b**) degenerate photon pairs and (**c,d**) nondegenerate photon pairs. In the case of the TSSA state, the two photons are in two different spatial modes. For the TSSB state, the two photons are in the same spatial mode.

$$|\Psi\rangle_{\text{TSSA}} = \frac{1}{\sqrt{2}}[|H\rangle_T|V(\Delta x_1)\rangle_R + |V\rangle_R|H(\Delta x_1)\rangle_T],$$
$$|\Psi\rangle_{\text{TSSB}} = \frac{1}{\sqrt{2}}[|H\rangle_T|H(\Delta x_1)\rangle_T + \exp(i2\phi)|V\rangle_R|V(\Delta x_1)\rangle_R], \qquad (6)$$

and for nondegenerate photon pairs

$$|\Psi\rangle_{\text{TSSA}} = \frac{1}{2}\begin{bmatrix}|\omega_1, H\rangle_T|\omega_2, V(\Delta x_1)\rangle_R + |\omega_1, V\rangle_R|\omega_2, H(\Delta x_1)\rangle_T \\ +|\omega_2, H\rangle_T|\omega_1, V(\Delta x_1)\rangle_R + |\omega_2, V\rangle_R|\omega_1, H(\Delta x_1)\rangle_T\end{bmatrix},$$
$$|\Psi\rangle_{\text{TSSB}} = \frac{1}{2}\begin{bmatrix}|\omega_1, H\rangle_T|\omega_2, H(\Delta x_1)\rangle_T + \exp(i2\phi)|\omega_1, V\rangle_R|\omega_2, V(\Delta x_1)\rangle_R \\ +|\omega_2, H\rangle_T|\omega_1, H(\Delta x_1)\rangle_T + \exp(i2\phi)|\omega_2, V\rangle_R|\omega_1, V(\Delta x_1)\rangle_R\end{bmatrix}, \qquad (7)$$

where the subscripts T (transmission) and R (reflection) denote the two spatial modes of the PBS output and $\omega_{1,2}$ represent the central frequencies of the two correlated photons. Here, it is worth noting that the two-photon amplitudes in Eq. (7), i.e., $|\omega_2, H\rangle_T|\omega_1, H(\Delta x_1)\rangle_T$ and $|\omega_2, V\rangle_R|\omega_1, V(\Delta x_1)\rangle_R$ for the TSSB state and $|\omega_2, H\rangle_T|\omega_1, V(\Delta x_1)\rangle_R$ and $|\omega_2, V\rangle_R|\omega_1, H(\Delta x_1)\rangle_T$ for the TSSA state, can be ignored when the two input photons are well separated from each other compared with the coincidence resolving time window T_R $(\Delta x_1/c \gg T_R)$. Similar to the MZI scenario, these two kinds of two-photon states are probabilistically coexistent, whether the two photons are in the same spatial mode (TSSB) and have the same polarization or if they are in two different spatial modes (TSSA) and are orthogonally polarized.

In the experiment, we set the Δx_1 of the vertically polarized photons to 3.2 mm. Figure 5a,d show the two-photon interference fringes measured by varying Δx_2 with 1-μm step size. The square symbols are the measured coincidence counts and are plotted as functions of Δx_2, while the gray are correspond to the phase-sensitive oscillatory fringe patterns. The solid lines denote the envelope curves obtained from Eq. (5). To observe the phase super-resolved fringe, we scanned one of the mirrors (M2) using the PZT actuator, for $\Delta x_2 \approx 0$. Figure 5b,e show the two-photon interference fringes with 26.21±0.58% (30.97±1.41%) visibility measured for degenerate (nondegenerate) photon pairs. If we randomized the relative phase between the two interferometer arms, we could then extract the interference fringe of the TSSA state from that of the TSSB state; this is because the TSSB state is very sensitive to the relative phase difference, while the TSSA state has no phase-sensitive interference fringe32,33. Figure 5c,f show the measured TSSA two-photon interference fringes as functions of Δx_2 that were obtained when a DC voltage value of 0~90 V was applied to the PZT actuator with a frequency of 10 Hz. The fringe visibility was found to be 39.11±3.38% (43.82±2.03). In the experiment, slightly lower visibility is mainly due to the imperfect alignment, instability in the interferometer including PZT actuator, and imperfection of the polarization optical components such as the PBS and wave plates.

DISCUSSION

Although various kinds of two-photon interference experiments involving correlated photons and using specific interferometers have been performed over the past three decades, the interference phenomena considered in this work, which arise from two-photon states composed of temporally separated photons, have not been fully analyzed. Moreover, a two-photon interference experiment with temporally separated photons in a Michelson interferometer has not yet been reported. Note that studies related to our MZI experiment have been conducted previously, using a coherently recurrent pump mode as the SPDC pump34,35. Those studies have presented some mathematical analyses of the interference phenomena. However, the present study provides fully generalized mathematical analyses (in Methods section), and qualitative and more comprehensive explanations of both the interference fringe shapes in the case of simultaneous inputs of the TSSA and TSSB states, and of the phase-insensitive side peak fringes arising from ordinary HOM-state inputs. In addition, the dispersion cancellation effects observed in two-photon interference experiments involving frequency-anticorrelated photon pairs in a HOM scheme can also be explained in the context of TSSA state interference, as depicted in Fig. 1b32,36. Indeed, many of the two-photon quantum interference phenomena can be more clearly understood by employing probabilistically coexisting states such as the TSSA and TSSB two-photon states.

In conclusion, we have demonstrated two-photon interference experiments involving temporally separated photons in an MZI and a PMI. We have introduced the concept of TSSA and TSSB two-photon states in order to distinguish from the conventional two-photon state with no time delay between the two constituent photons. By introducing a large time delay in the input stage of the interferometer, we successfully prepared two kinds of symmetrically superposed states of TSSA and TSSB two-photon states. The two-photon interference fringes measured for the two different kinds of two-photon states revealed the interferometric properties of both the single- and the two-photon wavepackets simultaneously, within a single interferometric setup. Further experimental investigation and related analysis can further clarify the origin of the two-photon interference effects in both the MZI and PMI. We believe that the present results will enable a more comprehensive understanding of the interference phenomena involving correlated quantum particles.

METHODS

Theoretical description

The quantum state of the photon pair source in Fig. 1 can be descripted as $|\Psi\rangle = \iint d\omega_1 d\omega_2 \Phi(\omega_1, \omega_2)\hat{a}_1^\dagger(\omega_1)\hat{a}_2^\dagger(\omega_2)|0\rangle$, where $\Phi(\omega_1, \omega_2)$ denotes the two-photon wave function, $\hat{a}_i^\dagger(\omega_j)$ is the creation operator of frequency ω_i at path j, and $|0\rangle$ is vacuum state. The two-photon coincidence counting rate between path 5 and 6 in Fig. 1 is proportional to the time-averaged value of the photon detection probability defined as $P_{5,6}(t_1, t_2) = |\langle 0|\hat{E}_5^{(+)}(t_1)\hat{E}_6^{(+)}(t_2)|\Psi\rangle|^2$, where $\hat{E}_k^{(+)}(t_l)$ denotes the positive part of the electric field operator at time t_l in path k. The electric field operators for paths 5 and 6 are superposed of them for path 1 and 2 in Fig. 1. If we assume that paths 2 and 4 have optical delay lines denoted τ_1 and τ_2, respectively, the normalized coincidence counting probability is calculated as

$$P = \frac{1}{2} + \frac{1}{8} Re \left\{ \begin{array}{l} 2\iint \Phi^*(\omega_1, \omega_2)\Phi(\omega_1, \omega_2)[e^{i\tau_2(-\omega_1+\omega_2)} + e^{i\tau_2(\omega_1+\omega_2)}]d\omega_1 d\omega_2 \\ + \iint \Phi^*(\omega_2, \omega_1)\Phi(\omega_1, \omega_2)[2e^{i\tau_1(\omega_1-\omega_2)+i\tau_2(\omega_1+\omega_2)} - e^{i(\tau_1+\tau_2)(\omega_1-\omega_2)} \\ - e^{i(\tau_1-\tau_2)(\omega_1-\omega_2)}]d\omega_1 d\omega_2 \end{array} \right\}. \tag{8}$$

For a comprehensive understanding, we assume that the two-photon wave function is symmetric, $\Phi(\omega_1, \omega_2) = \Phi(\omega_2, \omega_1)$, and the time delay τ_1 is zero or much larger than two-photon coherence time $\tau_{coh.}$. When τ_1 is zero, Eq. (8) can be simplified as

$$P = \frac{1}{2}\left\{1 + Re\left[\iint |\Phi(\omega_1, \omega_2)|^2 e^{i\tau_2(\omega_1+\omega_2)} d\omega_1 d\omega_2\right]\right\}. \tag{9}$$

It represents phase sensitive two-photon interference as shown in Fig. 3a 37. The envelope function of the interference patterns is related with Fourier transform of two-photon wave function for the direction of $\omega_1 = \omega_2$, so that the width of the interference pattern is decided by the spectral bandwidth of photon pair source22. If τ_1 is much larger than $\tau_{coh.}$, the interference patterns are revealed only when τ_2 is around 0 or $\pm\tau_1$. For the region of $|\tau_2 = \Delta\tau| < \tau_{\text{coh.}}$, Eq. (8) is simplified as

$$P = \frac{1}{2}\left\{1 + \frac{1}{2} Re\left[\iint |\Phi(\omega_1, \omega_2)|^2 (e^{i\Delta\tau(\omega_1-\omega_2)} + e^{i\Delta\tau(\omega_1+\omega_2)}) d\omega_1 d\omega_2\right]\right\}, \tag{10}$$

which includes phase insensitive HOM peak and phase sensitive two-photon interferences due to the terms of $(\omega_1 - \omega_2)$ and $(\omega_1 + \omega_2)$, respectively, and their amplitudes are reduced to a half. The width of HOM interference is related with the single-photon coherence time and the Fourier transform of two-photon wave function for the direction of $\omega_1 = -\omega_2$. It is decided by phase matching condition of the SPDC process and filter bandwidth used in the experiment. For the region of $|\tau_2 \pm \tau_1 = \Delta\tau| < \tau_{\text{coh.}}$, Eq. (8) is simplified as

$$P = \frac{1}{2}\left\{1 - \frac{1}{4} Re\left[\iint |\Phi(\omega_1, \omega_2)|^2 e^{-i\Delta\tau(\omega_1-\omega_2)} d\omega_1 d\omega_2\right]\right\}. \tag{11}$$

It only shows HOM dip interference with quarter amplitude. Equation. (9, 10, 11) clearly describe the experimental results in Fig. 3, and show the definitions of envelope functions $f(\Delta x_2)$ and $g(\Delta x_2)$ in Eq. (5). We note that Eq. (8) is not simplified and shows complicated interference patterns under a condition of $0 < \tau_1 < \tau_{\text{coh.}}$. For the case of non-degenerated photon pairs (frequency-entangled states), the assumption of the symmetry, $\Phi(\omega_1, \omega_2) = \Phi(\omega_2, \omega_1)$ is still satisfied. The only difference from degenerated case is that the beating fringes also arise in HOM interference due to the wavelength difference between two photons, as shown in Fig. 5f.

ACKNOWLEDGEMENTS

This work was supported by the Basic Science Research Program through the National Research Foundation of Korea funded by the Ministry of Education, Science and Technology (No. 2015R1A2A1A05001819, No. 2014R1A1A2055488, and No. 2014R1A1A2054719). Also, this work was supported by the Measurement Research Center (MRC) Program for Korea Research Institute of Standards and Science.

CONTRIBUTIONS

H.K. and H.S.M. conceived the project and designed the experiment. H.K. performed the experiment. S.M.L. contributed to the theoretical analysis. All authors discussed the results and contributed to writing the manuscript.

REFERENCES

1. Hong, C. K., Ou, Z. Y. & Mandel, L. Measurement of subpicosecond time intervals between two photons by interference. Phys. Rev. Lett. 59, 2044–2046 (1987).
2. Ou, Z. Y., Zou, X. Y., Wang, L. J. & Mandel, L. Experiment on nonclassical fourth-order interference. Phys. Rev. A 42, 2957–2965 (1990).
3. Lee, H., Kok, P. & Dowling, J. P. A quantum Rosetta stone for interferometry. J. Mod. Opt . 49, 2325–2338 (2002).
4. Yuan, Z. S. et al. Entangled photons and quantum communication. Phys. Rep . 497, 1–40 (2010).
5. Pan, J. W. et al. Multiphoton entanglement and interferometry. Rev. Mod. Phys. 84, 777–838 (2012).
6. Mandel, L. Quantum effects in one-photon and two-photon interference. Rev. Mod. Phys. 71, S274–S282 (1999).
7. Jaeger, G. & Sergienko, A. V. Multi-photon quantum interferometry, in E. Wolf, Progress in Optics 42, Elsevier Science B.V, pp. 277–324 (2001).
8. Pittman, T. B. et al. Can two-photon interference be considered the interference of two photons? Phys. Rev. Lett. 77, 1917–1920 (1996).
9. Strekalov, D. V., Pittman, T. B. & Shih, Y. H. What we can learn about single photons in a two-photon interference experiment. Phys. Rev. A 57, 567–570 (1998).
10. Kim, Y.-H. & Grice, W. P. Quantum interference with distinguishable photons through indistinguishable pathways. J. Opt. Soc. Am. B 22, 493–498 (2005).
11. Silverstone, J. W. et al. On-chip quantum interference between silicon photon-pair sources. Nature Photon . 8, 104–108 (2014).
12. Martino, G. D. et al. Observation of quantum interference in the plasmonic Hong-Ou-Mandel effect. Phys. Rev. Applied 1, 034004 (2014).
13. Toyoda, K., Hiji, R., Noguchi, A. & Urabe, S. Hong-Ou-Mandel interference of two phonons in trapped ions. Nature 527, 74–77 (2015).
14. Lopes, R. et al. Atomic Hong–Ou–Mandel experiment. Nature 520, 66–68 (2015).

15. Kobayashi, T. et al. Frequency-domain Hong-Ou-Mandel interference. Nature Photon . 10, 441–444 (2016).
16. Boto, A. N. et al. Quantum interferometric optical lithography: exploiting entanglement to beat the diffraction limit. Phys. Rev. Lett. 85, 2733–2736 (2000).
17. Edamatsu, K., Shimizu, R. & Itoh, T. Measurement of the photonic de Broglie wavelength of entangled photon pairs generated by spontaneous parametric down-conversion. Phys. Rev. Lett. 89, 213601 (2002).
18. Resch, K. J. et al. Time-reversal and super-resolving phase measurements. Phys. Rev. Lett. 99, 069902 (2007).
19. Nagata, T., Okamoto, R. O'Brien, J. L., Sasaki, K. & Takeuchi, S. Beating the standard quantum limit with four entangled photons. Science 316, 726–729 (2007).
20. Afek, I., Ambar, O. & Silberberg, Y. High-NOON states by mixing quantum and classical light. Science 328, 879–881 (2010).
21. Jin, X.-M. et al. Sequential path entanglement for quantum metrology. Sci. Rep. 3, 1779 (2013).
22. Kim, H., Lee, S. M. & Moon, H. S. Generalized quantum interference of correlated photon pairs. Sci. Rep . 5, 9931 (2015).
23. Rarity, J. G. et al. Two-photon interference in a Mach-Zehnder interferometer. Phys. Rev. Lett. 65, 1348–1351 (1990).
24. Larchuk, T. S. et al. Interfering entangled photons of different colors. Phys. Rev. Lett. 70, 1603–1606 (1993).
25. Brendel, J., Mohler, E. & Martienssen, W. Time-resolved dual-beam two-photon interferences with high visibility. Phys. Rev. Lett. 66, 1142–1145 (1991).
26. Odate, S., Wang, H. & Kobayashi, T. Two-photon quantum interference in a Michelson interferometer. Phys. Rev. A 72, 063812 (2005).
27. Lopez-Mago, D. & Novotny, L. Coherence measurements with the two-photon Michelson interferometer. Phys. Rev. A 86, 023820 (2012).
28. Wiegner, R., Thiel, C.,. Zanthier, J. & Agarwal, G. S. Quantum interference and entanglement of photons that do not overlap in time. Opt. Lett. 36, 1512–1514 (2011).
29. Megidish, E., Halevy, A., Shacham, T., Dvir, T., Dovrat, L. & Eisenberg, H. S. Entanglement swapping between photons that have never coexisted. Phys. Rev. Lett. 110, 210403 (2013).

30. Kim, H., Lee, H. J., Lee, S. M. & Moon, H. S. Highly efficient source for frequency-entangled photon pairs generated in a 3rd-order periodically poled MgO-doped stoichiometric LiTaO3 crystal. Opt. Lett. 40, 3061–3064 (2015).

31. Lee, S. M., Kim, H., Cha, M. & Moon, H. S. Polarization-entangled photon-pair source obtained via type-II non-collinear SPDC process with PPKTP crystal. Opt. Express 24, 2941–2953 (2016).

32. Halder, M. et al. Photon-bunching measurement after two 25-km-long optical fibers. Phys. Rev. A 71, 042335 (2005).

33. Huy. K. P., Nguyen, A. T., Brainis, E., Haelterman, M. & Emplit, P. Photon pair source based on parametric fluorescence in periodically poled twin-hole silica fiber. Opt. Express 15, 4419–4426 (2007).

34. Kwon, O., Ra, Y.-S. & Kim, Y.-H. Coherence properties of spontaneous parametric down-conversion pumped by a multi-mode cw diode laser. Opt. Express 17, 13059 (2009).

35. Kwon, O., Ra, Y.-S. & Kim, Y.-H. Observing photonic de Broglie waves without the maximally-path-entangled |N,0 > + |0,N > state. Phys. Rev. A 81, 063801 (2010).

36. Steinberg, A. M., Kwiat, P. G. & Chiao, R. Y. Dispersion cancellation in a measurement of the single-photon propagation velocity in glass. Phys. Rev. Lett. 68, 2421–2424 (1992).

37. Abouraddy, A. F., Yarnall, T. M. & Giuseppe, G. D. Phase-unlocked Hong-Ou-Mandel interferometry. Phys. Rev. A 87, 062106 (2013).

CHAPTER 15

SQUEEZED-LIGHT-DRIVEN FORCE DETECTION WITH AN OPTOMECHANICAL CAVITY IN A MACH–ZEHNDER INTERFEROMETER

ChangWoo Lee[1], Jae Hoon Lee[2] & Hyojun Seok[1]

[1] Department of Physics Education, Kongju National University, Gongju 32588, South Korea

[2] Korea Research Institute of Standards and Science, Daejeon 34113, South Korea

ABSTRACT

We analyze the performance of a force detector based on balanced measurements with a Mach–Zehnder interferometer incorporating a standard optomechanical cavity. The system is driven by a coherent superposition of coherent light and squeezed vacuum field, providing quantum correlation along with optical coherence in order to enhance the measurement sensitivity beyond the standard quantum limit. We analytically find the optimal measurement strength, squeezing direction, and squeezing strength at which the symmetrized power spectral density for the measurement noise is minimized below the standard quantum limit. This force detection

Citation: (APA): Lee, C. W., Lee, J. H., & Seok, H. (2020). Squeezed-light-driven force detection with an optomechanical cavity in a Mach–Zehnder interferometer. Scientific Reports, 10(1), 1-16. (16 pages)

scheme based on a balanced Mach–Zehnder interferometer provides better sensitivity compared to that based on balanced homodyne detection with a local oscillator in the low frequency regime.

INTRODUCTION

Quantum mechanics hinders the ability to measure a physical observable with arbitrarily high precision in continuous measurements. The corresponding quantum object being measured is not only agitated by its environment, but also disturbed by a quantum probe[1]. Therefore, the quantum limit of the measurement sensitivity is associated with the quantum characteristics of the object, the probe, and their mutual interactions. Efforts to optically measure the mechanical motion with high sensitivity triggered the research field of cavity optomechanics[2,3] in which the intensity of a cavity field is coupled to the motion of a mechanical oscillator via radiation pressure force[4,5]. This type of optomechanical measurement first became relevant in research where feeble forces act on macroscopic quantum objects[6] and in gravitational wave detection[7]. It has been known that the noise resulting from individual photons randomly arriving at the detectors for absorption, and the noise due to random scattering events between intracavity photons and the mechanical oscillator limit the measurement sensitivity of the displacement or the external force[6,8,9]. These two types of noises, namely the photon counting noise and the radiation pressure backaction noise, have opposite scaling behavior with respect to the input laser power: increasing the input power decreases the photon counting noise but increases the radiation pressure backaction noise, and vice versa. Balancing these two noises gives rise to the so-called standard quantum limit (SQL) that describes the limit of sensitivity for conventional interferometric measurements[6,8,9]. Routes to enhancing the sensitivity beyond the SQL thus requires evading the photon counting noise or the radiation pressure backaction noise.

Nonclassical states of light are possible resources for enhancing the sensitivity of displacement or force sensors. Particularly, squeezed states of light are one of the most prominent nonclassical probes of which the quantum fluctuations in one of the quadratures are less than the vacuum noise level[10,11]. Employing squeezed states of light was first proposed to demonstrate a reduction of the photon counting noise in the context of gravitational wave detection, yet undesirably accompanying an increase in the radiation pressure backaction noise[12]. This proposal prompted much theoretical effort on overcoming the SQL for gravitational wave detection, for example, exploiting quantum correlation in the input field[13], frequency-dependent squeezing of the input field[14,15,16,17], and variational measurements of the output field[18,19,20]. Such suppression of photon counting noise by squeezing has been realized in various types of table-top interferometers including Mach–Zehnder[21], polarization[22], Sagnac[23], and Michelson interferometers[24] and recently demonstrated with gravitational wave interferometers[25]. It has also been experimentally shown that squeezing the input field can modify the radiation pressure backaction noise[26].

Another approach to surpassing the SQL is using quantum nondemolition (QND) measurements in which the observable being measured is not dynamically coupled to its conjugate observable[27,28,29]. In QND measurements, such dynamical decoupling restrains backaction noise of the conjugate observable from being fed back into the observable being measured, allowing for backaction-evading measurements. A simple example of QND measurements in cavity optomechanics is to measure a static observable, for instance, only one of the quadratures of the mechanical motion[27,28,29,30]. Such backaction-evading measurements have been experimentally demonstrated by modulating the amplitude of the driving field in single-mode mechanical systems[31,32,33,34] and in a collective quadrature of multimode mechanical systems[35,36]. Recently, dynamical decoupling between an observable and its conjugate was proposed in the so-called quantum–mechanics-free subsystem (QMFS) by introducing an anti-backaction noise path in the system dynamics for coherent cancellation of the backaction noise[37,38]. Such observables in the QMFS are dynamical QND observables in that they evolve in time but are not coupled to their conjugate observables[38]. Backaction-evading measurements of motion based on QMFS have been proposed in a variety of optomechanical systems including an optomechanical cavity incorporating Kerr medium[39], Bose–Einstein condensates[40], an optomechanical cavity with an auxiliary cavity[41], and hybrid atom-optomechanical systems[42,43]. Furthermore, it has recently been reported that the backaction noise is somewhat suppressed in atomic ensembles that are subject to external magnetic fields[44].

In this paper, we study schemes employing quantum correlations in squeezed states of light and further utilize the notion of coherent quantum-noise cancellation developed in QND measurements to suppress the photon counting noise in an optomechanical force sensor operating in the low frequency regime. Specifically, we analyze the performance of a force detector based on a balanced Mach–Zehnder interferometer (MZI) incorporating a standard optomechanical cavity. The optomechanical system is driven by a coherent superposition of coherent light and squeezed vacuum field, providing quantum correlation along with optical coherence in order to enhance the measurement sensitivity beyond the SQL. We analytically find the optimal measurement strength, squeezing direction, and squeezing strength at which the symmetrized power spectral density (sPSD) for the measurement noise is minimized below the SQL. This force detection scheme based on a balanced MZI provides better sensitivity compared to that based on a balanced homodyne detection with a local oscillator field in the low frequency regime.

The remainder of this paper is structured as follows. In "Results" section we provide a description of the model system of interest and discuss the quantum characteristics of the optomechanical cavity output field. We then compare the measurement sensitivity of the two detection schemes, specifically, the balanced homodyne detection scheme and the balanced MZI scheme. In "Methods" section we present detailed derivation of the sPSD of the measurement noise, and subsequently show the optimization processes for suppressing the sPSD with respect to the squeezing direction and squeezing strength. Finally, we give our summary and conclusions in "Conclusions" section.

RESULTS

Model system

We consider a Mach–Zehnder interferometer in which an optomechanical cavity of resonant frequency ωcωc with a harmonically bound mirror of mass m and mechanical frequency Ω is integrated into one of the optical paths depicted in Fig. 1a. A coherent mixture of monochromatic coherent light and squeezed vacuum is created with a 50/50 beam splitter (BS1). The coherent light oscillates at frequency ω_L while the broadband squeezed vacuum is characterized by squeezing strength r, squeezing direction θ, and central frequency ω_L. The optomechanical system is optically driven by this coherently superposed field, and the output field from the optomechanical cavity is combined with the reference field at a second 50/50 beam splitter (BS2). In order to analyze the quantum noise introduced by this measurement scheme, the difference in photon flux between the output ports of BS2 is measured, i.e., balanced detection. The Hamiltonian governing the dynamics of the optomechanical system in a frame rotating at the driving frequency reads

$$\hat{H} = -\hbar\Delta\hat{a}^{\dagger}\hat{a} + i\hbar\sqrt{\kappa_{\text{in}}}\alpha_{\text{in}}(\hat{a}^{\dagger} - \hat{a}) + \frac{\hbar\Omega}{2}(\hat{q}^2 + \hat{p}^2) - \hbar g_0\sqrt{2}\hat{a}^{\dagger}\hat{a}\hat{q} + \hat{H}_{\text{diss}} + \hat{H}_{\text{ext}}. \tag{1}$$

The first two terms describe the cavity field driven by a pump field of detuning $\Delta=\omega_L-\omega_c$ with rate $\sqrt{\kappa_{\text{in}}}\alpha_{\text{in}}$ where κ_{in} is the loss rate through the cavity input port mirror, and $\alpha_{\text{in}} = \sqrt{\mathcal{P}/(\hbar\omega_L)}$ is the coherent amplitude of the input field with PP being the continuous wave optical power into the cavity. The bosonic annihilation operator of the cavity field a^ satisfies the relation [a^,a^†]=1. The third and fourth terms describe the mechanical oscillator driven by the intracavity radiation pressure, where g0g0 is the single-photon optomechanical coupling coefficient. The position and momentum operators of the mechanical oscillator are normalized by the natural length $q_0 = \sqrt{\hbar/(m\Omega)}$ and momentum $p_0 = \sqrt{\hbar m\Omega}$ so that the dimensionless position and momentum operators of the mechanics q^ and p^ satisfy $[\hat{q}, \hat{p}] = i$. $\hat{H}_{\text{diss}}$ which accounts for dissipation and decoherence represents the incoherent interaction of the optomechanical system with the optical reservoir at zero temperature and the mechanical bath at finite temperature *T*. Finally, $H\hat{}_{ext}$ describes the coupling between the momentum of the mechanical oscillator and an external force to be measured.

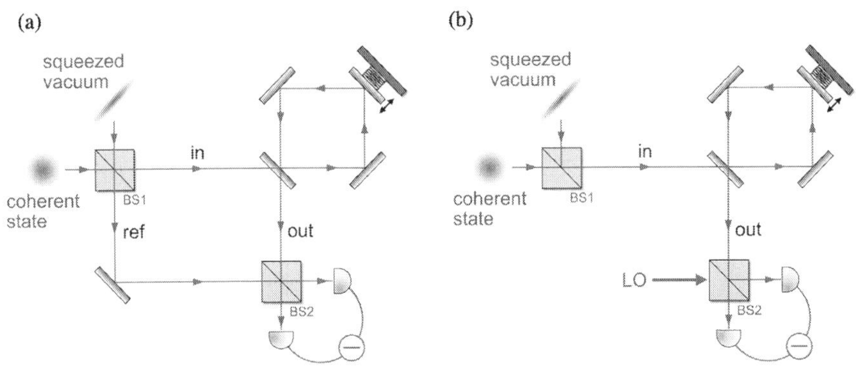

Figure 1: (**a**) Schematics of the Mach–Zehnder interferometer incorporating an optomechanical cavity for quantum enhanced force detection. The coherent field and squeezed vacuum field are combined at the first 50/50 beam splitter (BS1) producing an input field for the optomechanical cavity and a correlated reference field. The field exiting the optomechanical cavity is then combined with the reference beam at the second 50/50 beam splitter (BS2) for balanced detection. (**b**) Schematics of a standard homodyne optomechanical force sensing setup. The output field of the optomechanical cavity is combined with a local oscillator field at BS2. In both schemes, the photon flux difference between the two output ports of BS2 is measured in order to analyze the noise spectrum.

In this setup, the external force changes the dynamic variables for the mechanics, and subsequently the phase of the output optical field through the optomechanical interaction. In this manuscript, we assume a weak external force where the effects on the optomechanical system is similar to other noise levels. Hence, we focus only on the generic quantum noise generated by the force measurement process.

Output field from the optomechanical cavity

Mean fields

The semiclassical equation of motion for the cavity field derived from Eq. (1) can be written as

$$\dot{\alpha} = (i\Delta_{\text{eff}} - \kappa/2)\alpha + \sqrt{\kappa_{\text{in}}}\alpha_{\text{in}}, \quad (2)$$

where the classical mean field amplitude for the cavity field is denoted by $\alpha = \langle \hat{a} \rangle$ and the effective detuning is defined by $\Delta_{\text{eff}} = \Delta + \sqrt{2}g_0\langle q \rangle$. We assume for simplicity that there are no other loss channels except that from the cavity input port. Therefore, the total dissipation rate for the cavity field is simply the loss rate of the input port mirror, $\kappa=\kappa_{\text{in}}$. In the case where there are other loss channels, such as cavity internal loss, one can expect limited quantum enhancement of the measurement sensitivity45·46. Then the steady-state solution for the mean cavity field amplitude in the long-time limit can be obtained as

$$\alpha_{\text{ss}} = \frac{\sqrt{\kappa}}{-i\Delta_{\text{eff}} + \kappa/2}\alpha_{\text{in}}. \quad (3)$$

The steady-state position for the mechanics can be similarly obtained as $\langle q \rangle_{\text{ss}} = -\sqrt{2}g_0|\alpha_{\text{ss}}|^2/\Omega$. Invoking the input-output relation, $\hat{a}_{\text{out}} = \hat{a}_{\text{in}} - \sqrt{\kappa_{\text{in}}}\hat{a}$, results in the steady-state coherent amplitude for the cavity output field as

$$\alpha_{\text{out}} = \frac{-i\Delta_{\text{eff}} - \kappa/2}{-i\Delta_{\text{eff}} + \kappa/2}\alpha_{\text{in}}. \quad (4)$$

Notice that the magnitude of the coherent amplitude is the same for the cavity input and output fields, but there is a difference in phase depending on the optomechanical interaction. This phase shift of the optical field accounts for various static optomechanical effects, for example, the static optomechanical spring effect and the optical bistability47·48. In this paper, we assume that the effective detuning is zero so that the steady-state intracavity field amplitude is simply $\alpha_{ss} = 2\alpha_{in}/\sqrt{\kappa}$ and the classical amplitude of the input and output field is 180∘180∘ out of phase, αout=−αinαout=−αin. As is well-known, zero effective detuning does not lead to various dynamic optomechanical effects including optomechanical cooling or amplification of the mechanical oscillator[49,50,51], entanglement between the mechanical motion and cavity fields[52], and optomechanically induced transparency[53], to mention just a few.

Quantum fluctuations

So as to better capture the steady-state features of the quantum fluctuations in the optical and mechanical fields, we transform to a displaced frame where the expectation values of the optical field operator and the mechanical oscillator position operator are zero. Performing the displacement operation, $\hat{a} \to \alpha_{ss} + \hat{d}, \hat{q} \to \langle q \rangle_{ss} + \hat{Q}$, the Heisenberg–Langevin equations of motion for the quantum fluctuations of the optical field and the mechanical oscillator can be approximated as[46]

$$\dot{\hat{X}} = -\frac{\kappa}{2}\hat{X} + \sqrt{\kappa}\hat{X}_{in}, \tag{5a}$$

$$\dot{\hat{Y}} = -\frac{\kappa}{2}\hat{Y} + 2g\hat{Q} + \sqrt{\kappa}\hat{Y}_{in}, \tag{5b}$$

$$\dot{\hat{Q}} = \Omega\hat{P}, \tag{5c}$$

$$\dot{\hat{P}} = -\Omega\hat{Q} - \Gamma\hat{P} + 2g\hat{X} + \sqrt{2\Gamma}\hat{P}_{in}, \tag{5d}$$

where the intrinsic nonlinear optomechanical coupling is neglected in the weak coupling regime, $g=g_0\alpha_{ss}$ is the field-amplified optomechanical coupling strength, and Γ is the mechanical dissipation rate. The operators $\hat{X} = \frac{1}{\sqrt{2}}(\hat{d}^\dagger + \hat{d})$ and $\hat{Y} = \frac{i}{\sqrt{2}}(\hat{d}^\dagger - \hat{d})$ represent the quantum fluctuations of the amplitude and phase quadrature for the intracavity field, and $\hat{X}_{in}$ and $\hat{Y}_{in}$ are the input field noise operators for the two quadratures of the cavity field. Finally, $\hat{P}_{in} = \hat{F}_{th} + \hat{F}_{ext}$ represents the force acting on the mechanical oscillator where $\hat{F}_{th}$ and $\hat{F}_{ext}$ are the operators for the thermal force and the external force of interest, i.e., force to be detected, with units of √Hz.

Now that the Heisenberg equations of motion for the quantum fluctuations are linear with respect to the field operators, it is convenient to solve them in the frequency domain exploiting the Fourier transform for the operators,

$$\hat{O}(\omega) = \frac{1}{\sqrt{2\pi}} \int dt\, e^{i\omega t} \hat{O}(t). \tag{6}$$

With the input-output relation in the frequency domain, $\hat{d}_{\text{out}}(\omega) = \hat{d}_{\text{in}}(\omega) - \sqrt{\kappa}\hat{d}(\omega)$, the fluctuations in the quadrature for the output field are obtained as

$$\hat{X}_{\text{out}}(\omega) = -\frac{\chi_o^*}{\chi_o}\hat{X}_{\text{in}}(\omega), \tag{7}$$

$$\hat{Y}_{\text{out}}(\omega) = -\frac{\chi_o^*}{\chi_o}\hat{Y}_{\text{in}}(\omega) + 4G\chi_o^{*2}\chi_m\hat{X}_{\text{in}}(\omega) - 2\sqrt{2\Gamma G}\chi_o^*\chi_m\hat{P}_{\text{in}}(\omega). \tag{8}$$

Here, we have defined the measurement strength

$$G = g^2\kappa = 4g_0^2\mathcal{P}/(\hbar\omega_L), \tag{9}$$

where we used the definition of the amplified optomechanical strength $g=g_0\alpha_{ss}$, the steady-state intracavity field amplitude $\alpha_{ss}=2\alpha_{in}/\kappa$ for resonant driving, and the coherent amplitude of the input field $\alpha_{\text{in}} = \sqrt{\mathcal{P}/(\hbar\omega_L)}$. Note that while the intracavity photon number is inversely proportional to the cavity linewidth, the measurement strength G is independent of the cavity linewidth since it is directly related to the photon number in the output field. The optical and mechanical susceptibilities are given by

$$\chi_o = \frac{1}{i\omega + \kappa/2}, \tag{10}$$

$$\chi_m = \frac{\Omega}{\Omega^2 - \omega^2 - i\omega\Gamma}. \tag{11}$$

As can be inferred from Fig. 1, the quantum noise in the input field of the optomechanical system derives from the coherent superposition of the vacuum fluctuations d^v and squeezed vacuum fluctuations d^sq. Since the squeezed vacuum undergoes a reflection at BS1 there is a relative phase of 90° between the two fields, i.e., $\hat{d}_{\text{in}} = (\hat{d}_{\text{v}} + i\hat{d}_{\text{sq}})/\sqrt{2}$. The quantum fluctuations in the quadrature of the output field then become

$$\hat{X}_{\text{out}}(\omega) = -\frac{1}{\sqrt{2}}\frac{\chi_o^*}{\chi_o}\left[\hat{X}_{\text{v}}(\omega) - \hat{Y}_{\text{sq}}(\omega)\right], \tag{12}$$

$$\hat{Y}_{\text{out}}(\omega) = -\frac{1}{\sqrt{2}}\frac{\chi_o^*}{\chi_o}\left[\hat{Y}_{\text{v}}(\omega) + \hat{X}_{\text{sq}}(\omega)\right] + \frac{1}{\sqrt{2}}4G\chi_o^{*2}\chi_m\left[\hat{X}_{\text{v}}(\omega) - \hat{Y}_{\text{sq}}(\omega)\right] - 2\sqrt{2\Gamma G}\chi_o^*\chi_m\hat{P}_{\text{in}}(\omega). \tag{13}$$

It is important to note that the external force is imprinted only on the phase quadrature of the output field. This consequence can be attributed to having zero effective detuning which leads to the amplitude quadrature of the output field containing no information of the mechanics54. Thus, in order to capture information with regard to the external forces of interest one must measure $\hat{Y}_{\mathrm{out}}(\omega)$. The additional quantum fluctuations are inevitably introduced in the measurement of the phase quadrature of the output field.

Balanced homodyne detection with a local oscillator field

We first explore a scheme based on a balanced homodyne detection with a local oscillator (LO) field which may be considered more prototypical, as depicted in Fig. 1b in order to better appreciate the performance of our proposed balanced Mach–Zehnder measurement scheme which optimally utilizes quantum interference effects. As shown in Fig. 1b, the output field from the optomechanical cavity is coherently mixed with a LO field at BS2. The optical fields $\hat{a}_{+}$, $\hat{a}_{-}$ at the two output ports of BS2 can be described as

$$\begin{pmatrix} \hat{a}_{+} \\ \hat{a}_{-} \end{pmatrix} = \frac{1}{\sqrt{2}} \begin{pmatrix} 1 & i \\ i & 1 \end{pmatrix} \begin{pmatrix} \hat{a}_{\mathrm{out}} \\ \hat{a}_{\mathrm{LO}} \end{pmatrix}, \tag{14}$$

where $\hat{a}_{\mathrm{LO}}$ is the bosonic annihilation operator for the LO field. If the amplitude of the LO field is sufficiently larger than that of the cavity output field, the LO field can be treated classically. The quantum fluctuations in the photon flux difference between the two output ports $\hat{I} = \hat{a}_{-}^{\dagger}\hat{a}_{-} - \hat{a}_{+}^{\dagger}\hat{a}_{+}$, can now be approximated as

$$\delta\hat{I} = -\sqrt{2}|\alpha_{\mathrm{LO}}|(\hat{Y}_{\mathrm{out}}\cos\varphi + \hat{X}_{\mathrm{out}}\sin\varphi), \tag{15}$$

where $\alpha_{\mathrm{LO}} = |\alpha_{\mathrm{LO}}|e^{i\varphi}$ is the complex amplitude of the LO field. Since the force we want to detect is only dependent on $\hat{Y}_{\mathrm{out}}$, see Eq. (13), setting $\varphi=\pi$ for maximum amount of information gives rise to

$$\delta\hat{I} = \sqrt{2}|\alpha_{\mathrm{LO}}|\hat{Y}_{\mathrm{out}}. \tag{16}$$

It should be kept in mind that $\delta\hat{I}$ does not contain interference between $\hat{Y}_{\mathrm{out}}$ and additional quantum fluctuations since the LO is treated classically. The role of the LO field in this scheme is to choose the appropriate quadrature for measuring the output field, $\hat{Y}_{\mathrm{out}}$, without adding additional quantum fluctuations.

In this situation, the unbiased estimator for the external force of interest reads

$$\hat{F}_{\text{est}}(\omega) = -\frac{\delta I(\omega)}{4|\alpha_{\text{LO}}|\sqrt{\Gamma G}\chi_o^*\chi_m} = \hat{F}(\omega) + \hat{F}_{\text{ext}}(\omega), \tag{17}$$

where $\hat{F}(\omega)$ is the quantum noise introduced in this force measurement scheme and can be written as

$$\hat{F}(\omega) = c_{Y_v}\hat{Y}_v(\omega) + c_{X_{sq}}\hat{X}_{sq}(\omega) + c_{X_v}\hat{X}_v(\omega) + c_{Y_{sq}}\hat{Y}_{sq}(\omega) + \hat{F}_{th}(\omega) \tag{18}$$

with

$$c_{Y_v} = \frac{1}{4\sqrt{\Gamma G}\chi_o\chi_m}, \quad c_{X_{sq}} = c_{Y_v}, \quad c_{X_v} = -\sqrt{\frac{G}{\Gamma}}\chi_o^*, \quad c_{Y_{sq}} = -c_{X_v}. \tag{19}$$

It is apparent from Eq. (19) that the quantum fluctuations in the phase quadrature of the vacuum field are in phase with those in the amplitude quadrature of the squeezed vacuum field. Contrarily, the quantum fluctuations in the amplitude quadrature of the vacuum field are perfectly out of phase with those in the phase quadrature of the squeezed vacuum field. Additionally, we can deduce that in the weak-driving regime the quantum noise in the phase quadrature for the vacuum field and the amplitude quadrature for the squeezed vacuum field will be the dominant sources of detection noise, since c_{Yv} and c_{Xsq} are inversely proportional to the square root of the measurement strength. On the other hand, since both c_{Xv} and c_{Ysq} are proportional to the square root of the measurement strength, the quantum noises in the amplitude quadrature for the vacuum field and the phase quadrature for the squeezed vacuum field will be the dominant sources of detection noise in the strong-driving regime. Finally, the thermal fluctuations of the mechanical oscillator are independent of the measurement strength, as expected.

We are now in a position to calculate the sPSD of F^(ω) relevant to the homodyne detection

$$\bar{S}_{FF}(\omega) = \frac{1}{2}[S_{FF}(\omega) + S_{FF}(-\omega)], \tag{20}$$

where SFF(ω) is the self-PSD of F^(ω) and is defined by

$$S_{FF}(\omega) = \int_{-\infty}^{\infty} d\omega' \langle \hat{F}^\dagger(-\omega')\hat{F}(\omega)\rangle. \tag{21}$$

Since F^(ω) has the dimension of 1 $1/\sqrt{\text{Hz}}$, $\bar{S}_{FF}(\omega)$ is thus dimensionless in the unit system used in this manuscript. Note that $2\Gamma m\hbar\Omega\bar{S}_{FF}(\omega)$ with units of newtons squared per hertz should be used when comparing with experimental measurements in the laboratory. Assuming that the vacuum fluctuations, the squeezed vacuum fluctuations, and the thermal fluctuations of the mechanics have no memory effect and are uncorrelated with respect to each other, the derivation of the sPSD of F^(ω) is rather straightforward and is presented in "Methods" section. The resulting sPSD of F^(ω) can be decomposed into contributions from the vacuum, squeezed vacuum, and thermal fluctuations as

$$\bar{S}_{FF}(\omega) = \bar{S}_{\mathrm{v}}(\omega) + \bar{S}_{\mathrm{sq}}(\omega) + \bar{S}_{\mathrm{th}}(\omega), \tag{22}$$

where the contribution from the vacuum noise is

$$\bar{S}_{\mathrm{v}}(\omega) = \frac{1}{2}\left[|c_{X_{\mathrm{v}}}|^2 + |c_{Y_{\mathrm{v}}}|^2\right], \tag{23}$$

the contribution from the squeezed vacuum noise

$$\bar{S}_{\mathrm{sq}}(\omega) = \frac{1}{2}\cosh(2r)\left[|c_{X_{\mathrm{sq}}}|^2 + |c_{Y_{\mathrm{sq}}}|^2\right] + \frac{1}{2}\sinh(2r)\left[\cos\theta\left(|c_{X_{\mathrm{sq}}}|^2 - |c_{Y_{\mathrm{sq}}}|^2\right) - \sin\theta\,\mathrm{Re}[2c_{X_{\mathrm{sq}}}c^*_{Y_{\mathrm{sq}}}]\right], \tag{24}$$

and the contribution from thermal fluctuations of the mechanics

$$\bar{S}_{\mathrm{th}}(\omega) = \frac{\omega}{\Omega}\left[\bar{n}_{\mathrm{th}}(\omega) + \frac{1}{2}\right], \tag{25}$$

with $\bar{n}_{\mathrm{th}}(\omega) = [\exp(\hbar\omega/k_B T) - 1]^{-1}$ being the mean phonon occupation number in thermal equilibrium with the mechanical reservoir at finite temperature T.

Standard quantum limit

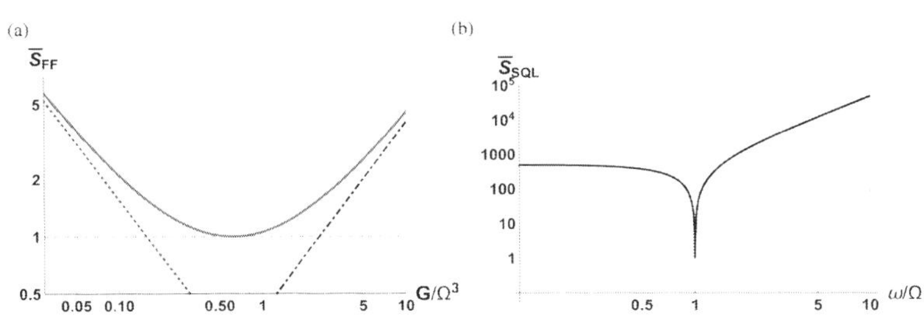

Figure 2: (**a**) sPSD of F^(ω) as a function of scaled measurement strength G/Ω^3 (red solid curve) along with the photon counting noise (blue dotted curve), radiation pressure backaction noise (black dot-dashed curve) and the standard quantum limit (gray dashed curve). (**b**) Standard quantum limit as a function of the scaled detection frequency. The parameters used are $\kappa/\Omega=10^2$, $\Gamma/\Omega=10^{-3}$.

In the case when the squeezing strength is zero, $\bar{S}_{FF}(\omega)$ is given by

$$\bar{S}_{FF}(\omega) = \frac{1}{16\Gamma G|\chi_o|^2|\chi_m|^2} + \frac{G|\chi_o|^2}{\Gamma} + \frac{\omega}{\Omega}\left[\bar{n}_{\text{th}}(\omega) + \frac{1}{2}\right]. \tag{26}$$

The first term, stemming from $Y_v(\omega)$ and $X_{sq}(\omega)$, is known as the photon counting noise. The second term is known as the radiation pressure backaction noise and derives from $X_v(\omega)$ and $Y_{sq}(\omega)$. Notice that the optical susceptibility affects the photon counting noise and the radiation pressure backaction noise in an inversed manner. Roughly speaking, a large cavity linewidth decreases the intracavity photon number, and therefore increases the photon counting noise while decreasing the radiation pressure backaction noise. Figure 2a shows the sPSD of F^(ω) as a function of the scaled measurement strength at ω=Ω. This plot depicts the common behavior that the photon counting noise is the dominant source of fluctuations at weak measurement strengths while the backaction noise becomes dominant at strong measurement strengths. For this reason, the sPSD can be minimized with respect to the measurement strength where one balances between the photon counting noise and radiation pressure backaction noise to reach the SQL

$$\bar{S}_{\text{SQL}}(\omega) = \frac{1}{2\Gamma|\chi_m|} + \frac{\omega}{2\Omega}, \tag{27}$$

with the mechanical bath at zero temperature and the measurement strength

$$G_{\text{SQL}} = \frac{1}{4|\chi_o|^2|\chi_m|}. \tag{28}$$

The first term in Eq. (27) comes from the quantum nature of the optical field and the second term results from the zero-point fluctuations of the mechanics. It is interesting to point out that the radiation pressure backaction gives rise to the dependence of the response function of the mechanics in the first term. Figure 2b shows the SQL as a function of the scaled detection frequency ω/Ω with a fixed mechanical damping rate. Since the mechanical susceptibility is maximized at the mechanical frequency, $\chi_m(\Omega)=i/\Gamma$, the sPSD for the measurement noise reaches the lower bound which is unity in the unit system used in this manuscript.

Beating the SQL with squeezed vacuum

In the case where r≠0, one can exploit the quantum correlation in the squeezed vacuum fluctuations to increase the sensitivity beyond the SQL. It is obvious from Eq. (24) that the sPSD resulting from the squeezed vacuum fluctuations has two contributions. The first term in Eq. (24) represents the photon counting noise and the radiation pressure backaction noise which increase with the squeezing strength while are independent of the direction of squeezing. The second term is associated with the quantum correlation of the squeezed vacuum fluctuations and is subject to both the direction and strength of squeezing. Therefore, the sPSD can be minimized with respect to the angle of squeezing as well as the squeezing strength. With broadband and frequency-dependent squeezing55, the optimization procedure is rather straightforward and is presented in "Methods"* section. By applying the optimal squeezing strength and squeezing angle at a given frequency, the minimized sPSD can be found as

$$\bar{S}_{FF}(\omega) = \frac{1}{32\Gamma G|\chi_o|^2|\chi_m|^2} + \frac{G|\chi_o|^2}{2\Gamma} + \frac{\omega}{4\Omega} + \frac{\omega}{\Omega}\left[\bar{n}_{\text{th}}(\omega) + \frac{1}{2}\right]. \tag{29}$$

The first and second terms introduced by the vacuum fluctuations characterize the photon counting noise and radiation pressure backaction noise, respectively. Notice that the sPSD due to the photon counting noise and radiation pressure backaction noise are halved compared to those in Eq. (26) since the photon counting and radiation pressure backaction noise in Eq. (29) result from only the vacuum fluctuations. The third term results from the squeezed vacuum fluctuations and is optimized with respect to both the squeezing strength and direction, but surprisingly depends only on the detection frequency scaled to the mechanical frequency. Finally, by balancing the photon counting and backaction noise the sPSD can reach an optimized lower bound

$$\bar{S}_{FF}^{\text{opt}}(\omega) = \frac{1}{4\Gamma|\chi_m|} + \frac{3\omega}{4\Omega} \tag{30}$$

with the mechanical bath at zero temperature, the optimum measurement strength

$$G_{\text{opt}} = G_{\text{SQL}}, \tag{31}$$

the optimum squeezing direction

$$\theta_{\text{opt}} = \frac{\pi}{2}, \tag{32}$$

and the optimum squeezing strength

$$r_{\text{opt}} = \frac{1}{2}\text{sech}^{-1}\left[\frac{\Gamma\omega|\chi_{\text{m}}|}{\Omega}\right]. \tag{33}$$

Notice that the optimum measurement strength is the same as that for the SQL in that the optimized photon counting and radiation backaction noise scale in the same manner as that for the SQL. In addition, θ_{opt} is independent of the physical parameters of the optomechanical system as well as the detection frequency. This underlies the fact that the optimized squeezing is along the π/4 direction in phase space so as to lead to a maximum negative correlation between quantum fluctuations in the amplitude and phase quadratures13. In contrast to the squeezing direction, roptropt relies on mechanical parameters such as the quality factor of the mechanical oscillator and the detection frequency.

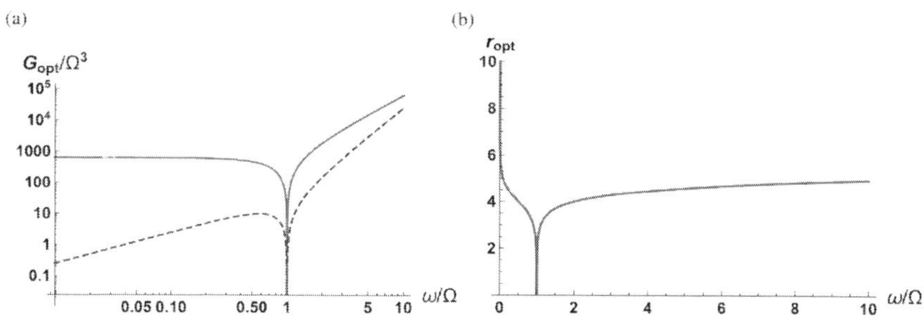

Figure 3: (**a**) Optimal measurement strength as a function of scaled detection frequency ω/Ω for the case where the output field interferes with a classical LO field (red solid curve) and a MZI reference field (blue dashed curve). (**b**) Optimized squeezing strength as a function of the scaled detection frequency. The parameters used are $\kappa/\Omega=10^2$ and $\Gamma/\Omega=10^{-3}$.

Figure 3a,b show the scaled optimal measurement strength and the optimal squeezing strength as a function of scaled detection frequency ω/Ω with a fixed cavity linewidth and mechanical damping rate (red solid curve), respectively. When the detection frequency is the same as the mechanical frequency, the mechanical susceptibility is maximized and both the optimum measurement strength and squeezing strength are minimized. In this on-resonance case, the optimal squeezing strength is zero, indicating that the squeezed vacuum field can not assist in beating the SQL for enhanced sensitivity. As the detection frequency moves off resonant from the mechanical frequency, the mechanical susceptibility decreases so that G_{opt} and r_{opt} increase in a monotonic fashion. It should be kept in mind that force measurements of dc fields or frequencies much larger than other system parameters are out of scope in this manuscript since the optimum squeezing strength approaches infinity. This condition breaks the essential assumption for the linearization of the optomechanical interaction, i.e., the amplitudes of the quantum fluctuations are much smaller than the relevant classical expectation values4·5.

Figure 4a shows the optimum sPSD along with the SQL as a function of the scaled detection frequency, displaying that the optimum sPSD is always lower than the SQL except on resonance. Figure 4b presents the optimum sPSD normalized with the SQL, indicating that $\bar{S}_{FF}^{opt}(\omega)$ is one half of the SQL when the detection frequency is far off-resonance. In the case of far off-resonance, the contribution from the squeezed vacuum fluctuations can be considered to be negligible leading to the vacuum field contribution in Eq. (30) to be halved compared to that in Eq. (27).

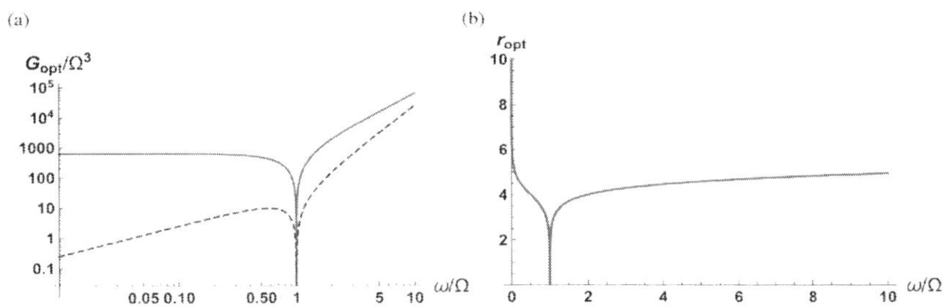

Figure 4: (**a**) Optimum sPSD of F^(ω) as a function of scaled detection frequency ω/Ω (red solid curve) along with the SQL (black dashed curve). (**b**) Optimum sPSD of F^(ω) normalized to the SQL as a function of the scaled detection frequency. The parameters used are $\Gamma/\Omega=10^{-3}$.

Balanced Mach–Zehnder interferometer

In this section, we present the sPSD corresponding to the quantum noise introduced in the measurement process described in Fig. 1a. In this scheme the coherent light experiences a reflection at BS1 where the quantum fluctuations in the reference field can be written as $\hat{d}_{\rm ref} = (i\hat{d}_{\rm v} + \hat{d}_{\rm sq})/\sqrt{2}$. Subsequently, the output field from the optomechanical cavity is coherently mixed with the reference field at BS2. The optical fields $\hat{a}_+$, $\hat{a}_-$ at the two output ports of BS2 can be written as

$$\begin{pmatrix} \hat{a}_+ \\ \hat{a}_- \end{pmatrix} = \frac{1}{\sqrt{2}} \begin{pmatrix} 1 & i \\ i & 1 \end{pmatrix} \begin{pmatrix} \hat{a}_{\text{out}} \\ i\hat{a}_{\text{ref}} \end{pmatrix}, \tag{34}$$

where the additional *i* in front of the field operator for the reference field accounts for a reflection at BS1. The quantum fluctuations of the balanced signal between the two output ports $\hat{I}$ can be approximated as

$$\delta\hat{I} = \sqrt{2}\alpha_{\text{in}}\left(\hat{Y}_{\text{out}} - \hat{X}_{\text{ref}}\right), \tag{35}$$

where we have used the fact that the reference field is ahead of the input field in relative phase by 90°, $\alpha_{ref}=i\alpha_{in}$, and the output field is 180° out of phase relative to the input field, $\alpha_{\text{out}} = -\alpha_{\text{in}}$. $\hat{X}_{\text{ref}} = (-\hat{Y}_{\text{v}} + \hat{X}_{\text{sq}})/\sqrt{2}$ represents the quantum fluctuations in the amplitude quadrature of the reference field. Notice that the quantum fluctuations in the phase quadrature of the output field interfere with those in the amplitude quadrature of the reference field destructively. Owing to the fact that the term for optical shot noise in the phase quadrature of the output field is given by $-\frac{1}{\sqrt{2}}\frac{\chi_o^*}{\chi_o}\left(\hat{Y}_{\text{v}} + \hat{X}_{\text{sq}}\right)$ from Eq. (13), the resulting optical shot noise in $\delta\hat{I}$ can be described by $-\alpha_{\text{in}}\chi_o^*\left(2i\omega\hat{Y}_{\text{v}} + \kappa\hat{X}_{\text{sq}}\right)$. Overall, the unbiased estimator for the external force of interest can be obtained as

$$\hat{F}_{\text{est}}(\omega) = -\frac{\delta I(\omega)}{4\alpha_{\text{in}}\sqrt{\Gamma G}\chi_o^*\chi_m} \equiv \hat{F}(\omega) + \hat{F}_{\text{ext}}(\omega), \tag{36}$$

where $\hat{F}(\omega)$ is the quantum noise concomitant in this force measurement scheme and can be written in the form of Eq. (18) with the relevant coefficients

$$c_{Y_{\text{v}}} = \frac{i\omega}{2\sqrt{\Gamma G}\chi_m}, \quad c_{X_{\text{sq}}} = \frac{\kappa}{2i\omega}c_{Y_{\text{v}}}, \quad c_{X_{\text{v}}} = -\sqrt{\frac{G}{\Gamma}}\chi_o^*, \quad c_{Y_{\text{sq}}} = -c_{X_{\text{v}}}. \tag{37}$$

While $c_{Yv}=c_{Xsq}$ in Eq. (19), the destructive interference between $\hat{Y}_{out}(\omega)$ and $\hat{X}_{ref}(\omega)$ at BS2 results in a different scaling between $Y_v(\omega)$ and $X_{sq}(\omega)$. Keep in mind that the contribution from the squeezed vacuum fluctuations manifests itself in the regime where the cavity linewidth is much larger than the detection frequency, therefore, we limit ourselves to the regime where $\omega \ll \kappa$. Comparing this balanced MZI based scheme with the conventional balanced homodyne scheme, we observe that the optical shot noise now displays quantum interference between the cavity output field and the reference field. On the other hand, the coefficients associated with the fluctuations in the amplitude quadrature of the vacuum field and the phase quadrature of the squeezed vacuum field are the same for the two schemes, so that the contribution from the radiation pressure backaction noise and the thermal noise remain the same. In other words, the quantum interference between the cavity output field and the reference field alters only the photon counting noise which differentiates the two schemes.

Before exploiting the quantum correlation in the squeezed vacuum field, it is helpful to consider the case where r=0 for completeness. In this case, the sPSD of F^(ω) is given by

$$\bar{S}_{FF}(\omega) = \frac{1}{8\Gamma G|\chi_o|^2|\chi_m|^2} + \frac{G|\chi_o|^2}{\Gamma} + \frac{\omega}{\Omega}\left[\bar{n}_{\text{th}}(\omega) + \frac{1}{2}\right]. \quad (38)$$

As expected, the radiation pressure backaction noise contribution and thermal noise contribution are the same as those in Eq. (26). However, the photon counting noise is doubled compared to that in Eq. (26) due to the presence of optical shot noise in the reference field. This result entails that we should not expect any additional benefits by exploiting quantum interference in the case of zero squeezing strength.

Beating the SQL based on squeezed vacuum and quantum interference

We now turn to the case where r≠0 to explore the effects of quantum interference and quantum correlation in the squeezed vacuum field on the detected quantum noise. As mentioned before, minimizing the sPSD of F^(ω) with respect to the squeezing strength and direction leads to

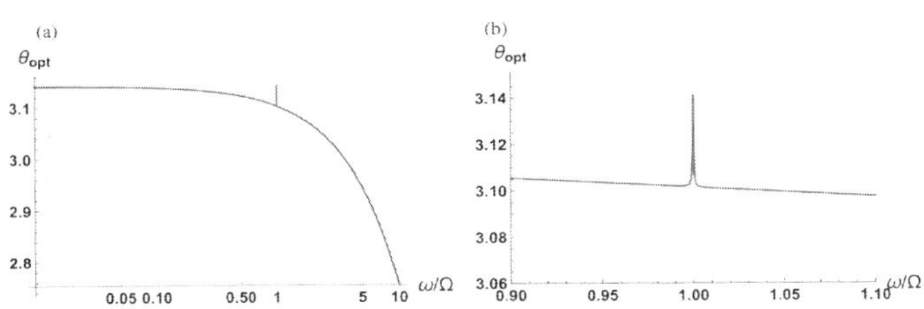

Figure 5: Optimum squeezing direction as a function of the scaled detection frequency at the optimum measurement strength (**a**) across the low frequency regime and (**b**) enlarged in the vicinity of the mechanical resonance. The parameters used are $\kappa/\Omega=10_2$, $\Gamma/\Omega=10^{-3}$.

$$\bar{S}_{FF}(\omega)=\frac{\omega^2}{8\Gamma G|\chi_m|^2}+\frac{G|\chi_o|^2}{2\Gamma}+\frac{\omega\kappa|\chi_o|^2}{4\Omega\Gamma}\left|\omega^2-\Omega^2-\frac{\Gamma\kappa}{2}\right|+\frac{\omega}{\Omega}\left[\bar{n}_{\text{th}}(\omega)+\frac{1}{2}\right]. \tag{39}$$

It is instructive to discuss the key features following the quantum interference in Eq. (39), which we note is absent in Eq. (29). First, notice that while the radiation pressure backaction noise and the thermal noise are the same, the photon counting noise is different when compared to Eq. (29). A large cavity linewidth compared to the detection frequency leads to increased photon counting noise and decreased radiation pressure backaction noise in Eq. (29). However, in Eq. (39), a relatively large cavity linewidth decreases the radiation pressure backaction noise but does not alter the photon counting noise. Second, the term due to the squeezed vacuum fluctuations, i.e., third term in Eq. (39), is minimized at frequency

$$\Omega_s=\sqrt{\Omega^2+\frac{\Gamma\kappa}{2}}, \tag{40}$$

whereas in Eq. (29), this term is independent of the optical cavity parameters. As before, balancing between the photon counting noise and radiation pressure backaction noise can suppress the sPSD of F^(ω) to reach a lower bound

$$\bar{S}^{\text{opt}}_{FF}(\omega)=\frac{\omega|\chi_o|}{2\Gamma|\chi_m|}+\frac{\omega\kappa|\chi_o|^2}{4\Omega\Gamma}\left|\omega^2-\Omega^2-\frac{\Gamma\kappa}{2}\right|+\frac{\omega}{2\Omega}, \tag{41}$$

with the mechanical bath at zero temperature, the optimum measurement strength

$$G_{\text{opt}}=\frac{\omega}{2|\chi_o||\chi_m|}, \tag{42}$$

the optimum squeezing direction

$$\theta_{\text{opt}}=\cos^{-1}\left[\frac{\omega^2-\kappa^2/4}{\left|\omega^2\frac{\chi_o^*(\omega)}{\chi_o(\omega)}+\frac{\kappa^2}{4}\frac{\chi_m^*(\omega)}{\chi_m(\omega)}\right|}\right], \tag{43}$$

and the optimum squeezing strength

$$r_{\text{opt}}=\frac{1}{2}\text{sech}^{-1}\left[\frac{\omega^2\kappa|\chi_o|^3|\chi_m|\left|\omega^2-\Omega^2-\frac{\Gamma\kappa}{2}\right|}{\Omega}\right]. \tag{44}$$

In this scenario, as shown in Fig. 3a, the optimal measurement strength is lower than that for the scheme using a classical LO, indicating that the optimum sPSD will be obtained with weaker input powers. Additionally, the optimal squeezing direction relies on both the detection frequency scaled to the cavity linewidth and the phase difference between the optical and mechanical susceptibilities. The optimal squeezing direction is displayed in Fig. 5 as a function of the scaled detection frequency in the low-frequency regime ($\omega \ll \kappa$). The optimum squeezing angle monotonically decreases from π with the detection frequency, except in the vicinity of the mechanical resonance. At resonance the optimum squeezing angle peaks to a value close to $\pi\pi$. In this case, the phase quadrature operator of the squeezed vacuum field is squeezed.

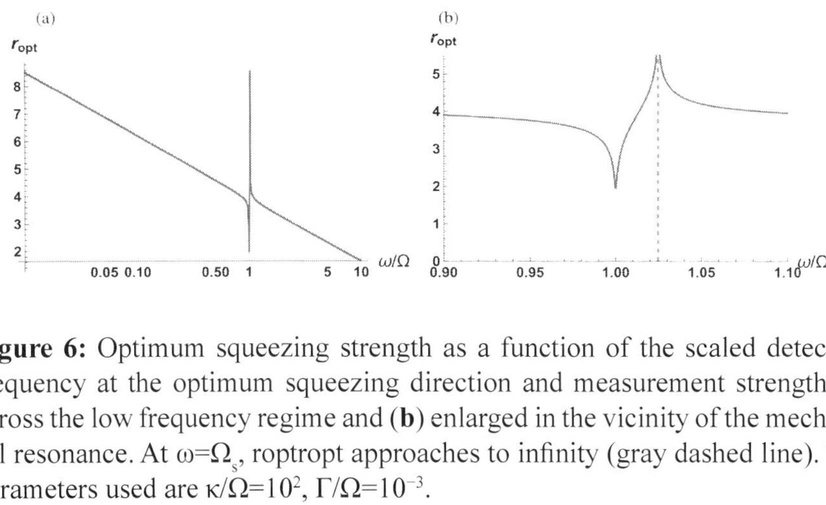

Figure 6: Optimum squeezing strength as a function of the scaled detection frequency at the optimum squeezing direction and measurement strength (**a**) across the low frequency regime and (**b**) enlarged in the vicinity of the mechanical resonance. At $\omega=\Omega_s$, roptropt approaches to infinity (gray dashed line). The parameters used are $\kappa/\Omega=10^2$, $\Gamma/\Omega=10^{-3}$.

Figure 6a,b show the optimal squeezing strength as a function of the scaled detection frequency in the low-frequency regime. Figure 6b clearly depicts the locally minimized optimal squeezing strength at $\omega=\Omega$, and that which approaches infinity at Ω_s (gray dashed line). Even though the squeezing strength for light in the laboratory is rapidly being advanced56, we consider cases when we want to measure external forces at frequencies near Ω_s to be outside the scope of this manuscript in addition to the extreme cases mentioned before, i.e., dc field external forces and forces oscillating at relatively high frequencies with respect to system parameters. When the detection frequency is close to Ω_s the amplitude of the quantum fluctuations is much larger than the relevant classical expectation values violating our initial assumptions.

Figure 7a shows the optimum sPSD of F^(ω) as a function of the scaled detection frequency for fixed optical and mechanical linewidths. The SQL is marked by a black dashed curve, and Fig. 7b illustrates a more detailed plot in the vicinity of the mechanical resonance with added curves representing contributions from vacuum fluctuations (orange solid curve) and squeezed vacuum fluctuations (blue dashed curve). When ω=Ω, the optimum sPSD becomes

$$\bar{S}_{FF}^{\text{opt}}(\Omega) = \frac{\Omega/2}{\sqrt{\Omega^2 + \kappa^2/4}} + \frac{\kappa^2/8}{\Omega^2 + \kappa^2/4} + \frac{1}{2}. \tag{45}$$

In the regime whereΩ≪κ, the contribution from the squeezed vacuum fluctuations becomes dominant, resulting in $\bar{S}_{FF}^{\text{opt}}(\Omega) \approx 1 + \Omega/\kappa$. Therefore, no added benefits will be observed by utilizing the quantum correlations of the squeezed vacuum with quantum interference when on resonance. However, for the case of off-resonance, a significant suppression of the noise can be observed when utilizing squeezed vacuum fluctuations in the balanced MZI scheme.

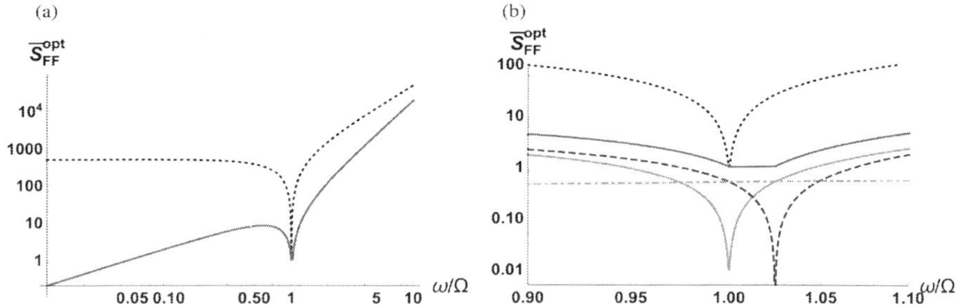

Figure 7: Optimum sPSD of F^(ω) as a function of scaled detection frequency ω/Ω (red solid curve) along with the SQL (black dashed curve) (**a**) across the low frequency regime and (**b**) enlarged in the vicinity of the mechanical resonance. The contribution from the vacuum fluctuations (orange solid curve), the squeezed vacuum fluctuations (blue dashed curve), and the thermal fluctuations (green dot-dashed curve). The parameters used are $\kappa/\Omega=10^2$ and $\Gamma/\Omega=10^{-3}$.

It is also found that the optimum sPSD of F^(ω) is flat across a frequency window of $\Omega<\omega<\Omega_s$ sustaining its minimum SQL value. This is well illustrated in Fig. 7b where the contribution from the vacuum fluctuations (orange solid curve) increases with the detection frequency on the blue side of the mechanical resonance, while the contribution from the squeezed vacuum fluctuations (blue dashed curve) decreases with detection frequency until reaching Ω_s. Indeed the two competing contributions are generally not exactly symmetric within this frequency window implying that the sPSD might not be completely flat with zero slope. However, when the contribution from thermal fluctuations (green dot-dashed curve) is added, the sPSD becomes perfectly flat at a constant minimum value. In practice, although the observation of the flat sPSD near Ωs is difficult even with state-of-the-art technology, the flat sPSD near Ω can be observed56. Note that the width of the flat sPSD window is determined by the product of the optical and mechanical linewidths, see Eq. (40).

Finally, in order to highlight the benefit of using the balanced MZI scheme, we present Fig. 8 showing the optimum sPSD normalized to the SQL. Recall that in the case where we use a classical LO field, the sPSD can be optimized to half of the SQL when off-resonance. On the other hand, as shown in Fig. 8, the sPSD can be optimized to be suppressed by multiple orders of magnitude below the SQL when off-resonance. In the later case, destructive interference engineers the photon counting noise in the balanced detection signal where its effect manifests itself in the regime when the detection frequency is much smaller than the cavity linewidth.

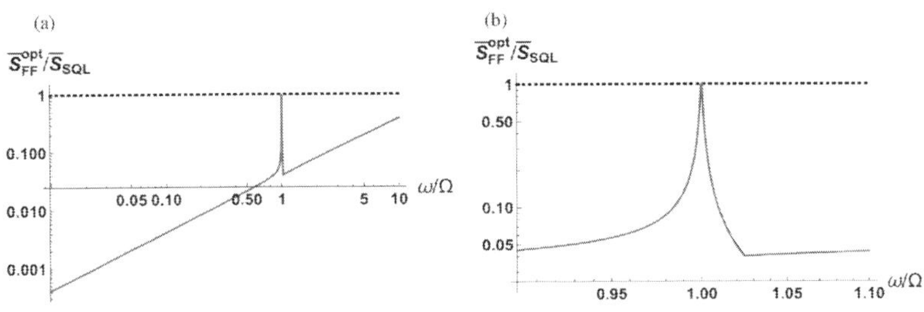

Figure 8: Optimum sPSD of F^(ω) normalized with the SQL as a function of scaled detection frequency ω/Ω (**a**) across the low frequency regime and (**b**) enlarged in the vicinity of the mechanical resonance. The parameters used are $\kappa/\Omega=10^2$ and $\Gamma/\Omega=10^{-3}$.

CONCLUSIONS

In summary, we have explored the quantum noise introduced to the force measurements based on a balanced MZI and standard optomechanical cavity. We have employed the coherent superposition of coherent light and squeezed vacuum field to enhance the measurement sensitivity beyond the SQL. The reference field destructively interferes with the output field of the optomechanical cavity so that the photon counting noise is suppressed in the low frequency regime. We analytically find the input parameters for which the sPSD of the detected noise is minimized below the SQL. The optimal parameters include the measurement strength, squeezing direction, and squeezing strength. The force detection scheme based on a balanced MZI introduced in this manuscript shows better sensitivity compared to that based on a balanced homodyne detection with a local oscillator field in the low frequency regime. The sensitivity of the optomechanical sensor introduced in this study is enhanced by quantum correlations in squeezed states of light as well as coherent quantum-noise cancellation developed in QND measurements. These results show the potential gains in sensitivity of MZI based optomechanical quantum sensors for a variety of weak forces and fields such as Casimir force, gravitational waves, and magnetic fields.

METHODS

Symmetrized power spectral density

This section presents the details of the procedure for obtaining Eq. (22). In order to calculate the self-PSD of the quantum noise introduced in the force measurement scheme, one needs the correlation functions for the vacuum field, the squeezed vacuum field, and the thermal fluctuations of the mechanics in the frequency domain. We first begin with the two-time correlation functions in the rotating frame at frequency ωL for the squeezed vacuum field of strength of the squeezing r and squeezing direction θ57

$$\langle \hat{d}_{sq}(t)\hat{d}_{sq}(t')\rangle = M^*\delta(t-t'), \tag{46a}$$

$$\langle \hat{d}^\dagger_{sq}(t)\hat{d}^\dagger_{sq}(t')\rangle = M\delta(t-t'), \tag{46b}$$

$$\langle \hat{d}^\dagger_{sq}(t)\hat{d}_{sq}(t')\rangle = N\delta(t-t'), \tag{46c}$$

$$\langle \hat{d}_{sq}(t)\hat{d}^\dagger_{sq}(t')\rangle = (N+1)\delta(t-t'), \tag{46d}$$

where $M = \cosh(r)\sinh(r)e^{i\theta}, N = \sinh^2(r)$. Transforming the two-time correlation functions into the frequency domain by Eq. (6), we have

$$\langle \hat{d}_{sq}(\omega)\hat{d}_{sq}(\omega')\rangle = M^*\delta(\omega+\omega'), \tag{47a}$$

$$\langle \hat{d}^\dagger_{sq}(\omega)\hat{d}^\dagger_{sq}(\omega')\rangle = M\delta(\omega+\omega'), \tag{47b}$$

$$\langle \hat{d}^\dagger_{sq}(\omega)\hat{d}_{sq}(\omega')\rangle = N\delta(\omega-\omega'), \tag{47c}$$

$$\langle \hat{d}_{sq}(\omega)\hat{d}^{\dagger}_{sq}(\omega')\rangle = (N+1)\delta(\omega-\omega'). \tag{47d}$$

Interchanging Ω and ω′ and using the even parity of the delta-function, we have the correlation functions for quantum fluctuations in the quadrature operators for the field, $\hat{X}_{sq}(\omega) = \frac{1}{\sqrt{2}}(\hat{d}^{\dagger}_{sq}(-\omega) + \hat{d}_{sq}(\omega))$ and $\hat{Y}_{sq}(\omega) = \frac{i}{\sqrt{2}}(\hat{d}^{\dagger}_{sq}(-\omega) - \hat{d}_{sq}(\omega))$, in the frequency domain as

$$\langle \hat{X}^{\dagger}_{sq}(-\omega')\hat{X}_{sq}(\omega)\rangle = \frac{1}{2}[\cosh(2r) + \sinh(2r)\cos\theta]\delta(\omega+\omega'), \tag{48a}$$

$$\langle \hat{X}^{\dagger}_{sq}(-\omega')\hat{Y}_{sq}(\omega)\rangle = \frac{1}{2}[i - \sinh(2r)\sin\theta]\delta(\omega+\omega'), \tag{48b}$$

$$\langle \hat{Y}^{\dagger}_{sq}(-\omega')\hat{X}_{sq}(\omega)\rangle = \frac{1}{2}[-i - \sinh(2r)\sin\theta]\delta(\omega+\omega'), \tag{48c}$$

$$\langle \hat{Y}^{\dagger}_{sq}(-\omega')\hat{Y}_{sq}(\omega)\rangle = \frac{1}{2}[\cosh(2r) - \sinh(2r)\cos\theta]\delta(\omega+\omega'). \tag{48d}$$

Therefore, the associated PSDs defined by $S_{OO'}(\omega) = \int_{-\infty}^{\infty} d\omega' \langle \hat{O}^{\dagger}(-\omega')\hat{O}'(\omega)\rangle$ for the squeezed vacuum field can be obtained as

$$S_{X_{sq}X_{sq}}(\omega) = \frac{1}{2}[\cosh(2r) + \sinh(2r)\cos\theta], \tag{49a}$$

$$S_{Y_{sq}Y_{sq}}(\omega) = \frac{1}{2}[\cosh(2r) - \sinh(2r)\cos\theta], \tag{49b}$$

$$S_{X_{sq}Y_{sq}}(\omega) = \frac{1}{2}[i - \sinh(2r)\sin\theta], \tag{49c}$$

$$S_{Y_{sq}X_{sq}}(\omega) = \frac{1}{2}[-i - \sinh(2r)\sin\theta]. \tag{49d}$$

where $S_{X_{sq}X_{sq}}$ and $S_{Y_{sq}Y_{sq}}$ are the self-PSDs of the amplitude and phase quadrature operators, respectively. $S_{X_{sq}X_{sq}}$ and $S_{Y_{sq}Y_{sq}}$ are the cross-PSDs between the two quadrature operators. As expected, the PSDs are independent of frequency since the associated two-time correlation functions are assumed to be delta-correlated. Furthermore, one can find the PSD of the vacuum field if the squeezing strength is zero and thus the associated PSDs for the vacuum field read

$$S_{X_vX_v}(\omega) = S_{Y_vY_v}(\omega) = \frac{1}{2}, \quad S_{X_vY_v}(\omega) = -S_{Y_vX_v}(\omega) = \frac{i}{2}. \tag{50}$$

Finally, if the two-time correlation function for F^th is58

$$\langle \hat{F}_{\text{th}}(t)\hat{F}_{\text{th}}(t') \rangle = \frac{1}{4\pi\Omega}\int_{-\infty}^{\infty} d\omega \, e^{-\omega(t-t')}\omega[1+\bar{n}_{\text{th}}(\omega)], \tag{51}$$

where n¯th(ω) is the mean phonon occupation number in thermal equilibrium, one can easily find the PSD of Fth in a similar way presented above. The PSDs of the thermal noise read

$$S_{F_{\text{th}}F_{\text{th}}}(-\omega) = \frac{\omega}{\Omega}\bar{n}_{\text{th}}(\omega), \quad S_{F_{\text{th}}F_{\text{th}}}(\omega) = \frac{\omega}{\Omega}[\bar{n}_{\text{th}}(\omega)+1]. \tag{52}$$

When F^(ω) is written as

$$\hat{F}(\omega) = c_{Y_v}\hat{Y}_v(\omega) + c_{X_{sq}}\hat{X}_{sq}(\omega) + c_{X_v}\hat{X}_v(\omega) + c_{Y_{sq}}\hat{Y}_{sq}(\omega) + \hat{F}_{\text{th}}(\omega), \tag{53}$$

The self-PSD of F^(ω) can be obtained as

$$\begin{aligned} S_{FF}(\omega) = {} & |c_{X_v}|^2 S_{X_vX_v}(\omega) + |c_{Y_v}|^2 S_{Y_vY_v}(\omega) + c^*_{Y_v}c_{X_v}S_{Y_vX_v}(\omega) + c^*_{X_v}c_{Y_v}S_{X_vY_v}(\omega) \\ & + |c_{X_{sq}}|^2 S_{X_{sq}X_{sq}}(\omega) + |c_{Y_{sq}}|^2 S_{Y_{sq}Y_{sq}}(\omega) + c^*_{X_{sq}}c_{Y_{sq}}S_{X_{sq}Y_{sq}}(\omega) + c^*_{Y_{sq}}c_{X_{sq}}S_{Y_{sq}X_{sq}}(\omega) + S_{F_{\text{th}}F_{\text{th}}}(\omega), \end{aligned} \tag{54}$$

where we have assumed that the vacuum, the squeezed vacuum, and thermal fluctuations are uncorrelated to each other and have used Eqs. (49a), (49b), (49c), (49d), (50), and (52). The sPSD relevant to the homodyne detection reads

$$\begin{aligned} \bar{S}_{FF}(\omega) = {} & \frac{1}{2}\left[|c_{X_v}|^2 + |c_{Y_v}|^2\right] + \frac{1}{2}\cosh(2r)\left[|c_{X_{sq}}|^2 + |c_{Y_{sq}}|^2\right] \\ & + \frac{1}{2}\sinh(2r)\left[\cos\theta(|c_{X_{sq}}|^2 - |c_{Y_{sq}}|^2) - \sin\theta \,\text{Re}[2c_{X_{sq}}c^*_{Y_{sq}}]\right] \\ & + \frac{\omega}{\Omega}\left[\bar{n}_{\text{th}}(\omega) + \frac{1}{2}\right]. \end{aligned} \tag{55}$$

Optimization of $\bar{S}_{sq}(\omega)$

It should be kept in mind for Eq. (24) that the sPSD stemming from the squeezed vacuum $\bar{S}_{sq}(\omega)$ relies on the measurement strength, squeezing strength as well as direction. Since $\bar{S}_{sq}(\omega)$ has a global minimum with respect to each of the independent parameters (G, r, and θ), the order of optimization procedure does not matter. However, the degree of complication regarding the calculation depends significantly on the order. We choose to first minimize $\bar{S}_{sq}(\omega)$ with respect to the angle of squeezing. The terms in the bracket in the second term can be minimized with respect to the angle of squeezing as

$$\left[\cos\theta(|c_{X_{sq}}|^2 - |c_{Y_{sq}}|^2) - \sin\theta \mathrm{Re}[2c_{X_{sq}} c^*_{Y_{sq}}]\right]_{\min} = -\left|c^2_{X_{sq}} + c^2_{Y_{sq}}\right|, \tag{56}$$

and the optimum direction of squeezing is

$$\theta_{\mathrm{opt}} = \cos^{-1}\left[\frac{-|c_{X_{sq}}|^2 + |c_{Y_{sq}}|^2}{\left|c^2_{X_{sq}} + c^2_{Y_{sq}}\right|}\right]. \tag{57}$$

Here, we have used the fact that

$$a\cos\theta - b\sin\theta \geq -\sqrt{a^2 + b^2}, \tag{58}$$

where

$$a = |c_{X_{sq}}|^2 - |c_{Y_{sq}}|^2, \quad b = \mathrm{Re}[2c_{X_{sq}} c^*_{Y_{sq}}]. \tag{59}$$

The resulting sPSD due to the squeezed vacuum fluctuations at the optimum direction of squeezing depends on the squeezing strength and the measurement strength and is thus given by

$$\bar{S}_{sq}(\omega)\Big|_{\theta_{\mathrm{opt}}} = \frac{1}{2}\cosh(2r)\left[|c_{X_{sq}}|^2 + |c_{Y_{sq}}|^2\right] - \frac{1}{2}\sinh(2r)\left|c^2_{X_{sq}} + c^2_{Y_{sq}}\right|, \tag{60}$$

where it follows from the minus sign of the second term in Eq. (60) that the contribution from the quantum correlation behaves in the opposite way to those of the photon counting and radiation pressure backaction noise. Therefore, the sPSD owing to the squeezed vacuum fluctuations can further be minimized with respect to the squeezing strength as

$$\bar{S}_{sq}(\omega)\Big|_{r_{\mathrm{opt}},\theta_{\mathrm{opt}}} = \left|\mathrm{Im}[c_{X_{sq}} c^*_{Y_{sq}}]\right| \tag{61}$$

at the optimum squeezing strength

$$r_{\mathrm{opt}} = \frac{1}{2}\cosh^{-1}\left[\frac{|c_{X_{sq}}|^2 + |c_{Y_{sq}}|^2}{2\left|\mathrm{Im}[c_{X_{sq}} c^*_{Y_{sq}}]\right|}\right]. \tag{62}$$

Here, we have used the fact that the inequality

$$a'\cosh(2r) - b'\sinh(2r) \geq \sqrt{a'^2 - b'^2}, \tag{63}$$

holds if $a' > b' \geq 0$, where

$$a' = \frac{1}{2}\left[|c_{X_{sq}}|^2 + |c_{Y_{sq}}|^2\right], \quad b' = \frac{1}{2}\left|c^2_{Y_{v}} + c^2_{Y_{sq}}\right|. \tag{64}$$

and the condition a'>b′ is guaranteed by the triangle inequality. Notice that the sPSD from the squeezed vacuum fluctuations at r_{opt} and θ_{opt} is independent of the measurement strength since the fluctuations in the amplitude and phase quadrature of the squeezed vacuum depends on the measurement strength in the opposite way, see Eqs. (19) and (37). The resulting sPSD can therefore be written as

$$\bar{S}_{FF}(\omega) = \frac{1}{2}\left[|c_{X_v}|^2 + |c_{Y_v}|^2\right] + \left|\mathrm{Im}[c_{X_{sq}} c^*_{Y_{sq}}]\right| + \frac{\omega}{\Omega}\left[\bar{n}_{\mathrm{th}}(\Omega) + \frac{1}{2}\right]. \tag{65}$$

ACKNOWLEDGEMENTS

This work is supported by the Institute of Information and communications Technology Planning and Evaluation (IITP) grant funded by the Korea Government (MSIT) (no. 2019-0-00720). C.-W. L. acknowledges the financial support of National Research Foundation of Korea (Grant no. NRF-2017R1D1A1B04032142). H. S. acknowledges the National Research Foundation of Korea grant funded by the Korea government (MSIT) (Grant no. NRF-2019R1F1A1042286).

CONTRIBUTIONS

C.-W.L. and H.S. conceived the study. J.H.L. and H.S. performed the theoretical analysis and numerical calculations. All authors wrote and reviewed the manuscript.

REFERENCES

1. Clerk, A. A., Devoret, M. H., Girvin, S. M., Marquardt, F. & Schoelkopf, R. J. Introduction to quantum noise, measurement, and amplification. *Rev. Mod. Phys.* **82**, 1155–1208. https://doi.org/10.1103/RevModPhys.82.1155 (2010).
2. Braginksy, V. B. & Manukin, A. B. Ponderomotive effects of electromagnetic radiation. *Sov. Phys. JETP.* **25**, 653 (1967).
3. Braginksy, V. B., Manukin, A. B. & Tikhonov, M. Y. Investigation of dissipative ponderomotive effects of electromagnetic radiation. *Sov. Phys. JETP.* **31**, 829 (1970).
4. Aspelmeyer, M., Kippenberg, T. J. & Marquardt, F. Cavity optomechanics. *Rev. Mod. Phys.* **86**, 1391–1452. https://doi.org/10.1103/RevModPhys.86.1391 (2014).
5. Meystre, P. A short walk through quantum optomechanics. *Ann. Phys.* **525**, 215–233. https://doi.org/10.1002/andp.201200226 (2013). https://onlinelibrary.wiley.com/doi/pdf/10.1002/andp.201200226.
6. Caves, C. M., Thorne, K. S., Drever, R. W. P., Sandberg, V. D. & Zimmermann, M. On the measurement of a weak classical force coupled to a quantum–mechanical oscillator. I. Issues of principle. *Rev. Mod. Phys.* **52**, 341–392. https://doi.org/10.1103/RevModPhys.52.341 (1980).
7. Thorne, K. S. Gravitational-wave research: Current status and future prospects. *Rev. Mod. Phys.* **52**, 285–297. https://doi.org/10.1103/RevModPhys.52.285 (1980).
8. Caves, C. M. Quantum-mechanical radiation-pressure fluctuations in an interferometer. *Phys. Rev. Lett.* **45**, 75–79. https://doi.org/10.1103/PhysRevLett.45.75 (1980).
9. Braginksy, V. B. & Khalili, F. Y. *Quantum Measurement* (Cambridge University Press, Cambridge, 1992).
10. Loudon, R. & Knight, P. Squeezed light. *J. Mod. Opt.* **34**, 709–759. https://doi.org/10.1080/09500348714550721 (1987).
11. Drummond, P. D. & Ficek, Z. *Quantum Squeezing* (Springer, New York, 2013).
12. Caves, C. M. Quantum-mechanical noise in an interferometer. *Phys. Rev. D* **23**, 1693–1708. https://doi.org/10.1103/PhysRevD.23.1693 (1981).

13. Yuen, H. P. Contractive states and the standard quantum limit for monitoring free-mass positions. *Phys. Rev. Lett.* **51**, 719–722. https://doi.org/10.1103/PhysRevLett.51.719 (1983).
14. Unruh, W. G. Quantum noise in the interferometer detector. In *Quantum Optics, Experimental Gravitation and Measurement Theory* (eds Meystre, P. & Scully, M. O.) 647 (Plenum, New York, 1983).
15. Bondurant, R. S. & Shapiro, J. H. Squeezed states in phase-sensing interferometers. *Phys. Rev. D* **30**, 2548–2556. https://doi.org/10.1103/PhysRevD.30.2548 (1984).
16. Luis, A. & Sánchez-Soto, L. L. Multimode quantum analysis of an interferometer with moving mirrors. *Phys. Rev. A* **45**, 8228–8234. https://doi.org/10.1103/PhysRevA.45.8228 (1992).
17. Pace, A. F., Collett, M. J. & Walls, D. F. Quantum limits in interferometric detection of gravitational radiation. *Phys. Rev. A* **47**, 3173–3189. https://doi.org/10.1103/PhysRevA.47.3173 (1993).
18. Vyatchanin, S. & Zubova, E. Quantum variation measurement of a force. *Phys. Lett. A* **201**, 269–274. https://doi.org/10.1016/0375-9601(95)00280-G (1995).
19. Kimble, H. J., Levin, Y., Matsko, A. B., Thorne, K. S. & Vyatchanin, S. P. Conversion of conventional gravitational-wave interferometers into quantum nondemolition interferometers by modifying their input and/or output optics. *Phys. Rev. D* **65**, 022002. https://doi.org/10.1103/PhysRevD.65.022002 (2001).
20. Khalili, F. Y. Quantum variational measurement in the next generation gravitational-wave detectors. *Phys. Rev. D* **76**, 102002. https://doi.org/10.1103/PhysRevD.76.102002 (2007).
21. Xiao, M., Wu, L.-A. & Kimble, H. J. Precision measurement beyond the shot-noise limit. *Phys. Rev. Lett.* **59**, 278–281. https://doi.org/10.1103/PhysRevLett.59.278 (1987).
22. Grangier, P., Slusher, R. E., Yurke, B. & LaPorta, A. Squeezed-light-enhanced polarization interferometer. *Phys. Rev. Lett.* **59**, 2153–2156. https://doi.org/10.1103/PhysRevLett.59.2153 (1987).
23. Eberle, T. *et al.* Quantum enhancement of the zero-area sagnac interferometer topology for gravitational wave detection. *Phys. Rev. Lett.* **104**, 251102. https://doi.org/10.1103/PhysRevLett.104.251102 (2010).

24. McKenzie, K., Shaddock, D. A., McClelland, D. E., Buchler, B. C. & Lam, P. K. Experimental demonstration of a squeezing-enhanced power-recycled michelson interferometer for gravitational wave detection. *Phys. Rev. Lett.* **88**, 231102. https://doi.org/10.1103/PhysRevLett.88.231102 (2002).
25. Aasi, J. *et al.* Enhanced sensitivity of the ligo gravitational wave detector by using squeezed states of light. *Nat. Photon.* **7**, 613–619. https://doi.org/10.1038/nphoton.2013.177 (2013).
26. Clark, J. B., Lecocq, F., Simmonds, R. W., Aumentado, J. & Teufel, J. D. Observation of strong radiation pressure forces from squeezed light on a mechanical oscillator. *Nat. Phys.* **12**, 683–687. https://doi.org/10.1038/nphys3701 (2016).
27. Thorne, K. S., Drever, R. W. P., Caves, C. M., Zimmermann, M. & Sandberg, V. D. Quantum nondemolition measurements of harmonic oscillators. *Phys. Rev. Lett.* **40**, 667 (1978).
28. Braginsky, V. B., Vorontsov, Y. I. & Thorne, K. S. Quantum nondemolition measurements. *Science* **209**, 547–557. https://doi.org/10.1126/science.209.4456.547 (1980). https://science.sciencemag.org/content/209/4456/547.full.pdf.
29. Braginsky, V. B. & Khalili, F. Y. Quantum nondemolition measurements: The route from toys to tools. *Rev. Mod. Phys.* **68**, 1–11. https://doi.org/10.1103/RevModPhys.68.1 (1996).
30. Clerk, A. A., Marquardt, F. & Jacobs, K. Back-action evasion and squeezing of a mechanical resonator using a cavity detector. *N. J. Phys.* **10**, 095010. https://doi.org/10.1088/1367-2630/10/9/095010 (2008).
31. Hertzberg, J. B. *et al.* Back-action-evading measurements of nanomechanical motion. *Nat. Phys.* **6**, 213–217. https://doi.org/10.1038/nphys1479 (2009).
32. Suh, J. *et al.* Mechanically detecting and avoiding the quantum fluctuations of a microwave field. *Science* **344**, 1262–1265. https://doi.org/10.1126/science.1253258 (2014). https://science.sciencemag.org/content/344/6189/1262.full.pdf.
33. Lecocq, F., Clark, J. B., Simmonds, R. W., Aumentado, J. & Teufel, J. D. Quantum nondemolition measurement of a nonclassical state of a massive object. *Phys. Rev. X* **5**, 041037. https://doi.org/10.1103/PhysRevX.5.041037 (2015).

34. Shomroni, I., Qiu, L., Malz, D., Nunnenkamp, A. & Kippenberg, T. J. Optical backaction-evading measurement of a mechanical oscillator. *Nat. Commun.* **10**, 2086. https://doi.org/10.1038/s41467-019-10024-3 (2019).
35. Woolley, M. J. & Clerk, A. A. Two-mode back-action-evading measurements in cavity optomechanics. *Phys. Rev. A* **87**, 063846. https://doi.org/10.1103/PhysRevA.87.063846 (2013).
36. Ockeloen-Korppi, C. F. *et al.* Quantum backaction evading measurement of collective mechanical modes. *Phys. Rev. Lett.* **117**, 140401. https://doi.org/10.1103/PhysRevLett.117.140401 (2016).
37. Tsang, M. & Caves, C. M. Coherent quantum-noise cancellation for optomechanical sensors. *Phys. Rev. Lett.* **105**, 123601. https://doi.org/10.1103/PhysRevLett.105.123601 (2010).
38. Tsang, M. & Caves, C. M. Evading quantum mechanics: Engineering a classical subsystem within a quantum environment. *Phys. Rev. X* **2**, 031016. https://doi.org/10.1103/PhysRevX.2.031016 (2012).
39. Bondurant, R. S. Reduction of radiation-pressure-induced fluctuations in interferometric gravity-wave detectors. *Phys. Rev. A* **34**, 3927 (1986).
40. Zhang, K., Meystre, P. & Zhang, W. Back-action-free quantum optomechanics with negative-mass bose-einstein condensates. *Phys. Rev. A* **88**, 043632. https://doi.org/10.1103/PhysRevA.88.043632 (2013).
41. Wimmer, M. H., Steinmeyer, D., Hammerer, K. & Heurs, M. Coherent cancellation of backaction noise in optomechanical force measurements. *Phys. Rev. A* **89**, 053836. https://doi.org/10.1103/PhysRevA.89.053836 (2014).
42. Bariani, F., Seok, H., Singh, S., Vengalattore, M. & Meystre, P. Atom-based coherent quantum-noise cancellation in optomechanics. *Phys. Rev. A* **92**, 043817. https://doi.org/10.1103/PhysRevA.92.043817 (2015).
43. Motazedifard, A., Bemani, F., Naderi, M. H., Roknizadeh, R. & Vitali, D. Force sensing based on coherent quantum noise cancellation in a hybrid optomechanical cavity with squeezed-vacuum injection. *N. J. Phys.* **18**, 073040. https://doi.org/10.1088/1367-2630/18/7/073040 (2016).
44. Møller, C. B. *et al.* Quantum back-action-evading measurement of motion in a negative mass reference frame. *Nature* **547**, 191–195. https://doi.org/10.1038/nature22980 (2017).

45. Kerdoncuff, H., Hoff, U. B., Harris, G. I., Bowen, W. P. & Andersen, U. L. Squeezing-enhanced measurement sensitivity in a cavity optomechanical system. *Annalen der Physik* **527**, 107–114. https://doi.org/10.1002/andp.201400171 (2015). https://onlinelibrary.wiley.com/doi/pdf/10.1002/andp.201400171.
46. Bowen, W. P. & Milburn, G. J. *Quantum Optomechanics* (CRC Press, Boca Raton, 2016).
47. Dorsel, A., McCullen, J. D., Meystre, P., Vignes, E. & Walther, H. Optical bistability and mirror confinement induced by radiation pressure. *Phys. Rev. Lett.* **51**, 1550–1553. https://doi.org/10.1103/PhysRevLett.51.1550 (1983).
48. Brennecke, F., Ritter, S., Donner, T. & Esslinger, T. Cavity optomechanics with a bose-einstein condensate. *Science* **322**, 235–238. https://doi.org/10.1126/science.1163218 (2008). https://science.sciencemag.org/content/322/5899/235.full.pdf.
49. Arcizet, O., Cohadon, P. F., Briant, T., Pinard, M. & Heidmann, A. Radiation-pressure cooling and optomechanical instability of a micromirror. *Nature* **444**, 71–74. https://doi.org/10.1038/nature05244 (2006).
50. Teufel, J. D. *et al.* Sideband cooling of micromechanical motion to the quantum ground state. *Nature* **475**, 359–363. https://doi.org/10.1038/nature10261 (2011).
51. Chan, J. *et al.* Laser cooling of a nanomechanical oscillator into its quantum ground state. *Nature* **478**, 89–92. https://doi.org/10.1038/nature10461 (2011).
52. Palomaki, T. A., Teufel, J. D., Simmonds, R. W. & Lehnert, K. W. Entangling mechanical motion with microwave fields. *Science* **342**, 710–713. https://doi.org/10.1126/science.1244563 (2013). https://science.sciencemag.org/content/342/6159/710.full.pdf.
53. Weis, S. *et al.* Optomechanically induced transparency. *Science* **330**, 1520–1523. https://doi.org/10.1126/science.1195596 (2010). https://science.sciencemag.org/content/330/6010/1520.full.pdf.
54. Kippenberg, T. J. & Vahala, K. J. Cavity optomechanics: Back-action at the mesoscale. *Science* **321**, 1172–1176. https://doi.org/10.1126/science.1156032 (2008). https://science.sciencemag.org/content/321/5893/1172.full.pdf.

55. Chelkowski, S. *et al.* Experimental characterization of frequency-dependent squeezed light. *Phys. Rev. A* **71**, 013806. https://doi.org/10.1103/PhysRevA.71.013806 (2005).
56. Vahlbruch, H., Mehmet, M., Danzmann, K. & Schnabel, R. Detection of 15 db squeezed states of light and their application for the absolute calibration of photoelectric quantum efficiency. *Phys. Rev. Lett.* **117**, 110801. https://doi.org/10.1103/PhysRevLett.117.110801 (2016).
57. Gardiner, C. & Zoller, P. *Quantum Noise* (Springer, Berlin, 2004).
58. Giovannetti, V. & Vitali, D. Phase-noise measurement in a cavity with a movable mirror undergoing quantum Brownian motion. *Phys. Rev. A* **63**, 023812. https://doi.org/10.1103/PhysRevA.63.023812 (2001).

CHAPTER 16

POINT DIFFRACTION INTERFEROMETRY

Daodang Wang[1] and Rongguang Liang[2]

[1] College of Metrology and Measurement Engineering, China Jiliang University, Hangzhou, China

[2] College of Optical Sciences, University of Arizona, Tucson, AZ, USA

ABSTRACT

The point diffraction interferometer (PDI) employs a point-diffraction spherical wavefront as ideal measurement reference, and it overcomes the accuracy limitation of reference optics in traditional interferometers. To overcome the limitation of measurement range either with pinhole (low light transmission) or with single-mode fiber (low NA), a single-mode fiber with narrowed exit aperture has been proposed to obtain the point-diffraction wavefront with both high NA and high power. It is a key issue to analyze the point-diffraction wavefront error in PDI, which determines the achievable accuracy of the system. The FDTD method based on the vector diffraction theory provides a powerful tool for the design and optimization of the PDI system. In addition, a high-precision method based on shearing interferometry can be applied to measure point-diffraction wavefront with

Citation: (APA): Wang, D., & Liang, R. (2017). Point Diffraction Interferometry. Optical Interferometry, 187. (25 pages)

high NA, in which a double-step calibration including three-dimensional coordinate reconstruction and symmetric lateral displacement compensation is used to calibrate the geometric aberration. The PDI is expected to be a powerful tool for high-precision optical testing. With the PDI method, a high accuracy with RMS value better than subnanometer can be obtained in the optical surface testing and submicron in the absolute three-dimensional coordinate measurement, demonstrating the feasibility and wide application foreground of PDI.

Keywords: point diffraction, optical testing, pinhole, optical fiber, wavefront sphericity

INTRODUCTION

The development of optical design and fabrication such as projection optics for extreme ultraviolet lithography (EUVL) and laser fusion, etc. places ultrahigh requirement on the optical testing precision and accuracy. In the EUVL operating at a wavelength of 13.5 nm [1–4], the projection optics is composed of 4–6 aspheric mirrors and each mirror needs to be finished with the figure error less than 0.2–0.3 nm (RMS), requiring a higher testing accuracy (e.g., 0.1 nm RMS).

The optical interferometers, including the Twyman-Green [5] and Fizeau interferometers [6], have been widely used in the testing lenses and mirrors for figure metrology, in which standard lenses are applied to produce the necessary reference wavefront. Due to the fabricating error of the reference optics, the achievable measurement accuracy of the traditional interferometers is generally limited within $\lambda/50$ (λ is the operating wavelength).

As a novel optical technique, the point-diffraction interferometer (PDI) [7–12] has been developed to overcome the accuracy limitation in traditional interferometers, and it can reach the measurement accuracy in the order of subnanometer. The PDI method employs point-diffraction spherical wavefront as ideal measurement reference, and it can achieve the accuracy better than $10^{-3}\lambda$.

The PDI does not require precise and costly standard parts, and it provides a feasible way to overcome the accuracy limitation of the reference optics in traditional interferometers. The process of point diffraction determines the reachable accuracy of PDI, thus, a good reproducibility of the measurement accuracy can be achieved with PDI. The ideal spherical wave can be generated from the point-diffraction source such as circle pinhole [13–15] and optical fiber [16–18]. Nikon realized that the high-precision testing of spherical and aspherical surfaces with the pinhole PDI [14, 15], and the Lawrence Livermore National Laboratory measured surface figure with the single-mode-fiber PDI [16]. The diffracted wave in both PDI systems serves as testing wave as well as reference wave.

To extend the aperture of the optics, a fiber PDI with two optical fibers serving as point sources was proposed for large optics measurement [18], in which the diffracted wave from one fiber acts as testing wave and that from the other fiber as reference wave. Considering the fact that the diffraction light power from pinhole is poor (transmittance <0.1‰) and the numerical aperture (NA) of diffracted wavefront with single-mode fiber is quite low (<0.20), both approaches limit the measurement range of the existing PDIs. A submicron-aperture (SMA) fiber with cone-shaped exit end has been proposed to obtain both the high diffraction light power and high-NA spherical wavefront [19–21], and it is considered as a feasible way to extend the measurement range of the system.

Due to the fact that the achievable testing accuracy of PDI is mainly determined by the sphericity of diffracted wavefront, the analysis of the diffracted wavefront error has become a fundamental way to evaluate the performance of PDI. In our research, both numerical analysis and experimental measurement have been performed to reconstruct the point-diffraction wavefront aberration.

Besides, various setups of PDI have been developed to realize the precise surface testing with adjustable contrast [12, 22] and three-dimensional (3D) coordinate measurement [20, 23, 24]. The new PDI for three-dimensional (3D) coordinate measurement avoids the measurement uncertainty introduced by the imperfect target in multilateration and allows the target (made of two point-diffraction sources) to take free movement within a volumetric space over NA of point-diffraction wave. Section 2 presents a brief review of the PDI for testing of wavefront and the optical surface. Section 3 presents the basic theory of the SMA fiber PDI for 3D coordinate measurement. Sections 4 and 5 explain the numerical method and the experimental measurement method for the evaluation of diffracted wavefront sphericity, as well as the corresponding analyzing results.

PDI FOR WAVEFRONT AND SURFACE TESTING

Wavefront testing with PDI

The PDI was first proposed by Smartt and Strong in 1972 [7]. The early version of the PDI was an interferometer using a PDI plate with partial transmission. Figure 1 shows the principle of the PDI with partially transparent PDI plate. The PDI plate consists of an absorbing metal coating mask on a clear substrate and a tiny pinhole placed near focus to divide the wave into two parts, namely the testing wave and reference wave. The pinhole picks off part of the incident focused light wave and generates the diffracted spherical wavefront as reference wavefront. The test wave passes through the PDI plate and then interferes with diffracted spherical wave. The PDI can be used to measure the wavefront error from imaging optics and flow field, etc. In the design of the PDI shown in Figure 1, it is a key issue to determine the pinhole size and plate transmittance [9]. Due to the reduction in light intensity incident on the tiny pinhole, the controllable reduction in the test wave intensity by mask attenuation is required to get maximum fringe contrast.

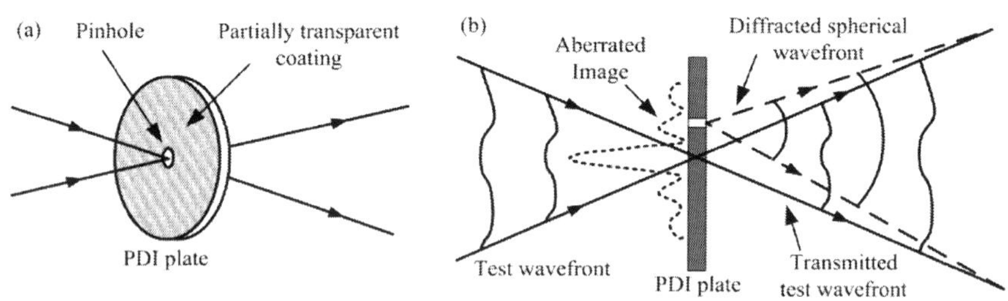

Figure 1. Principle of PDI with PDI plate. (a) PDI plate and (b) PDI operation.

Another PDI plate with double apertures has been applied in the PDI for the evaluation of wavefront error in imaging optics [25], as shown in Figure 2. A beam splitter such as the transmission grating is used to divide the test wave into two waves with a small angular separation. The PDI plate consists of one tiny pinhole and one large window on an opaque mask, and both of them are placed at the respect focal points of two beams. The pinhole is applied to generate the spherical reference wavefront by diffraction and the window transmits the test wave. The intensity of the diffracted wave relative to that of the conventional PDI is increased by several orders of magnitude.

With the application of double-aperture PDI plate, there is no need to further attenuation in the test wave to match the intensities of interfering waves. Besides, the beam division enables the potential to introduce the various phase shifts between the interfering waves.

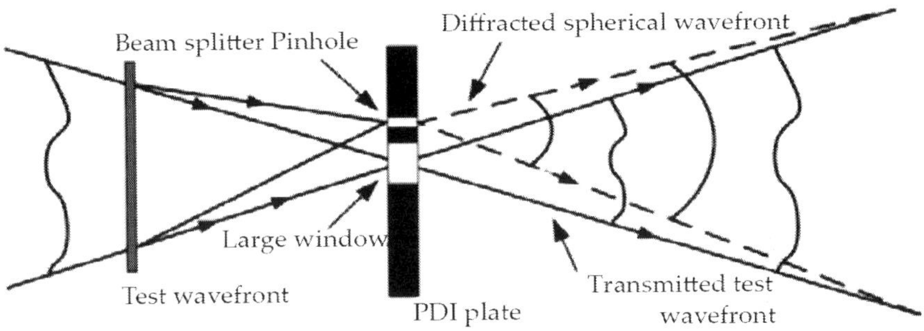

Figure 2. PDI plate with double aperture.

Optical surface testing with PDI

Figure 3 shows the typical configuration of PDI for optical surface testing. Either the pinhole (Figure 3a) [13–15] or the single-mode optical fiber (Figure 3b) [16–18] can be used as the point-diffraction source to generate the required spherical wavefront. The diffracted wavefront is separated to two parts, i.e., the test and reference wavefronts. The test wave travels toward the spherical surface under test, and the pinhole (exit end of fiber) is positioned at the curvature center of the test surface, so that the reflected wave from the test surface converges at the pinhole mirror (semitransparent metallic film on the output end of fiber) and then is reflected at the pinhole mirror (semitransparent metallic film on the output end of fiber). The test wave combines and interferes with the reference wave after reflection. By translating the test surface with a precise PZT scanner, the test surface error can be measured with the phase-shifting method. The selection of beam polarization state in the system can significantly influence the measurement. The polarization would affect the sphericity of the diffracted wavefront, and the reflection at mirror over a high NA can also introduce polarization-dependent phase shifts. In the PDI system, the polarization state of diffracted spherical wave is generally adjusted to be circularly polarized [26], in which the effect of polarization on the measurement precision is negligible.

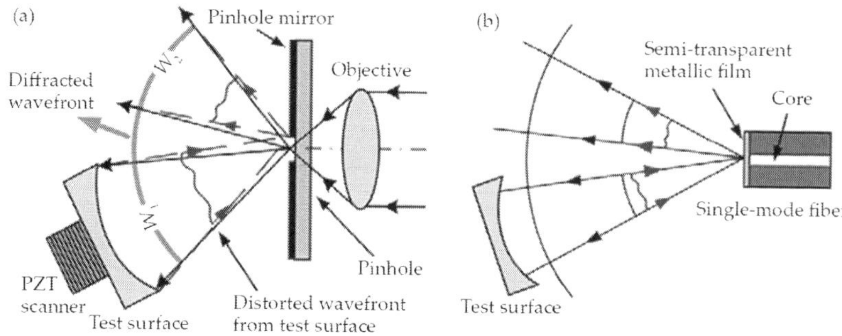

Figure 3. Configuration of PDI for optical surface testing. (a) Pinhole PDI [13], (b) single-mode fiber PDI [16].

In the PDI with pinhole method, the circular pinhole of submicron (or even smaller) diameter is adopted to obtain high measurable NA. Figure 4(a) shows a scanning-electron microscope (SEM) picture of the pinhole fabricated by etching the chromium film with the focused ion-beam etching (FIBE) method [22], in which the metallic layer is sputtered onto the silica substrate. A nearly perfect circular pinhole can be obtained with the FIBE method. The high measurable NA can be achieved with the pinhole method; however, the nonadjustable fringe contrast would limit the measurement accuracy in the testing of low-reflectivity surface, due to the poor fringe contrast and difficulty in fringe processing.

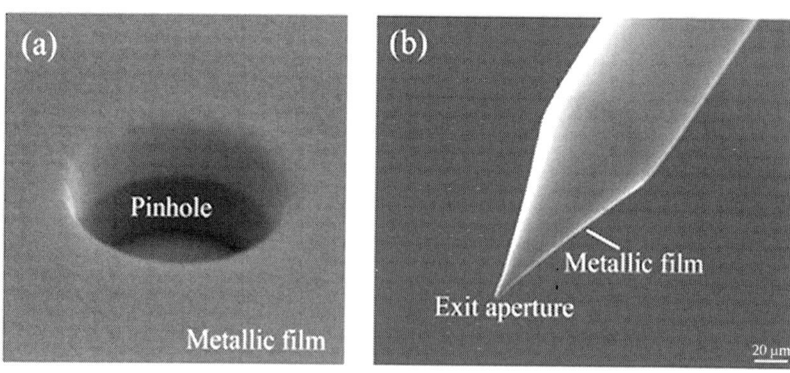

Figure 4. Point diffraction sources in PDI. (a) Pinhole [22], and (b) submicron-aperture fiber [21].

A pinhole PDI with adjustable fringe contrast can be adopted for the testing of high-NA spherical surfaces with low reflectivity [22]. The polarizing elements are applied to transform the polarization states and adjust the relative intensities of the interfering beams, by which the adjustable fringe contrast can be achieved. Figure 5 shows the optical configuration of the pinhole PDI with adjustable fringe contrast. A quarter-wave plate (QWP2) with special structure (consisting of a thin waveplate film and a plano-convex substrate) is placed at the test path, with the fast axis oriented at −45° to horizontal. The diffracted wave is adjusted to be circularly polarized. The test wave W_1 travels toward the test surface and passes through QWP2 twice, respectively, before and after the reflection at the test surface, then the test wave W_1 becomes opposite circularly polarized with respect to the reference wave W_2. The relative intensities of the test and reference beams can be adjusted by rotating the transmission axis of the analyzer, realizing the adjustable fringe contrast. Due to the fact that the diffracted wave is divergent, QWP2 in the test path would introduce different phase retardations in various directions. To minimize the effect of QWP2 on the divergent test wave, a true zero-order waveplate, which has advantages of less sensitive to variation in angle of incidence and wavelength, is employed in the system. In the true zero-order waveplate, a thin waveplate film is cemented on the glass substrate. To minimize the aberrations introduced by the glass substrate in the case of divergent waves, a plano-convex substrate is used in the true zero-order waveplate QWP2.

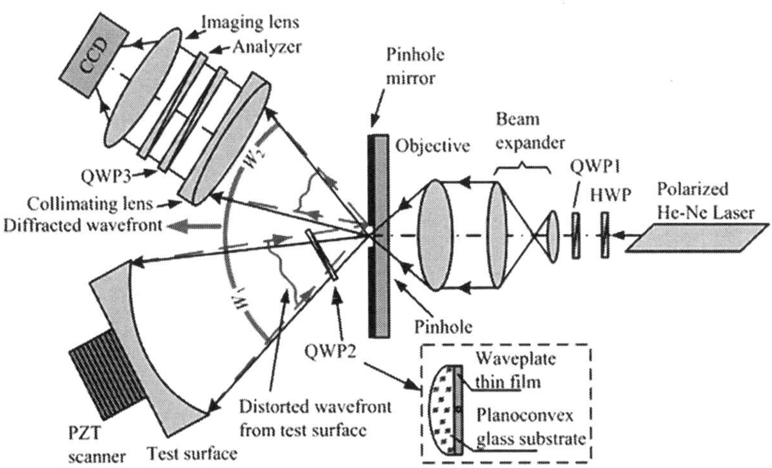

Figure 5. Configuration of pinhole PDI with adjustable fringe contrast [22].

Figure 6 shows the testing results about a spherical surface with the reflectivity 4%, NA 0.40 and aperture diameter 137.7 mm. Figure 6(a) is the surface error measured with the adjustable-contrast pinhole PDI system shown in Figure 5, and Figure 6(b) is that obtained with the ZYGO interferometer. According to Figure 6, a good agreement between the PDI result and that from the ZYGO interferometer is achieved, and the PV and RMS differences between the testing results are about 0.0214λ and 0.0025λ, respectively. Thus, high-accuracy testing of the surface with low reflectivity is realized with the adjustable-contrast PDI system.

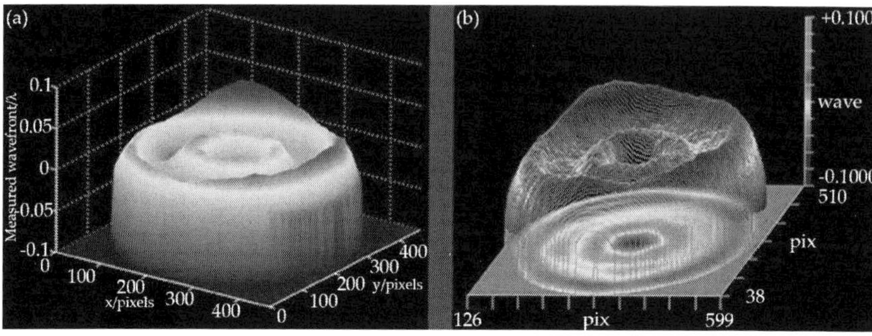

Figure 6. Measured surface error of test spherical surface with (a) the adjustable-contrast pinhole PDI and (b) the ZYGO interferometer [22].

The light transmission through the pinhole is quite low (<0.1‰), and it significantly limits the achievable measurement range of the pinhole PDI. In the PDI with single-mode fiber, the adjustable fringe contrast is easy to realize and high light transmittance (>10%) can be obtained; however, the measurable NA in the fiber method is limited by the NA of the fiber, which is commonly less than 0.2. Besides, the measurable aperture of the test surface is approximately half that of the diffraction wave, both for the pinhole PDI and single-mode fiber PDI shown in Figure 3. A PDI system with two optical fibers as point-diffraction sources was developed for full use of the diffracted wavefront [18], in which the diffracted wave from one fiber serves as reference wavefront and that from the other fiber as test wave. Besides, a novel submicron-aperture (SMA) fiber with cone-shaped exit end, as shown in Figure 4(b), has been proposed to obtain both the high diffraction light power and high-NA spherical wavefront [19–21]. The SMA fiber taper surface is coated with metallic film and the exit aperture is formed from the polished tip, it is formed with the same processing technology as manufacturing of fiber-based probes for the scanning near-field optical microscopy. With the SMA fiber, the measurable range can be almost extended to a half space, and the corresponding light transmittance over 50% is obtained [20]. Thus, it is considered as a feasible way to extend the measurement range of the system.

PDI FOR 3D COORDINATE MEASUREMENT

In this section, a fiber PDI with submicron aperture [20], which is based on the single-mode fiber with a narrowed exit aperture, for the absolute 3D coordinate measurement within large angle range is described, and the system configuration is shown in Figure 7. After passing through the half-wave plate (HWP1), the frequency-stabilized laser beam is separate into p-polarized and s-polarized beams by the polarized beam splitter (PBS), and they are coupled into the single-mode fibers SF1 and SF2, respectively. The exit ends of the fibers SF1 and SF2 are integrated into a target with certain lateral offset, and the output point-diffraction waves W_1 and W_2 interfere on the CCD detector. By translating the PZT scanner, the point-diffraction interference field can be retrieved with phase-shifting method, and the corresponding 3D coordinates of target can be obtained with numerical iterative reconstruction algorithm.

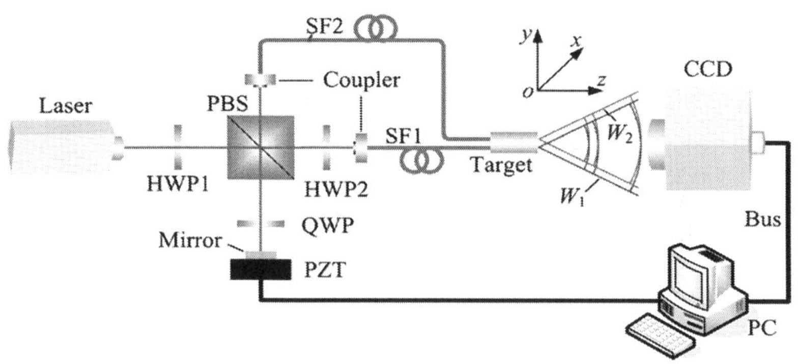

Figure 7. System configuration of the SMA fiber PDI for 3D coordinate measurement [20].

To increase the NA of diffracted wavefront with optical fiber, a single-mode fiber with submicron aperture (as shown in Figure 4b) is applied as point-diffraction source. Both the exit ends of two single-mode fibers are integrated in a target with certain lateral offset. According to the PDI system shown in Figure 7, the absolute 3D coordinates of target can be measured from the phase distribution in interference field corresponding to the optical path difference (OPD), as shown in Figure 8.

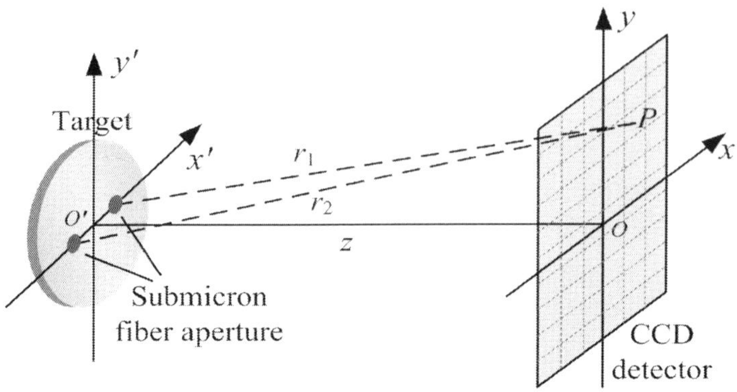

Figure 8. Model for 3D coordinate reconstruction [20].

Denoting the plane of CCD detector as the *xy* plane and the central position *o*, with the distances r_1 and r_2 between an arbitrary point P(x,y,z) on the CCD detector and the exit apertures of two fibers on the target, as shown in Figure 8, we have the phase difference φ(x,y,z),

$$\varphi(x,y,z) = k[r_1(x_1,y_1,z_1;x,y,z) - r_2(x_2,y_2,z_2;x,y,z)], \tag{1}$$

where k=2π/λ, (x,y,z) is the known 3D coordinate of the point P, (x_1,y_1,z_1) and (x_2,y_2,z_2) are those of two fiber apertures in the target under measurement. According to the one-to-one correspondence relationship of the phase φ at position P (that is the ith pixel on CCD) and 3D coordinates of two fiber apertures in the target, the phase difference can be written as follows:

$$f_i(\mathbf{\Phi}) = k(\varphi^i - \varphi^0) - \hat{\xi}_i, \tag{2}$$

where the vector Φ indicates the coordinates of two fiber apertures under measurement that is $\Phi=\{(x_1,y_1,z_1);(x_2,y_2,z_2)\}$; $\hat{\xi}_i$ is the measured phase difference. The phase differences φ^i and φ^0 can be written as follows:

$$\begin{cases} \varphi^i(x^i,y^i,z^i) = k[r_1(x_1,y_1,z_1;x^i,y^i,z^i) - r_2(x_2,y_2,z_2;x^i,y^i,z^i)], \\ \varphi^0(x^0,y^0,z^0) = k[r_1(x_1,y_1,z_1;x^0,y^0,z^0) - r_2(x_2,y_2,z_2;x^0,y^0,z^0)], \end{cases} \tag{3}$$

where (x^i,y^i,z^i) is the 3D coordinate of the ith pixel on CCD. From Eq. (2), various equations can be established corresponding to the pixel positions on the CCD, and we have the matrix equations,

$$f(\mathbf{\Phi}) = \{f_i(\mathbf{\Phi})\} = \begin{cases} k(\varphi^1 - \varphi^0) - \widehat{\xi}_1 \\ \vdots \\ k(\varphi^i - \varphi^0) - \widehat{\xi}_i \\ \vdots \\ k(\varphi^m - \varphi^0) - \widehat{\xi}_m \end{cases}, \quad (4)$$

where the subscript m is the number of the pixels selected for coordinate reconstruction. For the 6 unknowns in Eq. (2), at least 6 pixels are needed to obtain the 3D coordinates of two fiber apertures. To realize the accurate measurement of 3D coordinates with the PDI system, over 6 pixels could be applied to reconstruct the 3D coordinates, and a quadratic function corresponding to Eq. (4) can be obtained as:

$$F(\mathbf{\Phi}) = \frac{1}{2} f(\mathbf{\Phi})^T f(\mathbf{\Phi}) = \frac{1}{2} \sum_{i=1}^{m} f_i^2(\mathbf{\Phi}). \quad (5)$$

Thus, the space coordinates of two fiber apertures can be determined from the global minimum Φ^* of the function FF (that is the least-square solution of Eq. (4)), with a single true solution set.

An experimental SMA-fiber PDI system with aperture size about 0.5 μm, has been set up for the measurement of 3D coordinates. For comparison, the target is installed on the probe of a CMM with the positioning accuracy about 1.0 μm, and the 3D coordinates measured with CMM is taken as the nominal value. Figure 9 shows the measurement results about the 3D coordinate deviations in x and z directions corresponding to the initial target position (−0.375, 15, and 200) mm. According to Figure 9, a good agreement between the CMM results and those from the SMA-fiber PDI system is achieved, and the coordinate measurement errors in x and z directions are 0.47 μm and 0.68 μm RMS, respectively. Due to the decreasing in the light intensity and enlargement in fringe spacing, the measurement error also grows with the increase of measurement distance. Several hardware factors can limit the measurement range of the PDI system in practical application, including the detector size and light power, though there is no theoretical limit in principle. The PDI system can be further improved by increasing CCD detector size, lateral off set of two SMA fibers and source power. Compared to the traditional single-mode fiber, the SMA fiber provides a feasible way to obtain the ideal spherical reference wavefront, whose aperture angle almost covers half space, and the lateral measurement range is greatly extended.

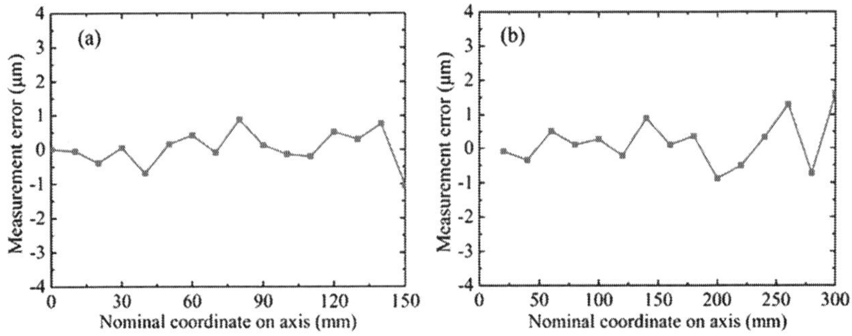

Figure 9. Measurement error in 3D coordinate measurement experiment. Measurement errors in (a) x direction and (b) z direction.

NUMERICAL ANALYSIS OF POINT-DIFFRACTION WAVEFRONT

As one of the most important elements in the PDI, the point source for the diffracted reference wavefront determines the achievable accuracy in the measurement. Purely empirical design of point source parameters is both time consuming and costly. The numerical method based on diffraction theory is a feasible way for the analyzing point-diffraction wavefront. The scalar diffraction theory is valid only when the pinhole size is several times larger than the operating wavelength. For the high-NA spherical wavefront emerging from a tiny aperture with the size comparable with or less than operating wavelength, the vector diffraction theory (that is a nonapproximate method) is required to realize the accurate estimation of diffracted wavefront error. In this section, numerical analysis based on finite difference time domain (FDTD) method (that is a vector diffraction theory) [27] is presented. Figure 10 shows the flow diagram for the simulation of point-diffraction wavefront based on FDTD method. Due to limitations of computer memory capacity and runtime, FDTD cannot be directly applied to calculate the far-field distribution of pinhole diffraction. In the first step, the near-field distribution of point diffraction is analyzed with the FDTD method, and then the near-to-far field transition based on Huygens' principle is performed to obtain the far-field distribution of point-diffraction wavefront at the position under study. Finally, the sphericity evaluation is carried out to get the departure of point-diffraction wavefront from an ideal sphere.

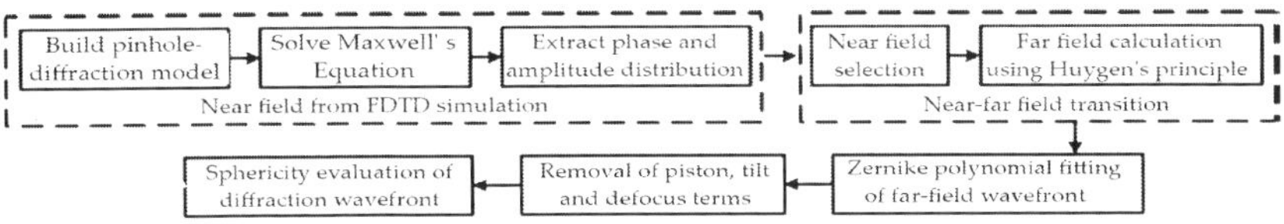

Figure 10. Procedure for simulation of point-diffraction wavefront based on FDTD method.

Simulation of pinhole diffraction

Figure 11 shows the amplitude and phase distribution in the near-field diffraction, respectively, corresponding to the pinholes of 0.5 μm and 1 μm diameters, respectively. According to Figure 11, the phase distribution within Airy disk range is close to an ideal sphere, and the angle of Airy disk increases with the decrease of pinhole diameter.

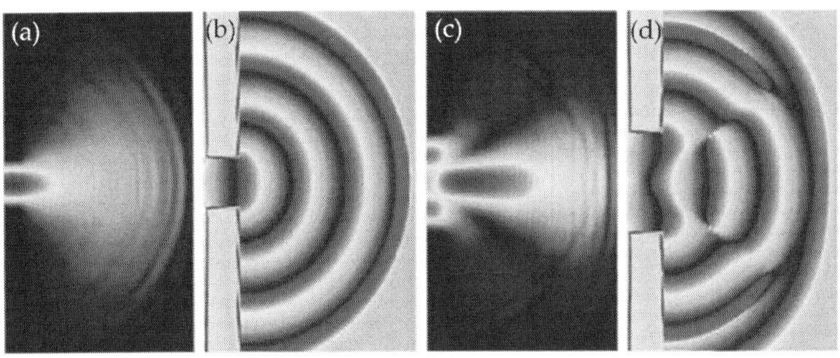

Figure 11. Near-field distribution of pinhole-diffraction wavefront. (a) Amplitude and (b) phase distribution with 0.5 μm-diameter pinhole; (c) amplitude and (d) phase distribution with 1 μm-diameter pinhole [27].

To analyze the wavefront error where the test surface with 500 mm curvature radius is placed, far field is positioned at 500 mm away from pinhole. The wavefront errors over various NAs of diffracted wavefronts corresponding to different pinhole diameters are shown in Figure 12(a). According to Figure 12(a), the point-diffraction wavefront error grows both with the pinhole size and NA range, and the wavefront error RMS for the 1 μm pinhole diameter over 0.35 NA is smaller than 0.53 nm. Thus, it can be taken as ideal reference wavefront and applied to realize the measurement accuracy reaching to RMS $\lambda/1000$ ($\lambda = 532$ nm). Besides, larger diffraction angle could be obtained with smaller pinhole, and it allows much higher measurable NA. The diffraction angle corresponding to 1 μm pinhole diameter is about 75°, and that for 3 μm pinhole diameter is about 30°. Thus, the small pinhole is required in the testing of surfaces with high NA. However, the small pinhole size would result in significant reduction of diffraction light intensity. The pinhole size could be optimized according to the requirement of testing accuracy and achievable diffracted wavefront precision corresponding to various pinhole diameters. The pinhole in PDI plays not only the role as a filter to remove the aberration in diffracted wavefront, but also as waveguide. Figure 12(b) shows the numerical results about the effect of film thickness on diffracted wavefront for different pinhole diameters, with the thickness ranging from 150 nm to 450 nm. The mean values of wavefront error RMS corresponding to 300 nm, 400 nm, and 500 nm pinhole diameters are about 0.02 nm, 0.08 nm, and 0.36 nm, respectively, and the corresponding fluctuation ranges are within 10%. Thus, the effect of variation in film thickness ranging from 150 nm to 450 nm is negligible.

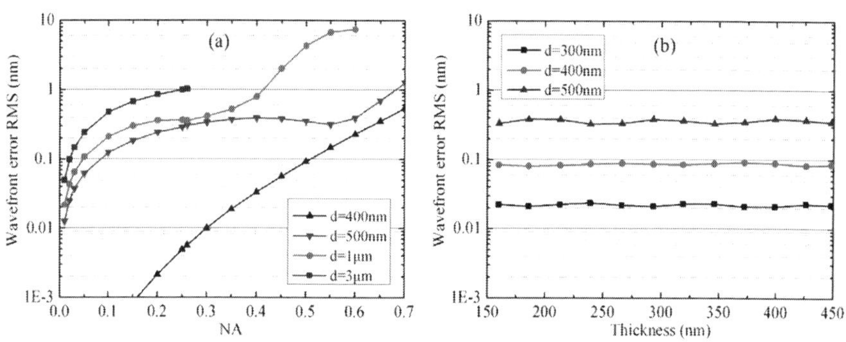

Figure 12. Simulation results for pinhole-diffraction wavefront. (a) Diffracted wavefront error over various NAs for different pinhole diameters, and (b) diffracted wavefront error under various film thicknesses [27].

Simulation of SMA fiber diffraction

Figure 13(a) shows the simulation results about the full wavefront aperture angle for the SMA fibers with various exit apertures [20]. According to Figure 13(a), the full aperture angle of diffracted wave obviously increases with the decreasing of fiber aperture, and that corresponding to the 0.5 μm fiber aperture are about 160°, providing the necessary conic boundary to extend the measurement range almost within a half space. Figure 13(b) shows the wavefront errors over various NAs of diffracted wavefronts corresponding to different SMA fiber apertures [20]. From Figure 13(b), the increase in NA range and fiber aperture could lead to the growth of diffracted wavefront error. The wavefront error over 0.70 NA range for the 0.5 μm fiber aperture is better than λ/1000 RMS, and it can be applied as ideal spherical reference wave in the PDI system.

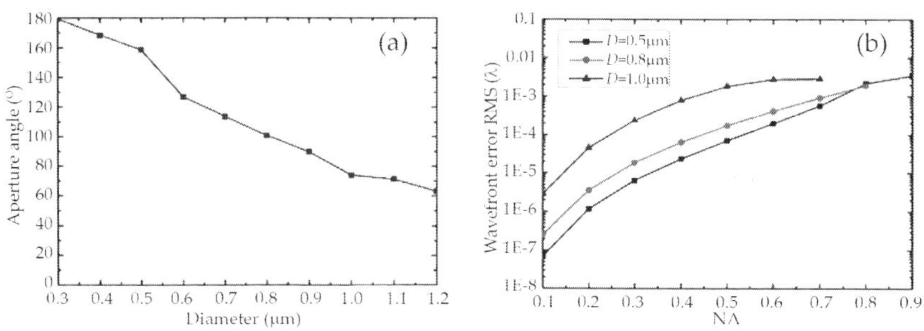

Figure 13. Simulation results for SMA-fiber-diffraction wavefront. (a) Aperture angle of diffracted wave and light transmittance for various fiber apertures, and (b) diffracted wavefront error within various NA ranges for different fiber apertures [20].

Figure 14 shows the analyzing result about light transmittance in the point diffraction both with the SMA fiber method and pinhole method. According to Figure 14, the light transmittance in both the SMA fiber diffraction and pinhole diffraction grows with the increase in aperture size. It can be seen from Figures 12(a) and 13(a) that the similar aperture angles can be obtained with the same diffraction aperture size, however, the light transmittance in the SMA fiber diffraction is far larger than that in pinhole diffraction according to Figure 14. The light transmittance corresponding to the 0.5 μm aperture in the SMA fiber method and pinhole method is about 67% and 6%, respectively. Thus, the SMA fiber provides a feasible way to obtain the high light intensity required in the optical testing, enabling the extension of measurement range with PDI system.

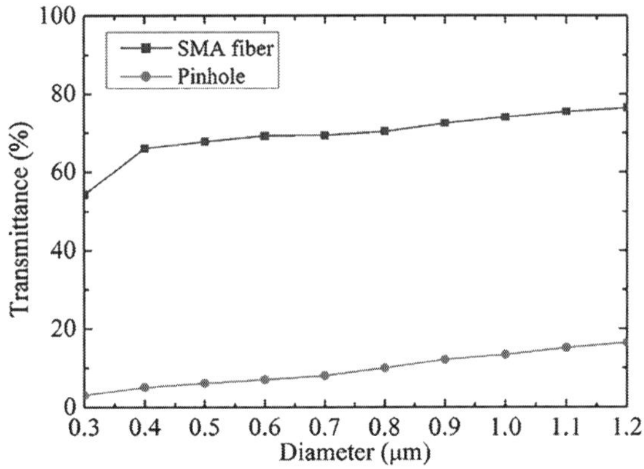

Figure 14. Simulation results for the light transmittance in pinhole diffraction and SMA fiber diffraction.

EXPERIMENTAL MEASUREMENT OF POINT-DIFFRACTION WAVEFRONT

The numerical simulation based on the vector diffraction theory such as FDTD method provides an easy and efficient way to estimate the point-diffraction wavefront error. However, both the computational accuracy and complex practical factors (such as environmental disturbance and performances of various optical parts) can introduce significant deviation from an ideal case. Due to the accuracy limitation of standard optics, the traditional interferometers fail to measure the point-diffraction wavefront error, which is expected to be in the order of subnanometer or even smaller. Various experimental testing methods have been proposed to measure the point-diffraction wavefront error, the majority of which are based on the hybrid method [15] and null test [28]. Typically, the hybrid method requires several measurements with the rotation and displacement of the optics under test, it is sensitive to environmental disturbance and cannot completely separate the systematic error. The null test is a self-reference method, and it is widely applied to reconstruct the point-diffraction wavefront and calibrate the PDI. However, it requires the foreknowledge about the system configuration to remove the high-order aberrations, which are introduced by the point-source separation and cannot be negligible in the case of high-NA wavefront. In this section, a high-precision measurement method, which is based on shearing interferometry, is presented to analyze the point-diffraction wavefront from SMA fiber [21].

Figure 15 shows the schematic diagram of SMA fiber projector used to evaluate the sphericity of point-diffraction wavefront. The SMA fiber can be applied to obtain both high diffraction light power and high-NA spherical wavefront. Four parallel SMA single-mode fibers S_1, S_2, S_3, and S_4 with coplanar exit ends are integrated in a metal tube. To obtain the identical wavefront parameters, the four SMA fibers with same aperture size and cone angle are carefully chosen in the experiment to minimize the measurement error. The coherent beams are coupled into the SMA fibers, and the point-diffraction waves from the fibers interfere on a detection plane, as shown in Figure 15(b). By alternatively switching on the waves from the fiber pairs G_1 (S_1 and S_2) and G_2 (S_3 and S_4), the shearing interferograms in *x* and *y* directions can be obtained and the corresponding shearing wavefronts can be measured with a phase-shifting method, respectively.

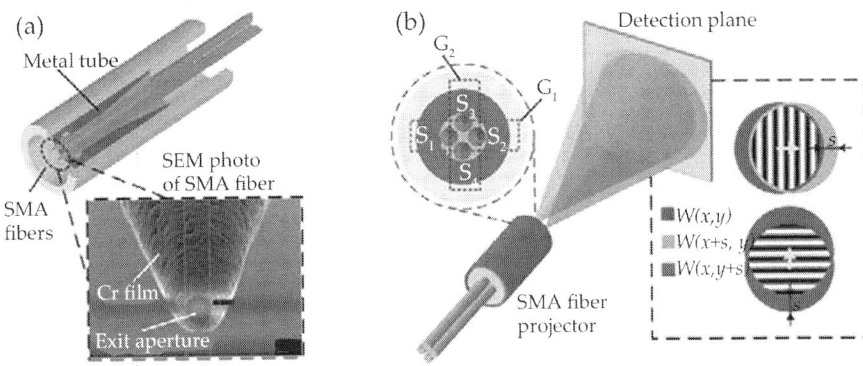

Figure 15. Schematic diagram of SMA fiber point-diffraction wavefront measurement system. (a) Structure of SMA fiber projector and (b) schematic diagram of SMA fiber point-diffraction wavefront measurement [21].

Point-diffraction wavefront retrieval method

With the shearing wavefronts obtained from the projector shown in Figure 15, the differential Zernike polynomials fitting method can be applied to retrieve the SMA fiber point-diffraction wavefront. Denoting the point-diffraction wavefront under test as W(x,y), we have the shearing wavefronts $\Delta W_x(x,y)$ and $\Delta W_y(x,y)$ in *x* and *y* directions,

$$\begin{cases} \Delta W_x(x,y) = W(x,y) - W(x+s,y) \\ \Delta W_y(x,y) = W(x,y) - W(x,y+s) \end{cases}, \tag{6}$$

where s is the shearing amount as shown in Figure 15(b). The wavefront W(x,y) can be described with Zernike polynomials and can be decomposed into a series (N terms) of orthonormal polynomials $\{Z_i(x,y)\}$ with the corresponding coefficients $\{a_i\}$,

$$W(x,y) = \sum_{i=1}^{N} a_i Z_i(x,y). \tag{7}$$

According to Eqs. (6) and (7), the shearing wavefronts $\Delta W_x(x,y)$ and $\Delta W_y(x,y)$ can be expressed as follows:

$$\begin{cases} \Delta W_x(x,y) = \sum_{i=1}^{N} a_i \Delta Z_{i,x}(x,y) \\ \Delta W_y(x,y) = \sum_{i=1}^{N} a_i \Delta Z_{i,y}(x,y) \end{cases}, \tag{8}$$

where the differential Zernike polynomials $\Delta Z_{i,x}(x,y)$ and $\Delta Z_{i,y}(x,y)$ can be written as follows:

$$\begin{cases} \Delta Z_{i,x}(x,y) = Z_i(x,y) - Z_i(x+s,y) \\ \Delta Z_{i,y}(x,y) = Z_i(x,y) - Z_i(x,y+s) \end{cases}. \tag{9}$$

Denoting the shearing wavefronts, differential Zernike polynomials and the coefficients in Eq. (8) as $\Delta W=(\Delta W_x,\Delta W_y)^T$, $\Delta Z=(\Delta Z_{i,x},\Delta Z_{i,y})^T$ and a, respectively, Eq. (8) can be transformed into a matrix form,

$$\Delta W = \Delta Z a. \tag{10}$$

Thus, the coefficients $\{a_i\}$ of Zernike polynomials in Eq. (7) can be obtained from the least-squares solution of Eq. (10),

$$\mathbf{a} = (\Delta \mathbf{Z}^T \Delta \mathbf{Z})^{-1} \Delta \mathbf{Z}^T \Delta \mathbf{W}. \tag{11}$$

According to Eq. (11), the retrieval of point-diffraction wavefront W(x,y) depends on the measurement precision of shearing wavefronts $\Delta W_x(x,y)$ and $\Delta W_y(x,y)$. Traditionally, the systematic error introduced by lateral displacement between SMA fibers can be calibrated by removing Zernike tilt and power terms. However, the residual high-order aberrations can significantly influence the measurement precision, especially in the case of high NA and large lateral displacement. Thus, a general and rigorous method for geometric error removal is required to realize the high-precision measurement of point-diffraction wavefront.

High-precision method for systematic error calibration

In the null test of pinhole diffraction wavefront and single-mode fiber diffraction wavefront, the high wavefront NA in a pinhole PDI and large lateral displacement between fibers in fiber PDI could introduce some high-order geometric aberrations, respectively. Different from traditional pinhole PDI and single-mode fiber PDI, the null test of SMA fiber diffraction wavefront involves both high NA and large lateral displacement, placing much higher requirement on the calibration of the systematic error introduced by lateral displacement between SMA fibers. A double-step calibration method based on three-dimensional coordinate reconstruction and symmetric lateral displacement compensation can be applied to completely remove systematic error. It should be noted that the possible longitudinal displacement between SMA fibers may also introduce certain systematic error; however, it can be well minimized with the fine adjusting mechanism. Besides, the error introduced by the longitudinal displacement is low-order aberration, and it can be well calibrated with traditional misalignment calibration method by subtracting the Zernike piston, tilt and power terms.

First-step calibration based on three-dimensional coordinate reconstruction

Without loss of generality, we take the displacement in x direction between SMA fibers as the geometric error calibration model to be analyzed, as shown in Figure 16.

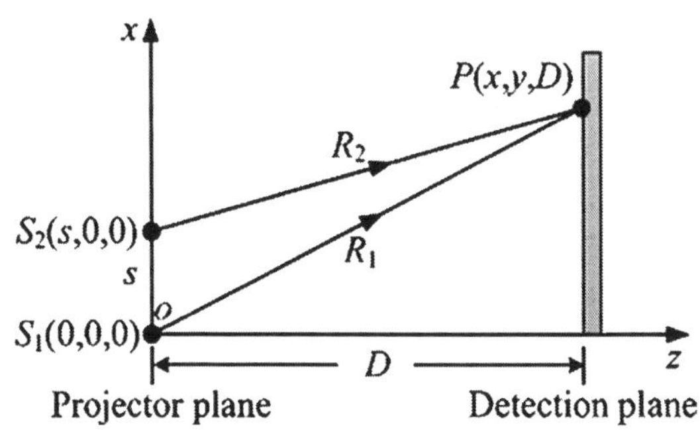

Figure 16. Geometry for systematic error analysis [21].

Supposing that the distance between exit apertures of fiber pairs G_1 (S_1 and S_2) and an arbitrary point P(x,y,z) on the detection plane are R_1 and R_2, we have the corresponding optical path difference OPD,

$$\text{OPD} = R_1(x_1, y_1, z_1; x, y, z) - R_2(x_2, y_2, z_2; x, y, z), \tag{12}$$

where (x_1,y_1,z_1) and (x_2,y_2,z_2) are the three-dimensional coordinates of S_1 and S_2, (x,y,z) is that of the arbitrary known point PP. To simplify the analysis, the origin of the coordinate system is located at S_1, the distance between the SMA fiber projector and CCD detector is D. Thus, the OPD in Eq. (12) can be simplified as follows:

$$\text{OPD} = \sqrt{x^2 + y^2 + D^2} - \sqrt{(x+s)^2 + y^2 + D^2}. \tag{13}$$

According to the one-to-one correspondence of the OPD distribution on the detection plane and the coordinate of fiber apertures, the 3D coordinate measurement method with PDI introduced in Section 3 can be applied to reconstruct the 3D coordinates in Eq. (12) that is the global minimum Φ* of the residual function F(Φ),

$$F(\mathbf{\Phi}) = \sum_k f_k^2(\mathbf{\Phi}) = \sum_k (\text{OPD}_k - \hat{\xi}_k)^2, \tag{14}$$

where the vector $\Phi=\{(x_1,y_1,z_1);(x_2,y_2,z_2)\}$ is the coordinates of SMA fibers under measurement, the subscript *k* indicates the point number on the detection plane, $\hat{\xi}_k$ is the measured OPD and OPD_k is the OPD reconstructed from coordinates Φ according to Eq. (13). With the reconstructed coordinates Φ*, the systematic error can be preliminarily calibrated,

$$\text{OPD}_1 = \text{OPD}(\mathbf{\Phi}) - \text{OPD}(\mathbf{\Phi}^*). \tag{15}$$

However, the reconstruction accuracy of fiber coordinates can only reach the order of submicron in the practical case, resulting in obvious residual error in the calibration result. The root mean square (RMS) value of the residual error in the preliminary calibration is 0.0077λ corresponding to the 0.5 μm coordinate reconstruction error for 0.60 NA fibers and 250 μm lateral displacement between two fibers. To further remove the residual systematic error, a second-step calibration, which is based on symmetric lateral displacement compensation, needs to be carried out.

Second-step calibration based on symmetric lateral displacement compensation

The expression for OPD in Eq. (12) can be simplified as follows:

$$\mathrm{OPD} \cong a_2 Z_2 + a_9 Z_9 + a_{19} Z_{19} + a_{33} Z_{33}, \quad (16)$$

where Z_2 refers to x tilt terms, Z_9, Z_{19}, and Z_{33} are Zernike primary, secondary, and tertiary x coma terms, respectively, a_2, a_9, a_{19}, and a_{33} are the corresponding coefficients, and

$$\begin{cases} a_2 = s[-t - ts^2/(2D^2) + t^3/3 + 3t^3s^2/(4D^2) - 3t^5/16 - 15t^5s^2/(16D^2) + t^7/8] \\ a_9 = s[t^3/6 + 3t^3s^2/(8D^2) - 3t^5/20 - 15t^5s^2/(20D^2) + t^7/8] \\ a_{19} = s[-3t^5/80 - 3t^5s^2/(16D^2) + 3t^7/56] \\ a_{33} = s(t^7/112) \end{cases} \quad (17)$$

Similarly, the reconstructed optical path difference OPD(Φ^*) (though with certain coordinate reconstruction error) follows the relationship given by Eq. (16). Thus, the residual systematic error OPD_1 after first-step calibration, according to Eqs. (15) and (16), can be written as follows:

$$\mathrm{OPD}_1 \cong \Delta a_2 Z_2 + \Delta a_9 Z_9 + \Delta a_{19} Z_{19} + \Delta a_{33} Z_{33}, \quad (18)$$

where $\{\Delta a_i\}$ ($i = 2, 9, 19, 33$) are the corresponding Zernike coefficients. With the implementation of the first-step calibration, the coordinate reconstruction error reaches the order of submicron, and the corresponding approximation error in Eq. (18) can be well restricted and is negligible. According to Eq. (16), the major systematic error introduced by the lateral displacement includes tilt and coma terms, they cannot be completely removed with traditional misalignment calibration method by subtracting the Zernike piston, tilt and power terms. From Eq. (16), the residual coma aberration due to lateral displacement depends on the lateral displacement s, NA, and distance DD, the corresponding Zernike coefficients are odd functions about s. Thus, the superposition of geometric errors $\mathrm{OPD}^{(s)}$ and $\mathrm{OPD}^{(-s)}$ corresponding to opposite shear directions can be expressed as follows:

$$\mathrm{OPD}^{(s)} + \mathrm{OPD}^{(-s)} \cong 0. \quad (19)$$

According to the analysis above, the geometric error can be further reduced by superposing two measurements with opposite shear directions, corresponding to lateral displacement s and −s, respectively. Denoting the preliminarily calibrated wavefront data corresponding to lateral displacement s and −s as $W_{m1}^{(s)}$ and $W_{m1}^{(-s)}$, and the true shearing wavefront under test ΔW, we have

$$\begin{cases} W_{m1}^{(s)} = \Delta W + \mathrm{OPD}_1{}^{(s)} \\ W_{m1}^{(-s)} = \Delta W + \mathrm{OPD}_1{}^{(-s)} \end{cases}. \tag{20}$$

According to Eqs. (19) and (20), the shearing wavefront ΔW can be obtained with

$$\Delta W \cong (W_{m1}^{(s)} + W_{m1}^{(-s)})/2. \tag{21}$$

To get the measurement data with the lateral displacement −s, one may rotate the SMA fiber projector by 180°. Figure 17(a) and (b) shows the residual systematic errors corresponding to various lateral displacement amounts with different NAs with traditional method and the double-step calibration method. The traditional method is not valid when the lateral displacement is over 50 μm because the residual error RMS is larger than $2.0\times10^{-3}\lambda$ with the 0.10 NA fiber. The residual error also increases significantly with the lateral displacement, especially for the case with high NA fiber. In Figure 17(b), the residual error is less than $1.0\times10^{-4}\lambda$ RMS within a 300 μm lateral displacement even for the 0.60 NA fiber, it confirms the feasibility of the double-step calibration in Section 5.2.

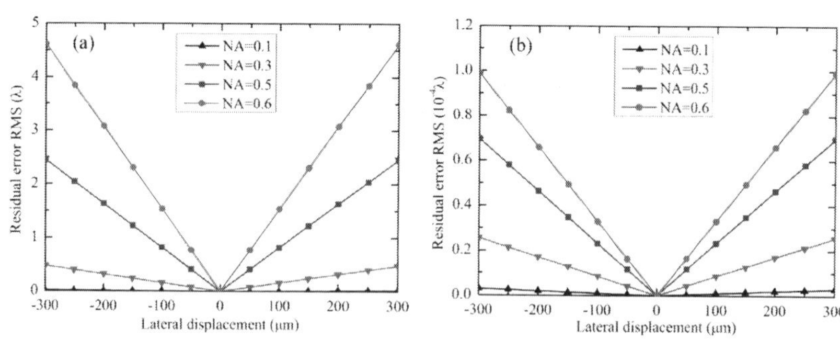

Figure 17. Residual errors in the calibration of systematic error under various lateral displacements and NAs in computer simulation. (a) RMS value with traditional method and (b) RMS value with the proposed double-step calibration method [21].

To obtain point-diffraction spherical wavefront in the experiment, the projector should be placed at the far-field zone that is the distance $D >> 2(2d)^2/\lambda$ from the fibers [29], where *d* is the aperture size. With the unwrapped phase distribution, the systematic error calibration is carried out to obtain the shearing wavefront with the high-precision double-step calibration method introduced in Section 5.2. The experimental measurement of SMA fiber point-diffraction wavefront has been carried out with the system diagram shown in Figure 15. The unwrapped original wavefront before and after 180° rotation in x and y directions is shown in Figure 18(a)–(d), and the corresponding precalibrated shearing wavefronts after the first-step calibration are shown in Figure 18(f)–(i), respectively. Figure 18(e) and (j) is the true shearing wavefronts in x and y directions obtained with the double-step calibration method. According to Figure 18, obvious residual errors can be seen in the obtained wavefronts after first-step calibration, the precalibrated shearing wavefront error RMS in x and y directions are 0.0254λ and 0.0194λ, respectively. The residual error is further removed after the second-step calibration, and true shearing wavefront error RMS in x and y directions are $1.38\times10^{-4}\lambda$ and $1.36\times10^{-4}\lambda$, respectively. Besides, there is no significant odd error in the finally retrieved wavefronts in Figure 18(e) and (j). Thus, the high-accuracy calibration is realized with the proposed method.

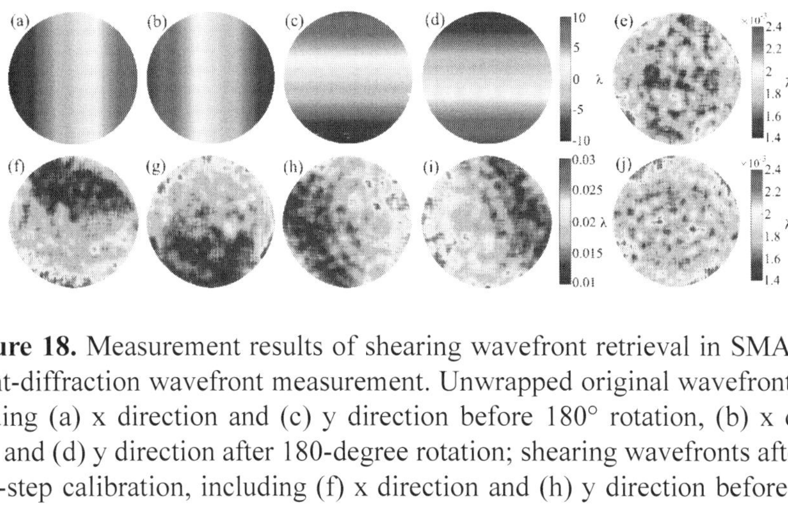

Figure 18. Measurement results of shearing wavefront retrieval in SMA fiber point-diffraction wavefront measurement. Unwrapped original wavefronts, including (a) x direction and (c) y direction before 180° rotation, (b) x direction and (d) y direction after 180-degree rotation; shearing wavefronts after the first-step calibration, including (f) x direction and (h) y direction before 180° rotation, (g) x direction and (i) y direction after 180-degree rotation; finally retrieved shearing wavefronts after the double-step calibration, including (e) x and (j) y directions [21].

With the true shearing wavefronts in x and y directions after systematic error calibration, the point-diffraction wavefront can be reconstructed based on the differential Zernike polynomials fitting method introduced in Section 5.1. The measured wavefront error compared with an ideal sphere is shown in Figure 19(a), PV and RMS are $9.20\times10^{-4}\lambda$ and $1.54\times10^{-4}\lambda$, respectively. Figure 19(b) and (c) shows the measured wavefronts after the 45° rotation and translation along z direction of projector, respectively. It can be seen from

Figure 19 that the measured wavefront shape in the original position agrees well with those obtained with projector rotation and translation, demonstrating the measurement repeatability and accuracy. The protective glass on CCD detector could introduce additional deformation on point-diffraction wavefront, and the measurement accuracy is expected to be further improved by removing the protective glass. Due to the smoothing effect of each pixel, the relatively large pixel acts as a low-pass filter and the measured wavefront should be taken as a macroparameter. By adopting the CCD with smaller pixel size and placing the SMA fiber projector at a position farther from CCD detector, the smoothing effect can be further decreased.

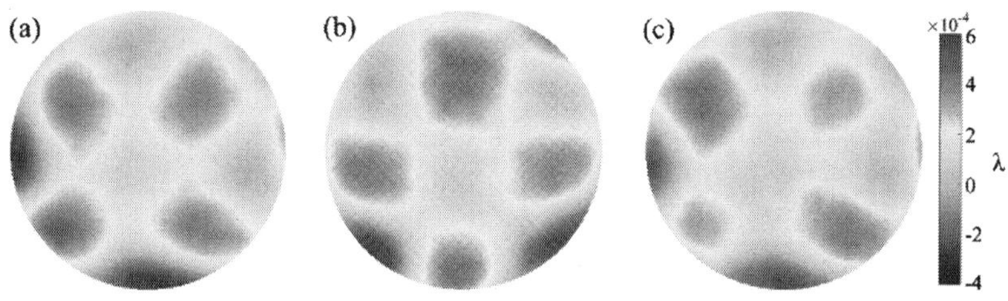

Figure 19. Experimental result of point-diffraction wavefront measurement. (a) Measured point-diffraction wavefront in original position, (b) measured wavefront after 45° projector rotation, and (c) measured wavefront after projector translation [21].

CONCLUSION

In this chapter, the application of PDI system in the wavefront and surface testing is introduced. Various PDI setups, including the fiber PDI and pinhole PDI, have developed for high-accuracy optical testing, and the measurement accuracy up to subnanometer was reported. Different from the pinhole PDI, the adjustable fringe contrast is easily realized with fiber PDI, and it can be applied to test the surface with various reflectivities. To realize the adjustable fringe contrast with pinhole method, the polarizing elements with special structure is needed to transform the polarization states and adjust the relative intensities of the interfering testing and reference beams. High measurable NA can be achieved with the pinhole; however, the light transmission is quite low; high light transmittance can be obtained with single-mode fiber; however, the measurable NA is limited by the NA of the fiber. To obtain both the high diffraction light power and high-NA spherical wavefront, a SMA fiber with cone-shaped exit end has been proposed, and it is considered as a feasible way to extend the measurement range of the system. Based on the

SMA fiber, a PDI system for the absolute 3D coordinate measurement was proposed. The system utilizes the SMA fiber to get spherical wave with both high NA and high light intensity, and large measurable range is achieved.

The numerical study of point-diffraction wavefront both with pinhole and SMA fiber is performed, which is based on the FDTD method. The aperture angle, light transmittance, and wavefront error are analyzed. According to simulation results, the ideal spherical wavefront with the accuracy better than subnanometer can be obtained both with pinhole and SMA fiber, and the increase in NA range and diffraction aperture size could lead to the growth of diffracted wavefront error. It can be seen from the comparison of pinhole diffraction and SMA fiber diffraction, the light transmittance in both the methods grows with the increase in aperture size, and the similar aperture angles can be obtained with the same diffraction aperture size. However, the light transmittance in SMA fiber diffraction is far larger than that in pinhole diffraction. Thus, the SMA fiber provides a feasible way to obtain the high light intensity required in the optical testing, enabling the extension of measurement range with PDI system.

The experimental measurement of SMA fiber point-diffraction wavefront is carried out to evaluate its accuracy in practical application, which is based on shearing interferometry. To realize the high-precision calibration of the systematic error introduced by the lateral displacement between SMA fibers, a double-step calibration method based on three-dimensional coordinate reconstruction and symmetric lateral displacement compensation is used to remove the geometric aberration from the shearing wavefront. Besides, the differential Zernike polynomials fitting method is used to reconstruct the point-diffraction wavefront. The spherical wavefront with the accuracy reaching the order of $10-4\lambda 10-4\lambda$ is obtained in the experiment. The calibration method can be performed without any preknowledge about the measurement system configuration. It realizes the high-accuracy measurement of point-diffraction wavefront and also provides a feasible method for geometric aberration calibration in interferometric system.

ACKNOWLEDGMENTS

This work was partially supported by the National Natural Science Foundation of China (11404312), Zhejiang Provincial Natural Science Foundation of China (LY17E050014), and Zhejiang Key Discipline of Instrument Science and Technology (JL150508). Authors are grateful to editors for giving them the chance to contribution to this book.

REFERENCES

1. Stulen R H, Sweeney D W. Extreme ultraviolet lithography. IEEE Journal of Quantum Electronics. 1999; 35(5): 694–699. DOI: 10.1109/3.760315
2. Goldberg K A, Naulleau P P, Denham P E, Rekawa S B, Jackson K, Anderson E H, et al. At-wavelength alignment and testing of the 0.3 NA MET optic. Journal of Vacuum Science & Technology B. 2004; 22(6): 2956–2961. DOI: 10.1116/1.1815303
3. Otaki K, Zhu Y, Ishii M, Nakayama S, Murakami K, Gemma T. Rigorous wavefront analysis of the visible-light point diffraction interferometer for EUVL. Proceeding of SPIE. 2004; 5193: 182–190. DOI: 10.1117/12.507046
4. Takeuchi S, Kakuchi O, Yamazoe K, Gomei Y, Decker T A, Johnson M A, et al. Visible light point-diffraction interferometer for testing of EUVL optics. Proceeding of SPIE. 2006; 6151: 61510E. DOI: 10.1117/12.656275
5. Malacara D. Twyman-Green Interferometer. In: Malacara D, editor. Optical Shop Testing. 3rd ed. New Jersey: Wiley; 2007. pp. 46–96.
6. Mantravadi M V, Malacara D. Newton, Fizeau, and Haidinger interferometers. In: Malacara D, editor. Optical Shop Testing. 3rd ed. New Jersey: Wiley; 2007. pp. 1–45.
7. Smartt R N, Steel W H. Theory and application of point-diffraction interferometers. Japanese Journal of Applied Physics. 1975; 14(S1): 351–356. DOI: 10.7567/JJAPS.14S1.351
8. Wang G Y, Zheng Y L, Sun A M, Wu S D, Wang Z J. Polarization pinhole interferometer. Optics Letters. 1991; 16(17): 1352–1354. DOI: 10.1364/OL.16.001352
9. Gong Q, Geary J M. Modeling point diffraction interferometers. Optical Engineering. 1996; 35(2): 351–356. DOI: 10.1117/1.600903
10. Millerd J E, Martinek S J, Brock N J, Hayes J B, Wyant J C. Instantaneous phase-shift, point-diffraction interferometer. Proceeding of SPIE. 2004; 5380: 422–429. DOI: 10.1117/ 12.557126
11. Gong Q, Eichhorn W. Alignment and testing of piston and aberrations of a segmented mirror. Proceeding of SPIE. 2005; 5869: 586912. DOI: 10.1117/12.613657
12. Neal R M, Wyant J C. Polarization phase-shifting point-diffraction interferometer. Applied Optics. 2006; 45(15): 3463–3476. DOI: 10.1364/AO.45.003463

13. Otaki K, Bonneau F, Ichihara Y. Absolute measurement of spherical surface by point diffraction interferometer. Proceeding of SPIE. 1999; 3740: 602–605. DOI: 10.1117/ 12.347755

14. Otaki K, Ota K, Nishiyama I, Yamamoto T, Fukuda Y, Okazaki S. Development of the point diffraction interferometer for extreme ultraviolet lithography: design, fabrication, and evaluation. Journal of Vacuum Science & Technology B. 2002; 20(6): 2449–2458. DOI: 10.1116/1.1526605

15. Otaki K, Yamamoto T, Fukuda Y, Ota K, Nishiyama I, Okazaki S. Accuracy evaluation of the point diffraction interferometer for extreme ultraviolet lithography aspheric mirror. Journal of Vacuum Science & Technology B. 2002; 20(1): 295–300. DOI: 10.1116/1.1445161

16. Sommargren G E. Phase shifting diffraction interferometry for measuring extreme ultraviolet optics. OSA Trends in Optics & Photonics. 1996; V4: 108–112.

17. Kihm H, Kim S W. Oblique fiber optic diffraction interferometer for testing. Optical Engineering. 2005; 44(12): 125601. DOI: 10.1117/1.2148367

18. Matsuura T, Udaka K, Oshikane Y, Inoue H, Nakano M, Yamauchi K, et al. Spherical concave mirror measurement by phase-shifting point diffraction interferometer with two optical fibers. Nuclear Instruments & Methods in Physics Research A. 2010; 616(2–3): 233–236. DOI: 10.1016/j.nima.2009.12.031

19. Chkhalo N I, Klimov A Y, Rogov V V, Salashchenko N N, Toropov M N. A source of a reference spherical wave based on a single mode optical fiber with a narrowed exit aperture. Review of Scientific Instruments. 2008; 79(3): 033107. DOI: 10.1063/1.2900561

20. Wang D, Chen X, Xu Y, Wang F, Kong M, Zhao J, et al. High-NA fiber point-diffraction interferometer for three-dimensional coordinate measurement. Optics Express. 2014; 22 (21): 25550–25559. DOI: 10.1364/OE.22.025550

21. Wang D, Xu Y, Liang R, Kong M, Zhao J, Zhang B, et al. High-precision method for submicron-aperture fiber point-diffraction wavefront measurement. Optics Express. 2016; 24(7): 7079–7090. DOI: 10.1364/OE.24.007079

22. Rhee H-G, Kim S-W. Absolute distance measurement by two-point-diffraction interferometry. Applied Optics. 2002; 41(28): 5921–5928. DOI: 10.1364/AO.41.005921

23. Rhee H-G, Chu J, Lee Y-W. Absolute three-dimensional coordinate measurement by the two-point diffraction interferometry. Optics Express. 2007; 15(8): 4435–4444. DOI: 10.1364/OE.15.004435
24. Wang D, Yang Y, Chen C, Zhuo Y. Point diffraction interferometer with adjustable fringe contrast for testing spherical surfaces. Applied Optics. 2011; 50(16): 2342–2348. DOI: 10.1364/AO.50.002342
25. Medecki H, Tejnil E, Goldberg K A, Bokor J. Phase-shifting point diffraction interferometer. Optics Letters. 1996; 21(19): 1526–1528. DOI: 10.1364/OL.21.001526
26. Johnson M A, Phillion D W, Sommargren G E, Decker T A, Taylor J S, Gomei Y, et al. Construction and testing of wavefront reference sources for interferometry of ultra-precise imaging systems. Proceeding of SPIE. 2005; 5869: 58690P. DOI: 10.1117/12.623185
27. Wang D, Wang F, Zou H, Zhang B. Analysis of diffraction wavefront in visible-light point-diffraction interferometer. Applied Optics. 2013; 52(31): 7602–7608. DOI: 10.1364/ AO.52.007602
28. Naulleau P, Goldberg K, Lee S, Chang C, Bresloff C, Batson P, et al. Characterization of the accuracy of EUV phase-shifting point diffraction interferometry. Proceeding of SPIE. 1998; 3331: 114–123. DOI: 10.1117/12.309563
29. Kong J A. Electromagnetic Wave Theory. 1st ed. New Jersey: Wiley; 1986. 667 p.

CHAPTER 17

RADIAL SHEARING INTERFEROMETER

Naiting Gu[1] and Qun Luo[2]

[1]Institute of Optics and Electronics, Chinese Academy of Sciences, China

[2]National Key Laboratory of Science and Technology on Blind Signal Processing, China

ABSTRACT

Radial shearing interferometer (RSI) is one of the most powerful tools in many domains, especially in optical testing. RSI has compact size and good vibration immunity, which is adaptive to various environments, due to its common-path configuration. Moreover, it is very convenient application because no plane referencing wavefront is needed. The disadvantages of the conventional RSIs are that the distorted wavefront is hard to extract quickly and accurately from one radial shearography due to the phase extract algorithm is complex. Fortunately, the new RSIs can receive benefits from the accuracy of the methods of phase-shifting interferometry, and phase-shifting shearography is more sensitive than simple digital shearography. There are two mainly trend to the RSIs based on phase-shifting technique, i.e. instantaneous phase-shifting and compact size. In this chapter, a development process of RSI will be introduced briefly firstly, and then the some new RSIs

Citation: (APA): Gu, N., & Luo, Q. (2018). Radial Shearing Interferometer. In Interferometry-Recent Developments and Contemporary Applications. IntechOpen. (22 pages)

based phase-shifting techniques in our work will be described in following parts, including initial RSI by using four-step polarization phase-shifting, modal wavefront reconstruction method for RSI with lateral shear and a new kind of compact RSI based micro-optics technique.

Keywords: radial shearing interferometer, phase-shifting tcchnique, instantaneous, vibration insensitive, modal wavefront reconstruction

INTRODUCTION

Radial shearing interferometer (RSI) was proposed firstly in 1961 [1]. After development of many years, the RSI has been used widely in optical testing [2, 3, 4], corneal topographic inspection [5, 6, 7], wavefront sensing [8, 9, 10, 11] and laser beam characterization [12, 13, 14]. Radial shear can be introduced by some classical optical components [15, 16] and the other different ways [17, 18, 19, 20, 21] including optical gratings [17, 18], zone plate [19], speckle interference [20, 21] and the other applications. It is very convenient application because no plane referencing wavefront is needed [24], especially comparing to the point diffraction interferometer (PDI) [22, 23]. RSI has compact size and good vibration immunity, which is adaptive to various environments, due to its common-path configuration. Recently, RSI has been becoming one of the most important tools for diagnosing the wavefront of laser beams and the other applications. The disadvantages of the conventional RSIs are that the distorted wavefront is hard to extract from only one radial shearography due to the complexity of the phase extract algorithm. However, fortunately, many of these applications can receive benefits from the high accuracy of phase-shifting interference methods, and phase-shifting shearography is more sensitive than simple radial shearography [25]. Thus some authors [8, 26, 27, 28] have proposed several kinds of RSIs by using temporal or spatial phase-shifting techniques. A cyclic RSI was developed in Ref. [26] for phase shifting interferometry with a polarization phase shifter. A RSI based on a Mach-Zehnder configuration by using liquid-crystal-device as the phase shifter was described in Ref. [27]. A Sagnac RSI with geometric phase-shifting technique was designed in Ref. [28]. However, all above these developed RSIs adopted the temporal phase shifting technique. In this case, the wavefront under test must be essentially stationary during the duration of acquiring several phase-stepping images, and they cannot be used to measure dynamic wavefront. In Ref. [8], a new RSI with spatial phase shifting techniques was developed for wavefront measurements and diagnosing dynamic measurements and the single-shot common-path phase-stepping configuration improved the performance of RSI especially for dynamic scenes. However, its complex configuration and extraction algorithm limits the testing accuracy, system stability, compactness and its application. Apparently, it is highly desirable to acquire compact one-shot vibration-insensitive RSIs. In Ref. [29], a new compact RSI with simultaneous phase-shifter based on a micro-retarder array (MRA) was proposed, which has some good features such as high speed, high accuracy and vibration immunity while decreasing the complexity. In Ref. [11], a new modal reconstruction algorithm was proposed to reconstruct accurately and quickly the wavefront under test in a RSI with or without lateral shear.

In this chapter, we will introduce the new RSIs with instantaneous phase-shifting and its corresponding modal wavefront reconstruction method. These works have been published earlier in [8, 11, 29], and they are re-arranged in this manuscript. This manuscript is organized as follows. In Section 2 a single-shot common-path phase-stepping RSI will be introduced. A new RSI with simultaneous phase-shifter based on a micro-retarder array will be described in Section 2.1. In Section 3 a modal wavefront reconstruction method based on Zernike polynomials in a RSI with or without lateral shear will be described. The conclusion of this chapter will be made in final section.

RSI BASED ON SIMULTANEOUS POLARIZATION PHASE-SHIFTER

RSIs in many domains can receive advantages from the accuracy of phase-shifting interference methods, and many new RSIs were proposed by utilizing temporal or spatial phase-shifting techniques. Comparatively, the temporal phase shifting technique requires stationary scene during the duration of acquiring several phase-stepping images, but the spatial phase-shifting technique not. Thus the RSIs with spatial phase-shifting technique are preferred to apply in several applications due to its ability for dynamic scene. In this part we will introduce a new RSI based on simultaneous polarization phase-shifter.

System design and basic principle

Figure 1 shows the schematic diagram and its experimental setup of the RSI based on simultaneous polarization phase-shifter. It is composed of three main individual parts, i.e. wavefront simulator (WS), Hartmann-Shack wavefront sensor (HS WFS) and common-path phase-shifting RSI, which is the core part of this system. The proposed RSI includes a four-channel polarization phase stepper (FCPPS) and a cyclic RSI, which comprise a polarizing beam splitter (PBS1), two lenses (L5 and L6) and two flat mirrors (M1 and M2). The cyclic RSI can be treated as a Keplerian telescope system comprised L5 and L6. P2 is used to filter the state of polarization (SoP) of incident light at a fixed filtering angle. The passing beam will be guided into the RSI system and divided into two beams according to its components of SoP. According to the property of PBS, the reflected and transmitted lights has vertical linear SoP and horizontal linear SoP respectively, they will transmit the cyclic RSI system along anticlockwise direction (i.e. PBS1 → L5 → M1 → M2 → L6 → PBS1) and clockwise direction (i.e. PBS1 → L6 → M2 → M1 → L5 → PBS1) respectively and are guided into FCCPS system. These two beams are magnified and de-magnified due to the different focal lengths of lenses L5 and L6.

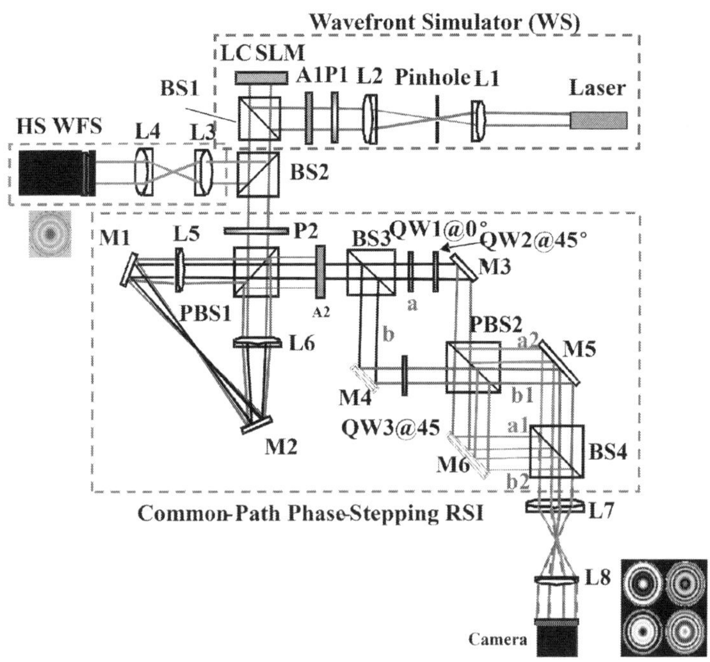

Figure 1. Concept design and experimental setup of the RSI with spatial phase-shifting method. L1–8, lenses; PBS1/2, polarization beam splitters; BS1–4, intensity beam splitters; M1–6, mirrors; QW1–3, quarter-wave plates; A1–2, diaphragms; P1–2, polarizers; WFS, wavefront sensor.

The complex amplitude of the incident light $E_0(x, y)$ can be presented as

$$E_0(x,y) = A_0(x,y)\exp\left[-i\varphi_0(x,y)\right] \quad (1)$$

where $A_0(x, y)$ and $\varphi_0(x, y)$ are the amplitude distribution and the phase distribution (i.e. wavefront) respectively, and (x, y) represents the spatial coordinates of amplitude and phase of the complex amplitude.

Assuming the focal length of L6 (i.e. f_6) larger than the focal length of L5 (i.e. f_5), the radial shearing ratio $s = f_6/f_5$ is larger than 1. Then $E_1(x, y)$ and $E_2(x, y)$, which are the complex amplitudes of two exiting beams from the cyclic RSI, can be presented by

$$\begin{cases} E_1(x,y) = A_0(x/s,y/s)\exp\left[-i\varphi_0(x/s,y/s)\right] \\ \qquad\quad = A_1(x,y)\exp\left[-i\varphi_1(x,y)\right] \\ E_2(x,y) = A_0(xs,ys)\exp\left[-i\varphi_0(xs,ys)\right] \\ \qquad\quad = A_2(x,y)\exp\left[-i\varphi_2(x,y)\right] \end{cases} \quad (2)$$

where $A_{1/2}(x, y)$ present the amplitude distributions of magnified and de-magnified beams respectively; $\varphi_1(x, y)$, $\Delta\varphi_2(x, y)$ presents the phase distributions of magnified and de-magnified beams respectively. Their definitions are shown as

$$\begin{cases} A_1(x,y) = A_0(x/s,y/s) \\ \varphi_1(x,y) = \varphi_0(x/s,y/s) \\ A_2(x,y) = A_0(xs,ys), \\ \varphi_2(x,y) = \varphi_0(xs,ys) \end{cases} \quad (3)$$

The aperture size of incident laser beam is limited as D by A1, and the corresponding beam size of exiting beams from cyclic RSI will be Ds and D/s respectively. The beam size is filtered again as same with their common area by A2 before they are introduced into FCCPS, as shown in Figure 1.

The FCPPS will generate the desired phase shifts of π/2 between each generated interferograms. Firstly, two beams, which generated from the cyclic RSI, are equally divided into two channels (i.e. channel a and channel b, shown in Figure 1) by the BS3. QW1 (angle distance of 0° from its fast axis to horizontal direction) and QW2 (angle distance of 45° from its fast axis to horizontal direction) are placed in channel *a*. QW1 generates a λ/4 phase delay between the Beam1 and the Beam2, and QW2 converts two beams into circularly polarized lights with levorotation and dextrorotation respectively. QW3 is placed into the channel b, and its fast axis keeps accordance to QW2. Another two circularly polarized beams with levorotation and dextrorotation are generated. The beams of channel a and b is split into four beam pairs by PBS2, i.e. a1, a2, b1 and b2 respectively, and it will introduces λ/2 phase delay into between channel 1 and 2. Four beam pairs are spatial arranged properly by mirrors M5 and M6 and beam splitter BS4, and they are projected onto the photosensitive plane of a CCD camera by the relay optics system composed of lenses L7 and L8. Finally, four interferograms are detected and recorded by the CCD camera.

The basic principle of the spatial phase-shifting can be explained by Jones formulas. The Jones matrices for a horizontal linear polarizer P_0, a vertical linear polarizer P_{90}, and a quarter-wave plate with horizontal fast axis Q_0, and 45° fast axis Q_{45}, are respectively

$$P_0 = \begin{bmatrix} 1 & 0 \\ 0 & 0 \end{bmatrix}, P_{90} = \begin{bmatrix} 0 & 0 \\ 0 & 1 \end{bmatrix}, Q_0 = \begin{bmatrix} 1 & 0 \\ 0 & -i \end{bmatrix}, Q_{45} = \frac{1}{\sqrt{2}} \begin{bmatrix} 1 & i \\ i & 1 \end{bmatrix} \tag{4}$$

The Jones matrix in the polarization state for each of the four channels can be represented by

$$a1 = P_0 Q_{45} Q_0 = \frac{1}{\sqrt{2}} \begin{bmatrix} 1 & 1 \\ 0 & 0 \end{bmatrix}, a2 = P_{90} Q_{45} Q_0 = \frac{1}{\sqrt{2}} \begin{bmatrix} 0 & 0 \\ i & -i \end{bmatrix},$$

$$b1 = P_0 Q_{45} = \frac{1}{\sqrt{2}} \begin{bmatrix} 1 & i \\ 0 & 0 \end{bmatrix}, b2 = P_{90} Q_{45} = \frac{1}{\sqrt{2}} \begin{bmatrix} 0 & 0 \\ i & 1 \end{bmatrix} \tag{5}$$

From Eq. (5), 0, 90, 180 and 270° phase shifting have been generated between each beam pair, which are the a1, b1, a2 and b2 respectively.

Wavefront difference $\Delta\varphi(x, y)$, which presents the difference between wavefront of Beam1 and Beam2 in their interference zone can be written as

$$\Delta\varphi(x,y) = \varphi_2(x,y) - \varphi_1(x,y) = \varphi_0(xs,ys) - \varphi_0(x/s,y/s) \tag{6}$$

The interference intensity of four interferograms, which are expressed by I_{a1}, I_{b1}, I_{a2} and I_{b2} respectively, can be represented by

$$\begin{aligned} I_{a1} &= A_1^2/2 + A_2^2/2 + A_1 A_2 \cos(\Delta\varphi) \\ I_{b1} &= A_1^2/2 + A_2^2/2 + A_1 A_2 \cos(\Delta\varphi + \pi/2) \\ I_{a2} &= A_1^2/2 + A_2^2/2 + A_1 A_2 \cos(\Delta\varphi + \pi) \\ I_{b2} &= A_1^2/2 + A_2^2/2 + A_1 A_2 \cos(\Delta\varphi + 3\pi/2) \end{aligned} \tag{7}$$

where I_{ai}, I_{bi}, A_i and $\Delta\varphi$ are the abbreviations of $I_{ai}(x, y)$, $I_{bi}(x, y)$, $A_i(x, y)$ and $\Delta\varphi(x, y)$, here i = 1,2.

Thus the wavefront difference $\Delta\varphi(x, y)$ can be calculated by classic four-frame phase shifting algorithm, which given by

$$\Delta\varphi(x,y) = \tan^{-1}\left[\frac{I_{b2}(x,y) - I_{b1}(x,y)}{I_{a1}(x,y) - I_{a2}(x,y)}\right] \tag{8}$$

Then, a suitable phase unwrapping algorithm is employed to acquire the absolute wavefront difference distribution.

Experimental results and analysis

As a comparison, a wavefront simulator (WS), which is composed mainly of an electrically addressed phase-only liquid crystal spatial light modulator (LC SLM), is adopt in experimental system to generate the candidate wavefront, and it will be measured by both a Hartmann-Shack wavefront sensor and the described RSI. For achieving this goal, the LC SLM should be adjusted in a phase-only modulation mode by keeping accordance of polarization state between polarizer and LC SLM. The generated wavefront by WS is measured separately, and compare with that obtained by the proposed RSI. The lenses L3 and L4 placed in HS WFS system relay the LC SLM to the HS WFS. The experiment is performed with a He-Ne laser, and its wavelength λ = 632.8 nm. The focal length of Lens1 and Lens2 are 250 mm and 300 mm respectively. Four interferograms are detected by one 8 bit, 576 × 768 pixels CCD camera. The full resolution of HS WFS is 32 × 32, and the aberrated wavefront are generated by a 512 × 512 pixel LC SLM.

Figure 2 shows the practical experimental setup of the proposed RSI. In experiments, we find the best angle for acquiring highest fringe visibility firstly. Then the LC SLM is controlled to generate a random aberration, which is a combination of 45-limit order Zernike polynomials. Four fringe patterns were detected simultaneously by the CCD camera, and they are shown in Figure 3.

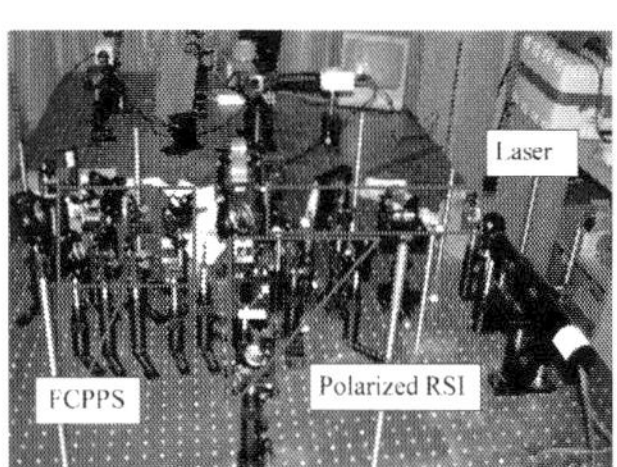

Figure 2. Experimental setup of the proposed RSI.

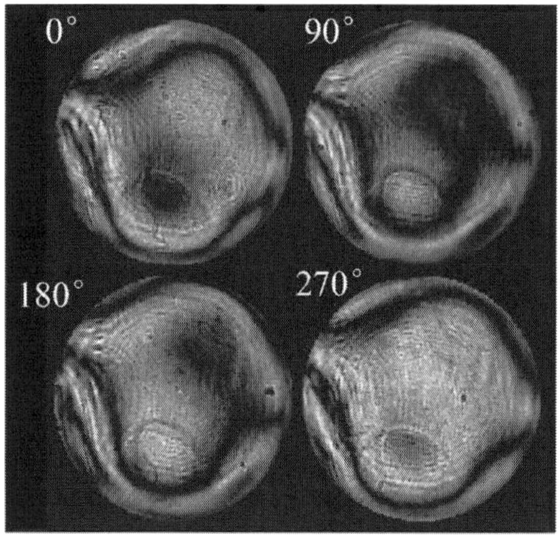

Figure 3. Interferograms recorded by the experimental setup.

The wrapped wavefront difference $\Delta\varphi(x, y)$ can be calculated according to Eq. (8). The unwrapped result is also shown in Figure 4(b). A modal wavefront reconstruction method is employed to reconstruct the wavefront under test from the unwrapped wavefront differences. The RMS and the PV of the reconstructed wavefront are 0.4753λ and 3.3222λ, respectively, and here RMS and PV represent the root-mean-square value and peak-to-valley value of the wavefront under test. When we normalize the RMS value of each order Zernike polynomial to 1.0λ, the Zernike polynomial coefficients of the wavefront under test can be obtained by using the least square method, and it is shown in Figure 5(b).

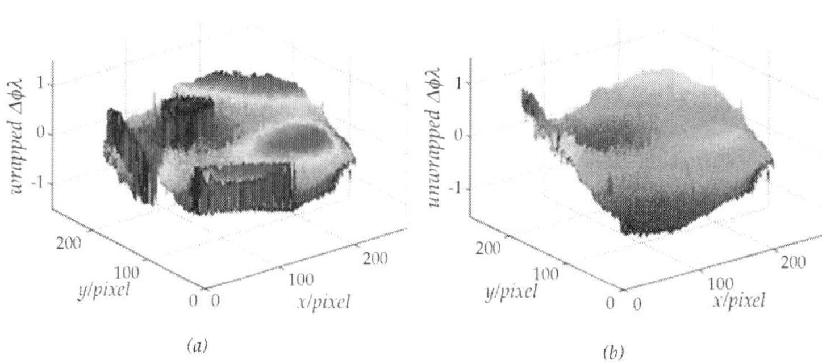

Figure 4. The wavefront difference calculated from the four interferograms (shown in Figure 3). (a) The wrapped wavefront difference and (b) the unwrapped wavefront difference.

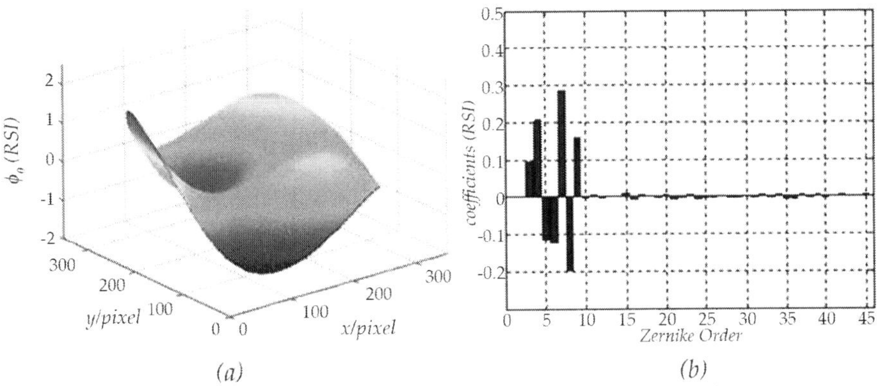

Figure 5. The wavefront under test measured by the proposed RSI. (a) The 3D plot of the wavefront under test reconstructed from the wavefront difference shown in Figure 4(b) and (b) the coefficients of the first 45 order Zernike polynomials.

The same aberrated wavefront also detected by HS WFS separately. A spot array is detected is split into HS WFS, and the corresponding gradient distributions along horizontal and vertical directions can be calculated. Finally the wavefront under test was reconstructed again by using a similar modal reconstruction method. Figure 6(a) shows the reconstructed result, and the RMS and PV are 0.4798λ and 3.5953λ, respectively. The residual error, which is the difference between these two wavefront under test measured by the proposed RSI and the HS WFS, is also drawn in Figure 6(b), and the corresponding RMS and PV are 0.0348λ and 0.3149λ, respectively.

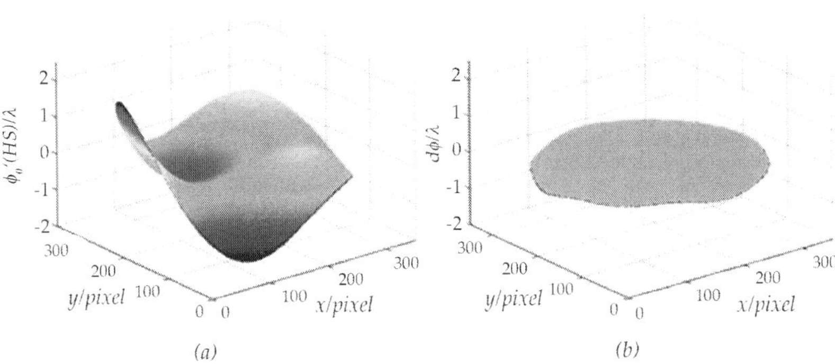

Figure 6. The measured results of HS WFS and the residual error. (a) Experimental result measured by HS WFS (RMS and PV are 0.4798λ and 3.5953λ respectively) and (b) residual errors between wavefront measured by RSI and HS-WFS (RMS and PV are 0.0348λ and 0.3149λ respectively).

From Figure 6, a small residual error is obtained, and the result measured by the proposed RSI keeping accordance with HS WFS. The difference is mainly due to the detecting noise of the CCD camera and the grating effect of the LC SLM. Better results would be acquired when a high performance CCD camera and a continuous surface wavefront simulator are used.

Figure 7 gives the other experimental results when the LC SLM generates some wavefront under test with single Zernike polynomial. Figure 8 shows the statistic results of RMS and PV of 3–35 order Zernike polynomials measured by the proposed RSI and HS WFS.

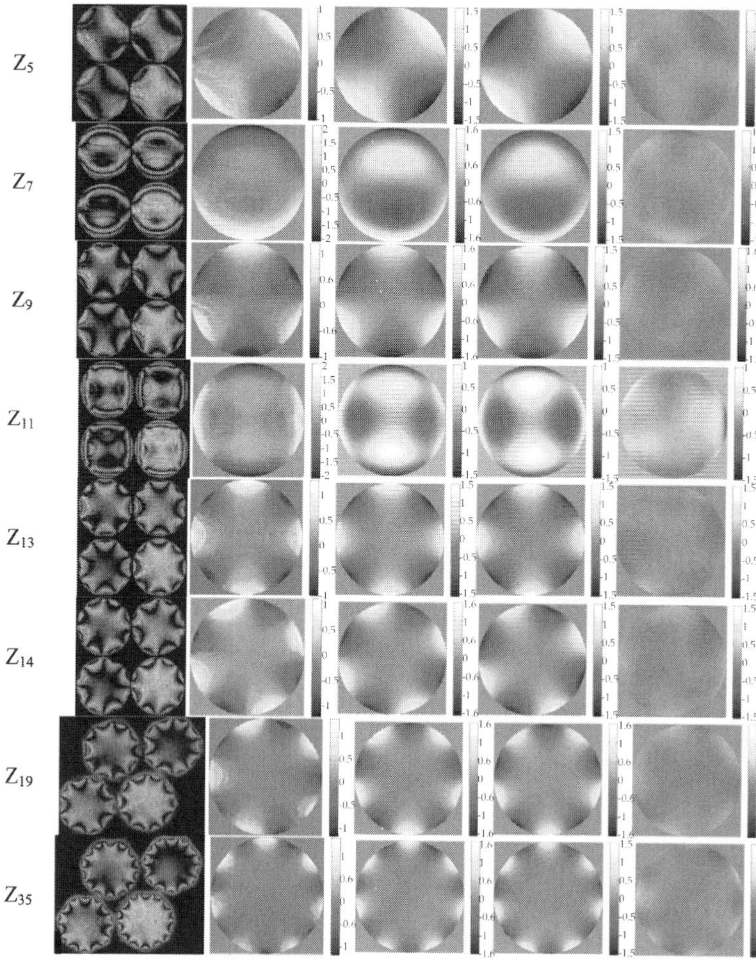

Figure 7. Experimental results when the LC SLM generates some wavefront under test with single Zernike polynomial.

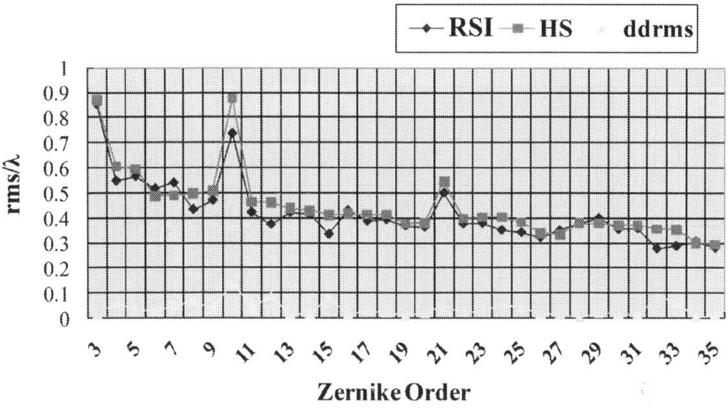

Figure 8. Experimental results when the LC SLM generates some wavefront under test with single Zernike polynomial.

COMPACT RSI BASED ON MICRO-RETARDER ARRAY

In RSIs, complex configuration limits the measurement accuracy, system stability, and its application. Apparently, it is highly desirable to design compact one-shot vibration-insensitive RSIs. In this part, we will introduce a compact RSI based on pixelated micro-retarder-array (MRA).

Design and theory of the compact RSI

The basic concept and the schematic layout of the new compact RSI based on MRA are shown in Figure 9. The cyclic RSI is a Keplerian telescope system, which is same as the counterpart of Figure 1. The complex amplitude $E_0(x, y)$ can also be presented by Eq. (1). The half-wave retarder (HWR) is to adjust the polarization of the introduced light beam. Let θ as the angle between the polarization orientation of the introduced beam and the fast axis of the HWR. The polarization of the beam will be rotated by 2θ after passing through the HWR. The PBS splits the incident light into two beams with perpendicular polarization. The complex amplitude distributions of the new beam are generated when they pass through L1 and L2 and L2 and L1, respectively.

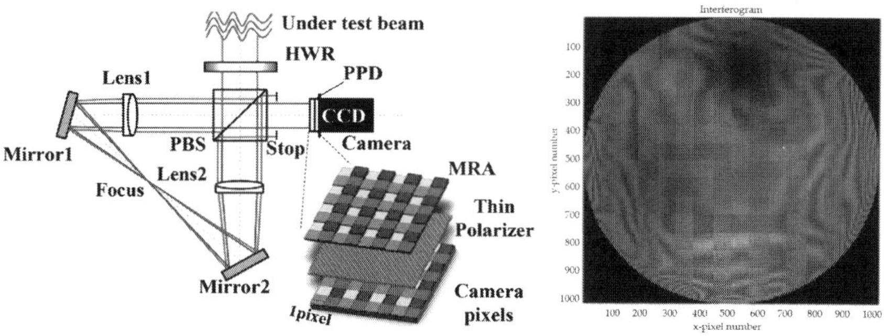

Figure 9. Concept design of the proposed compact RSI. PPD, pixelated polarization device; HWR, half-wave retarder; PBS, polarized beam splitter; MRA, micro-retarder array.

Figure 10 illustrates polarization conversion and complex amplitude after passing through different optical elements. $E_0(x, y)$ presents the optical field of the incident beam, and the variable α presents the angle of its SoP relative to x axis (i.e. horizontal direction). $E'_0(x, y)$ presents the complex amplitude of the exiting beam from HWR. The fast axis of HWR is presented by the dashed line. The complex amplitudes of reflection beam and transmission beam after passing through PBS can be presented as

$$E_p(x,y) = E'_0(x,y)\cos(\alpha - 2\theta),\ E_s(x,y) = E'_0(x,y)\sin(\alpha - 2\theta) \tag{9}$$

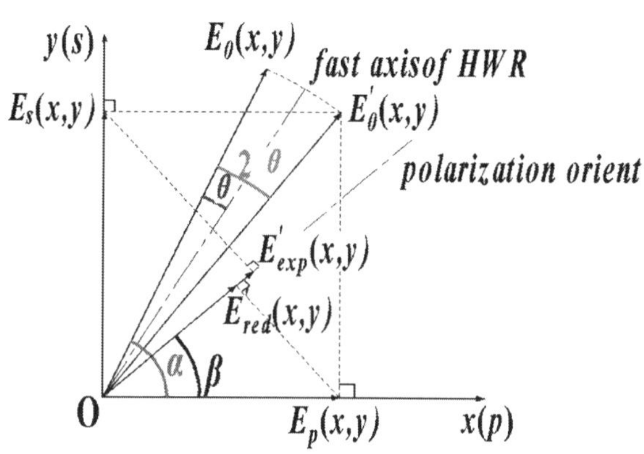

Figure 10. Sketch map of SoPs conversion.

In fact the $E'_0(x, y)$ can be treated same as the $E_0(x, y)$ in scalar field because they have the same amplitude and phase distribution when they passing through HWR, which just change the SoP of the incident beam under different fast axis. Similar to the cyclic RSI in Figure 1, there are also two beams, i.e. magnified (i.e. $E_{exp}(x, y)$) and de-magnified (i.e. $E_{red}(x, y)$) beams, are generated by the cyclic RSI, and they can be presented by

$$\begin{aligned} E_{\exp}(x,y) &= E_s(x/s, y/s)/s \\ E_{red}(x,y) &= sE_p(xs, ys) \end{aligned} \tag{10}$$

Finally the complex amplitude distribution of expanded and reduced beams can be presented as

$$\begin{aligned} E_{\exp}(x,y) &= \frac{1}{s}A_0'\left(\frac{x}{s},\frac{y}{s}\right)\exp\left[-j\varphi_0\left(\frac{x}{s},\frac{y}{s}\right)\right] \\ E_{red}(x,y) &= sA_0''(xs, ys)\exp\left[-j\varphi_0(xs, ys)\right] \end{aligned} \tag{11}$$

where $A'_0(x, y) = A_0(x, y)\sin(\alpha - 2\theta)$, and $A''_0(x, y) = A_0(x, y)\cos(\alpha - 2\theta)$.

Then the expanded beam and reduced beam with orthogonal polarization are introduced into the proposed phase-shifter, which consists of a micro-retarder array (MRA) and a piece of thin polarizer. Each macro pixel of MRA is composed of four neighboring pixels with different-thickness (as shown in Figure 11(a)). The MRA is made by a slice of birefringent crystal, and it can be quartz, calcite or the other kind of optical birefringent material. The fast axis of the selected birefringent crystal should be along or orthometric with x axis (i.e. horizontal direction) for keeping polarization status of incident light from changes when different phase-shifts are introduced. The schematic diagram of the MRA is drawn in Figure 11, and the basic principle of phase delay between two radial shearing beams is also introduced.

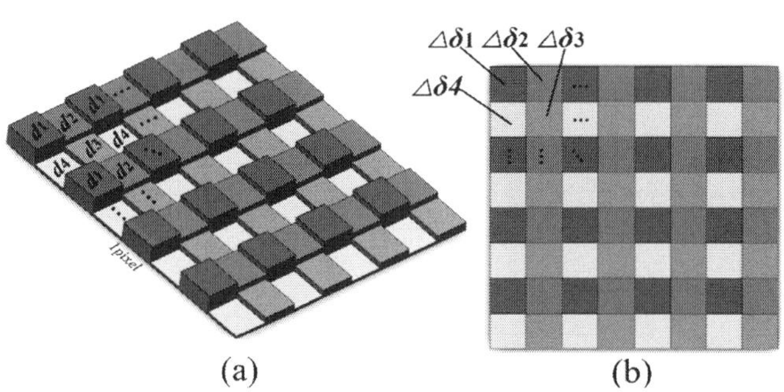

Figure 11. Schematic diagram of the MRA. (a) MRA composed of units with different-thickness pixels made by a slice of birefringence crystal, and (b) the phase delay of each pixel. d_1–d_4: thickness at four neighboring pixels, Δδ1–Δδ4: phase-shifting between s- and p-light for four neighboring pixels.

Pixel size of the MRA should be kept accordance with the pixel size of camera, and they should be integrated onto the photosensitive plane of the camera pixel by pixel. Here a negative single axis birefringent crystal is assumed, and let the fast axis of it along vertical direction. Thus the phase-delays of the four neighboring pixels is

$$\Delta\delta_i = (n_o - n_e)d_i \cdot 2\pi/\lambda, i = 1, 2, 3, 4 \tag{12}$$

where λ is the center wavelength of light, and *i* is the sequence number of the four neighboring pixels.

When these two beams exiting from the cyclic RSI transmit through MRA, Eq. (11) at each pixel is changed as

$$\begin{aligned} E_{\exp,i}(x,y) &= A_0'(x/s,y/s)\exp[-j(\varphi_0(x/s,y/s)+\delta_i+\Delta\delta_i)]/s \\ E_{red,i}(x,y) &= sA_0''(xs,ys)\exp[-j(\varphi_0(xs,ys)+\delta_i)] \end{aligned} \tag{13}$$

where δ_i are the phase-shifts when the o-light transmits through MRA.

After passing through the thin polarizer, Eq. (13) can be rewritten by

$$\begin{aligned} E'_{\exp,i}(x,y) &= E_{\exp,i}(x,y)\sin\beta \\ E'_{red,i}(x,y) &= E_{con,i}(x,y)\cos\beta \end{aligned} \tag{14}$$

here β presents the filtering angle of the thin polarizer relative to horizontal direction, as shown in Figure 10.

After passing through the thin polarizer, the beams will be interference with each other on the pixels of the camera, and the intensity at four adjacent pixels can be represented as

$$\begin{aligned} I_i(x,y) &= \left|E'_{\exp,i}(x,y)\right|^2 + \left|E'_{red,i}(x,y)\right|^2 + 2\left|E'_{\exp,i}(x,y)\right|\left|E'_{red,i}(x,y)\right|\cos[\Delta\varphi(x,y)+\Delta\delta_i(x,y)] \\ &= I_a(x,y) + I_b(x,y) + 2\sqrt{I_a(x,y)I_b(x,y)}\cos[\Delta\varphi(x,y)+\Delta\delta_i(x,y)] \end{aligned} \tag{15}$$

where $\Delta\varphi(x, y)$ represents the mean phase difference between the beams at (x, y) position. x and y presents the center position of each macro-pixel.

Combining the Eqs. (11), (13)–(15), a new equation can be obtained as

$$\begin{aligned} I_a(x,y) &= I_0\left(\frac{x}{s},\frac{y}{s}\right)\sin^2\beta\sin^2(\alpha-2\theta)/s^2, \\ I_b(x,y) &= s^2 I_0(xs,ys)\cos^2\beta\cos^2(\alpha-2\theta) \\ \Delta\varphi(x,y) &= \varphi_0(xs,ys) - \varphi_0\left(\frac{x}{s},\frac{y}{s}\right), \\ I_0(x,y) &= |E_0(x,y)|^2 = |A_0(x,y)|^2 \end{aligned} \tag{16}$$

Eq. (15) can be expressed in discrete form as

$$I_i(m,n) = I_a(m,n) + I_b(m,n) + 2\sqrt{I_a(m,n)I_b(m,n)}\cos[\Delta\varphi(m,n) + \Delta\delta_i(m,n)] \quad (17)$$

where (m, n) is a discrete form of (x, y), which presents the coordinate positions. Assuming the number of rows and columns of camera pixels are $2M_0$ and $2N_0$ respectively, the (m, n) should take a value from 1 to M_0 and 1 to N_0 respectively.

Then Eq. (17) can be rewritten as

$$I_i(m,n) = K(m,n) + L(m,n)\cos[\Delta\delta_i(m,n)] - R(m,n)\sin[\Delta\delta_i(m,n)] \quad (18)$$

where

$$K(m,n) = I_a(m,n) + I_b(m,n)$$

$$L(m,n) = 2\sqrt{I_a(m,n)I_b(m,n)}\cos[\Delta\varphi(m,n)]$$

$$R(m,n) = 2\sqrt{I_a(m,n)I_b(m,n)}\sin[\Delta\varphi(m,n)] \quad (19)$$

Eq. (19) can be expressed by matrix form, and it is

$$\begin{bmatrix} I_1(m,n) \\ I_2(m,n) \\ I_3(m,n) \\ I_4(m,n) \end{bmatrix} = \begin{bmatrix} 1 & \cos[\Delta\delta_1(m,n)] & -\sin[\Delta\delta_1(m,n)] \\ 1 & \cos[\Delta\delta_2(m,n)] & -\sin[\Delta\delta_2(m,n)] \\ 1 & \cos[\Delta\delta_3(m,n)] & -\sin[\Delta\delta_3(m,n)] \\ 1 & \cos[\Delta\delta_4(m,n)] & -\sin[\Delta\delta_4(m,n)] \end{bmatrix} \begin{bmatrix} K(m,n) \\ L(m,n) \\ R(m,n) \end{bmatrix} \quad (20)$$

Thus

$$\begin{bmatrix} K(m,n) \\ L(m,n) \\ R(m,n) \end{bmatrix} = \begin{bmatrix} A_1(m,n) & B_1(m,n) & C_1(m,n) & D_1(m,n) \\ A_2(m,n) & B_2(m,n) & C_2(m,n) & D_2(m,n) \\ A_3(m,n) & B_3(m,n) & C_3(m,n) & D_3(m,n) \end{bmatrix} \begin{bmatrix} I_1(m,n) \\ I_2(m,n) \\ I_3(m,n) \\ I_4(m,n) \end{bmatrix} \quad (21)$$

where

$$\begin{bmatrix} A_1(m,n) & B_1(m,n) & C_1(m,n) & D_1(m,n) \\ A_2(m,n) & B_2(m,n) & C_2(m,n) & D_2(m,n) \\ A_3(m,n) & B_3(m,n) & C_3(m,n) & D_3(m,n) \end{bmatrix} = \begin{bmatrix} 1 & \cos[\Delta\delta_1(m,n)] & -\sin[\Delta\delta_1(m,n)] \\ 1 & \cos[\Delta\delta_2(m,n)] & -\sin[\Delta\delta_2(m,n)] \\ 1 & \cos[\Delta\delta_3(m,n)] & -\sin[\Delta\delta_3(m,n)] \\ 1 & \cos[\Delta\delta_4(m,n)] & -\sin[\Delta\delta_4(m,n)] \end{bmatrix}^{i} \quad (22)$$

The matrix **K**, **L** and **R** can be solved for all units $M_0 \times N_0$ by

$$K = A_1. \times I_1 + B_1. \times I_2 + C_1. \times I_3 + D_1. \times I_4$$

$$L = A_2. \times I_1 + B_2. \times I_2 + C_2. \times I_3 + D_2. \times I_4$$

$$R = A_3. \times I_1 + B_3. \times I_2 + C_3. \times I_3 + D_3. \times I_4 \quad (23)$$

For constant phase-shifts $\Delta\delta_1$–$\Delta\delta_4$, A_1–D_3 are constant coefficients that can be calculated by Eq. (22) or calibrated by only one time. They can be used to extract random phase delay when the matrix of Eq. (22) is reversible.

Combining Eqs. (19) and (23), the phase difference between two shearing beams can be presented as

$$\Delta\varphi(m,n) = \tan^{-1}[R(m,n)/L(m,n)] \tag{24}$$

The wavefront under test $\varphi_0(x, y)$ can be reconstructed by Zernike mode [11] and iterative methods [30, 31].

The fringe visibility can be adjusted by changing the angle of HWR. Let ρ_0 as the mean power density of beam under test, the power density of the reduced beam ρ_p and the expanded beam ρ_s exiting the polarizer can be written by

$$\begin{aligned} \rho_p &= \rho_0 \cos^2(\alpha - 2\theta) \cdot s^2 \cdot \cos^2\beta \\ \rho_s &= \rho_0 \sin^2(\alpha - 2\theta)/s^2 \cdot \sin^2\beta \end{aligned} \tag{25}$$

For the best fringe visibility, ρ_p should be equal to ρ_s, and that means

$$\tan(\alpha - 2\theta)\tan\beta = s^2 \tag{26}$$

Generally, for a certain RSI system, the shear ratio s and the angle β are constants, so the fringe visibility can be adjusted by change of θ according to the polarization angle α of the beam under test.

Validation by numeric analysis

In this part, numeric analysis is made for validating the feasibility of the proposed compact RSI. The systematic parameters are listed as follows: aperture size Φ = 5 mm, random polarization angle of beam under test α = 139°, wavelength λ = 632.8 nm, focal lengths of Lens1 f_1 = 50 mm, focal lengths of Lens2 f_2 = 60 mm, β = 45°. And calcite retarder is chosen with vertical fast axis and (n_e, n_o) = (1.4852, 1.6557), thickness of four pixels per unit are [0.0591, 0.1536, 0.3072, 0.4785] μm. The pixel size of 1024 × 1024 camera is 9.8 μm × 9.8 μm. For the proposed RSI, the phase-shifts at four adjacent pixels per unit are [0.6283, 1.6336, 3.2673, 5.0894] rad, and θ is 41.89°for acquiring the best fringe visibility.

The beam under test is generated with random complex amplitude, which are shown in Figure 12(a) and (b), respectively. The interferogram generated by the proposed RSI is calculated from Eqs. (12), (15) and (16), as shown in Figure 12(a). The matrix **K**, **L** and **R** are calculated by Eqs. (22) and (23). The wavefront difference between two shearing beams, as shown in Figure 12(b), is extracted by Eq. (24).

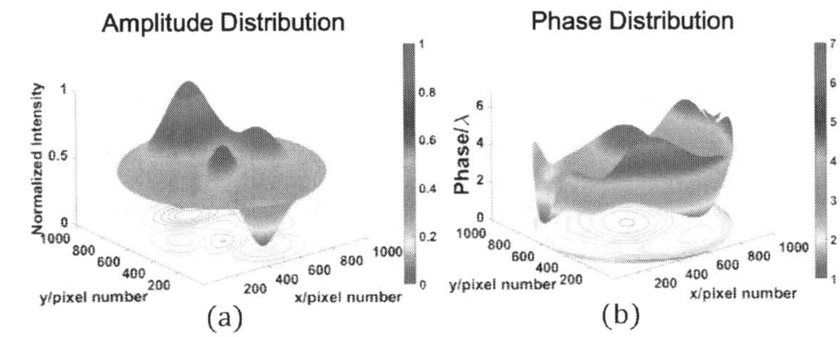

Figure 12. Complex amplitude of the beam under test. (a) Random amplitude distribution generated by combination of 3-Gaussi functions, and (b) random phase distribution generated by combination of 65-limit Zernike polynomials (RMS = 0.975λ, PV = 5.736 λ).

Figure 13 shows the wavefront difference between two radial shearing beams and the deviation relative to the ideal wavefront difference. As seen from Figure 13(b), the extracted error is less than 0.2%. The error is mainly caused by approximate treatment of intensity and phase values at four neighboring pixels. One should obtain a smaller deviation when a smoother intensity and phase distribution or large sampling number of camera is used.

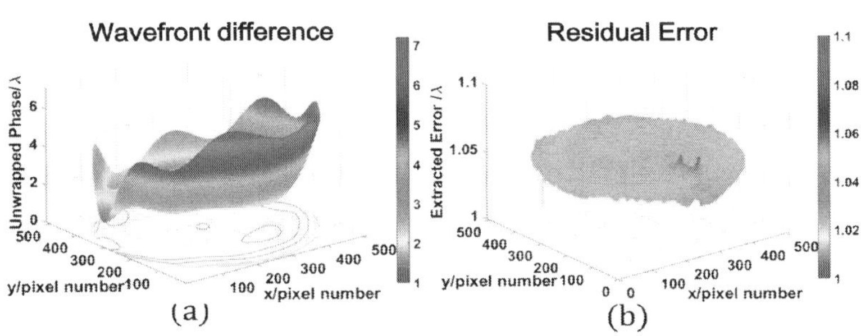

Figure 13. The wavefront difference and its residual error relative to ideal value. (a) The unwrapped wavefront difference (RMS = 0.972λ), and (b) the error of the extracted wavefront difference (RMS = 0.002λ).

Figure 14 shows the results of wavefront reconstruction. As seen from Figure 14, the RMS of the reconstructed wavefront is about 0.975λ, which is very close to the incident wavefront. The RMS of residual measurement error is about 0.008λ, which leads a relative deviation of 0.79%. In fact, if planar intensity amplitude and a smoother phase are employed in our numeric analysis, a better accuracy would be obtained based the method.

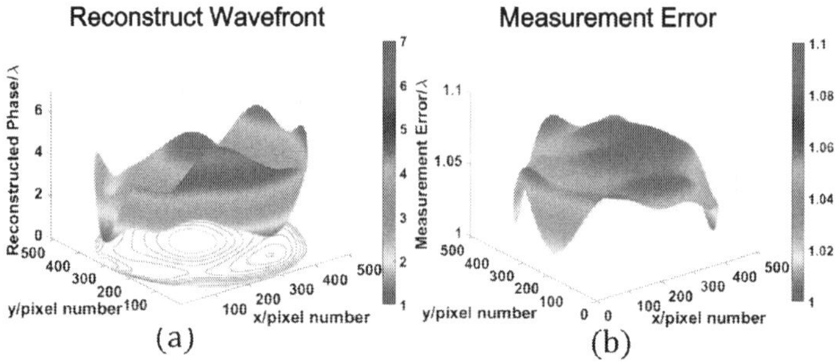

Figure 14. The reconstruction result and the residual error. (a) The reconstruction result (RMS = 0.975λ), and (b) the residual error (RMS = 0.008λ).

MODAL METHOD FOR RECONSTRUCTING WAVEFRONT IN RSI WITH OR WITHOUT LATER SHEAR

In RSI no extra referencing planar wavefront but a magnified wavefront can be used as a reference. Thus the magnified and de-magnified beams should be aligned accurately avoiding wrong wavefront reconstruction. Unfortunately, two beams in practical application are often misalignment, and a lateral shear is always exists in a RSI. Here we present a modal method to solve this problem.

Modal reconstruction method based Zernike polynomials

As shown in Figure 15, the wavefront difference at interference area can be described as

$$\Delta\varphi(x,y) = \varphi_1(x,y) - \varphi_2(x,y) \tag{27}$$

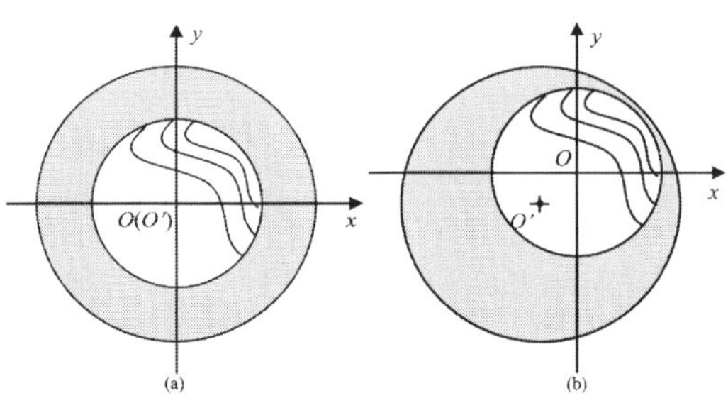

Figure 15. Radial shearing interferogram without and with lateral shear. (a) Shearogram without lateral shear and (b) shearogram with lateral shear.

The wavefront aberration $\varphi_1(x, y)$ is a de-magnified version of the wavefront under test $\varphi_0(x, y)$, and there no information is lost. So the wavefront aberration $\varphi_1(x, y)$ contains the same information with the wavefront under test $\varphi_0(x, y)$ if one normalize the diameter of the origin pupil and the de-magnified pupil. In other words, one can consider $\varphi_1(x, y)$ to be as the wavefront under test $\varphi_0(x, y)$. For simplicity, we define the interference area as *circle*(d), and in this area two versions of wavefront under test can be expressed as

$$\begin{cases} \varphi_1(x,y) = \varphi_0(x,y) \\ \varphi_2(x,y) = \varphi_0(x/s^2, y/s^2) \end{cases}, (x,y) \in circle(d) \tag{28}$$

Eq. (28) is under an ideal alignment condition as shown in Figure 15(a). However, in a practical RSI system lateral shear always exists and is inevitable. As shown in Figure 15(b), the smaller circle denotes the de-magnified beam, and it also is the interference area, i.e. *circle*(d). For convenience, the center of the contracted beam can be defined as the origin O of the Cartesian coordinate system, and the center position O′ of the magnified beam express the amount of lateral shear. Two variables, i.e. x_0 and y_0, are defined as the lateral shear at x direction and y direction respectively, and they can be calibrated by only one time measurement. Two versions of the wavefront under test can be rewritten as

$$\begin{cases} \varphi_1(x,y) = \varphi_0(x,y) \\ \varphi_2(x,y) = \varphi_0(x/s^2 - x_0, y/s^2 - y_0) \end{cases} \tag{29}$$

Suppose that the $\varphi_0(x, y)$ can be described by a N-limited Zernike polynomials, and thus $\varphi_0(x, y)$ can be expressed as

$$\varphi_0(x,y) = \sum_{k=0}^{N} a_k Z_k(x,y) \tag{30}$$

where $Z_k(x, y)$ is the kth order Zernike polynomial and a_k is the corresponding coefficient.

The magnified wavefront $\varphi_2(x, y)$ in *circle*(d) can be written by

$$\varphi_2(x,y) = \sum_{k=0}^{N} a_k P_k(x,y) \tag{31}$$

where $P_k(x, y)$ is a portion of Zernike polynomials $Z_k^D(x, y)$ in the expanded beam area (denoted as *circle*(D)), i.e. $P_k(x, y) = Z_k^D(x, y)$, $(x, y) \in circle(d)$.

According to the reference [32], provided an N-limited Zernike description of a wavefront on a large pupil, any circular portion inside it can be described by another Zernike ensemble, limited to the same N. So the function $P_k(x, y)$ can be expressed by

$$P_k(x,y) = \sum_{j=0}^{k} b_j^k Z_j(x,y) \tag{32}$$

Put the Eqs. (30)–(32) into Eq. (2), a new equation is generated, and it is

$$\Delta\varphi(x,y) = \sum_{k=0}^{N} a_k Z_k(x,y) - \sum_{k=0}^{N} a_k \left[\sum_{j=0}^{k} b_j^k Z_j(x,y) \right] \tag{33}$$

Eq. (33) describes the relationship between the wavefront difference and the Zernike polynomials. However, it is also indirect and not clear. Actually, Eq. (33) can be rewrite in its matrix form, and it is expressed by

$$\Delta\varphi = [a_1 \cdots a_k \cdots a_N] \left\{ \begin{bmatrix} 1 & & & & \\ & \ddots & & & \\ & & 1 & & \\ & & & \ddots & \\ & & & & 1 \end{bmatrix} - \begin{bmatrix} b_1^1 & & & & \\ \vdots & \ddots & & & \\ b_1^k & \cdots & b_k^k & & \\ \vdots & \vdots & \vdots & \ddots & \\ b_1^N & \cdots & b_k^N & \cdots & b_N^N \end{bmatrix} \right\} \begin{bmatrix} Z_1 \\ \vdots \\ Z_k \\ \vdots \\ Z_N \end{bmatrix} \tag{34}$$

where $\Delta\varphi$ and Z_k are the compact form of matrix formalism of $\Delta\varphi(x, y)$ and $Z_k(x, y)$, respectively.

Eq. (34) can be simplified as

$$\Delta\varphi = \mathrm{ABZ} \tag{35}$$

where $\mathbf{A} = \{a_1, \ldots, a_k, \ldots, a_N\}$ is a coefficient vector of the wavefront under test; $\mathbf{Z} = \{Z_1, \ldots, Z_k, \ldots, Z_N\}^T$ is a column vector which is composed by a set of Zernike polynomials; $\mathbf{B}$ is a coefficient matrix, and it is

$$\mathbf{B} = \begin{bmatrix} 1-b_1^1 & & & & \\ \vdots & \ddots & & & \\ -b_1^k & \cdots & 1-b_k^k & & \\ \vdots & \vdots & \vdots & \ddots & \\ -b_1^N & \cdots & -b_k^N & \cdots & 1-b_N^N \end{bmatrix} \tag{36}$$

On the other hand, the wavefront difference $\Delta\varphi(x, y)$ can be decomposed into a linear combination of the N-limited Zernike polynomials, and it is shown by

$$\Delta\varphi = \mathrm{CZ} \tag{37}$$

where $\mathbf{C} = \{c_1, \ldots, c_k, \ldots, c_N\}$ is the vector of coefficient.

There has only one unique decomposed result of Eq. (37) because the Zernike polynomials is orthogonal each other, and the solve can be presented as

$$C=AB \tag{38}$$

We can get

$$A=CB^{+} \tag{39}$$

where $\mathbf{B}^{+}$ is the generalized inverse of the matrix **B**.

The coefficient vector **C** can be calculated directly from Eq. (37), and the coefficient vector **A** can be determined by the matrix **B** finally. In fact, the matrix B can be determined beforehand by a numerical calculation or a calibration process for a certain radial ratio s and lateral shear x_0 and y_0. Then the wavefront difference $\Delta\varphi(x, y)$ can be expressed by a set of Zernike polynomials, and the corresponding coefficient vector **C** is calculated. Finally, the coefficient vector **A** can be determined by only one time matrix operation, and then the wavefront under test is reconstructed.

Numeric analysis

A random phase distribution is generated as wavefront under test for validating the proposed modal reconstruction method. A 45-limit Zernike polynomials are used, and the full resolution is 256 × 256. 1.2 of radial shear ratio s is assumed in a RSI, and $\eta_x = 10\%$ and $\eta_y = -18\%$ of relative lateral shear at x direction and at y direction respectively are used in our numeric analysis.

Here $\varphi_0(x, y)$ is presented by a random composition of 45-limited Zernike polynomials. Two variables, σ_{rms} and σ_{pv}, are used to express the root-mean-square (RMS) and peak-to-valley (PV) of the wavefront respectively. The 3D-plot of the simulated random wavefront $\varphi_0(x, y)$ is shown in Figure 16(a). The RMS and PTV of the simulated wavefront under test are 2.000λ and 12.285λ respectively, where λ is the wavelength of laser. According to the radial shear ratio s and the relative lateral shear η_x and η_y, the wavefront difference $\Delta\varphi(x, y)$ can be calculated easily, and it is shown in Figure 16(b).

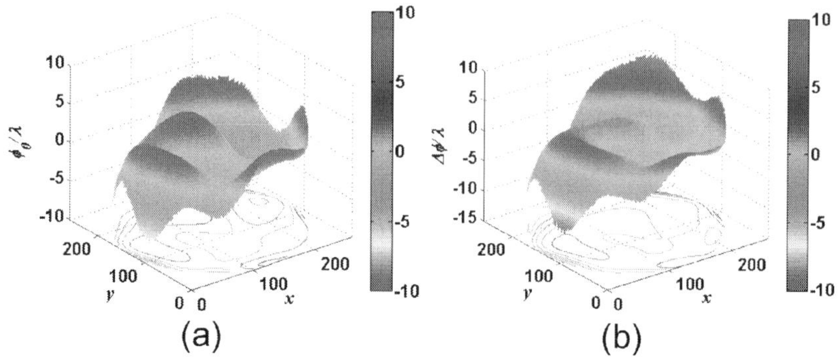

Figure 16. Wavefront under test and the corresponding wavefront difference. (a) Random wavefront under test $\varphi_0(x, y)$ and (b) The wavefront difference $\Delta\varphi(x, y)$ calculated from $\varphi_0(x, y)$; the RMS and PTV of $\Delta\varphi(x, y)$ are 2.091λ and 18.057λ respectively.

The coefficient matrix **B** can be calculated from Eqs. (32)–(36). According to the description of Eq. (39), the coefficient vector **A** is obtained. By applying Eq. (30), the wavefront of incident light can be reconstructed, and it is drawn in Figure 17(a). Figure 17(b) shows the residual error of wavefront reconstruction, and its corresponding RMS and PV are $7.003 \times 10^{-14}\lambda$ and $2.896 \times 10^{-13}\lambda$ respectively.

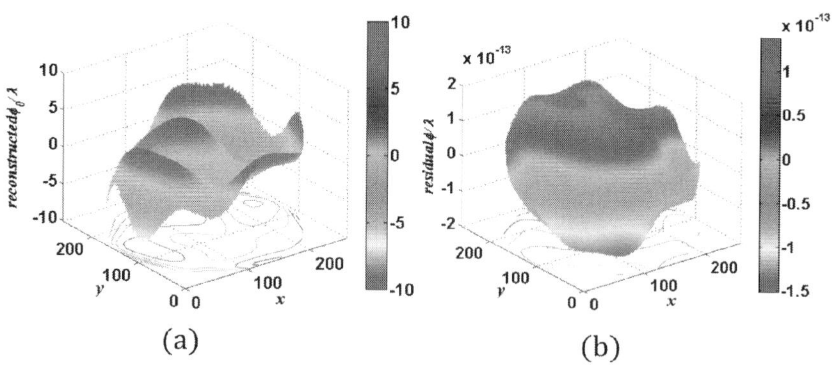

Figure 17. Wavefront reconstruction and the corresponding residual error. (a) The reconstructed wavefront under test by substituting the vector A into Eq. (30) and (b) the residual error between the origin wavefront and the reconstructed wavefront.

CONCLUSION

In this chapter, we have introduced briefly the history of RSI and the recent development. Two kinds of simultaneous RSI are also described in deep, including RSI based on simultaneous polarization phase-shifter and a new compact RSI based on micro-retarder array. Comparatively, the former one is suitable for wavefront measurements and diagnosing dynamic measurements, but its complex configuration limits the system stability, measurement accuracy and therefore its application. The new compact RSI based on micro-retarder array has several obvious advantages such as high-speed, high accuracy, vibration immunity, and compact size and so on, and it should be a kind of promising RSI in future. Complex extraction algorithm is another main barrier influencing the development of RSI. In this chapter a simple and useful modal reconstruction method is also given to extract the wavefront under test from the radial shearograms with or without lateral shear. Comparing with the previous method, the modal wavefront reconstruction method reduces effectively the noise accumulation and has good error propagation property. It is based on Zernike polynomials and its matrix formalism, and it should lead to an easier implementation in some practical situations.

ACKNOWLEDGMENTS

The authors are grateful to Prof. Wenhan Jiang for helpful discussions and suggestions. We would like to thank Prof. Changhui Rao, Prof. Linhai Huang, and Dr. Benxi Yao at Institute of Optics and Electronics (IOE), Chinese Academy of Science (CAS), for their cooperation and important help.

FUNDING

This work is supported by the National Natural Science Foundation of China under Grant Nos. 11643008 and 11727805.

REFERENCES

1. Hariharan P, Sen D. Radial shearing interferometer. Journal of Scientific Instruments. 1961;38:428-432. DOI: 10.1088/0950-7671/38/11/305
2. Liu D, Yang Y, Shen Y. System optimization of radial shearing interferometer for aspheric testing. Proceedings of SPIE. 2007;6834: 68340U-1-8. DOI: 10.1117/12.760132
3. Wang M, Zhang B, Shouping N. Radial shearing interferometer for aspheric surface testing. Proceedings of SPIE. 2002;4927:673-676. DOI: 10.1117/ 12.471690
4. Kohno T, Matsumoto D, Yazawa T. Radial shearing interferometer for inprocess measurement of diamond turning. Proceedings of SPIE. 1997;3173: 280-285. DOI: 10.1117/12.294519
5. Kowalik W, Garncarz B, Kasprzak H. Corneal topography measurement by means of radial shearing interference: Part I—Theoretical consideration. Optik. 2002;113:39-45. DOI: 10.1078/ 0030-4026-00113
6. Kowalik W, Garncarz B, Kasprzak H. Corneal topography measurement by means of radial shearing interference: Part II—experiment results. Optik. 2002;113:46-50. DOI: 10.1078/ 0030-4026-00112
7. Kowalik W, Garncarz B, Kasprzak H. Corneal topography measurement by means of radial shearing interference: Part III—measurement errors. Optik. 2003;114:199-206. DOI: 10.1078/ 0030-4026-00247
8. Gu N, Huang L, Yang Z. Changhui Rao: A single-shot common-path phasestepping radial shearing interferometer for wavefront measurements. Optics Express. 2011;19:4703-4713. DOI: 10.1364/ OE.19.004703
9. Cheung D, Barnes T, Haskell T. Feedback interferometry with membrane mirror for adaptive optics. Optics Communication. 2003;218:33-41. DOI: 10.1016/S0030-4018(03)01188-X
10. Shirai T, Barnes TH, Haskell TG. Adaptive wave-front correction by means of all-optical feedback interferometry. Optics Letters. 2000;25: 773-775. DOI: 10.1364/OL.25.000773
11. Gu N, Huang L, Yang Z, Luo Q, Rao C. Modal wavefront reconstruction for radial shearing interferometer with lateral shear. Optics Letters. 2011; 36:3693-3695. DOI: 10.1364/ OL.36.003693
12. Li D, Wang P, Li X, Yang H, Chen H. Algorithm for near-field reconstruction based on radial-shearing interferometry. Optics Letters. 2005;30: 492-494. DOI: 10.1364/OL.30.000492

13. Liu D, Yang Y, Wang L, Zhuo Y. Real time diagnosis of transient pulse laser with high repetition by radial shearing interferometer. Applied Optics. 2007;46:8305-8314. DOI: 10.1364/ AO.46.008305
14. Hernandez-Gomez C, Collier JL, Hawkes SJ, Danson CN, Edwards CB, Pepler DA, et al. Wave-front control of a large-aperture laser system by use of a static phase corrector. Applied Optics. 2000;39:1954-1961. DOI: 10.1364/ AO.39.001954
15. Murty M, Compact Radial A. Shearing interferometer based on the law of refraction. Applied Optics. 1964; 3:853-858. DOI: 10.1364/ AO.3.000853
16. Murty M, Shukla RP. Radial shearing interferometers using a laser source. Applied Optics. 1973;12: 2765-2767. DOI: 10.1364/ AO.12.002765
17. Ru Q-S, Ohyama N, Honda T, Tsujiuchi J. Constant radial shearing interferometry with circular gratings. Applied Optics. 1989;28:3350-3353. DOI: 10.1364/AO.28.003350
18. Silva DE. Talbot interferometer for radial and lateral derivatives. Applied Optics. 1972;11:2613-2624. DOI: 10.1364/AO.11.002613
19. Smartt RN. Zone plate interferometer. Applied Optics. 1974;13: 1093-1099. DOI: 10.1364/AO.13.001093
20. Mohanty RK, Joenathan CJ, Sirohi RS. High sensitivity tilt measurement by speckle shear interferometry. Applied Optics. 1986;25:1661-1664. DOI: 10.1364/AO.25.001661
21. Joenathan C, Torroba R. Simple electronic speckle-shearing-pattern interferometer. Optics Letters. 1990;15: 1159-1161. DOI: 10.1364/ OL.15.001159
22. Paterson C, Notaras J. Demonstration of closed-loop adaptive optics with a point-diffraction interferometer in strong scintillation with optical vortices. Optics Express. 2007;15:13745-13756. DOI: 10.1364/ OE.15.013745
23. Bai F, Rao C. Experimental validation of closed-loop adaptive optics based on a self-referencing interferometer wavefront sensor and a liquid-crystal spatial light modulator. Optics Communication. 2010;283: 2782-2786. DOI: 10.1016/j. optcom.2010.03.032
24. Bryngdahl O. Reversed-radialshearing interferometry. Journal of the Optical Society of America. 1970;60: 915-917. DOI: 10.1364/ JOSA.60.000915

25. Mihaylova E, Whelan M, Toal V. Simple phase-shifting lateral shearing interferometer. Optics Letters. 2004;29: 1264-1266. DOI: 10.1364/OL.29.001264
26. Kothiyal MP, Delisle C. Shearing interferometer for phase shifting interferometry with polarization phase shifter. Applied Optics. 1985;24: 4439-4447. DOI: 10.1364/ AO.24.004439
27. Chung CY, Cho KC, Chang CC, Lin CH, Yen WC, Chen SJ. Adaptive-optics system with liquid-crystal phase-shift interferometer. Applied Optics. 2006; 45:3409-3414. DOI: 10.1364/ AO.45.003409
28. Naik DN, Ezawa T, Miyamoto Y, Takeda M. 3-D coherence holography using a modified Sagnac radial shearing interferometer with geometric phase shift. Optics Express. 2009;17: 10633-10641. DOI: 10.1364/AO.17.010633
29. Gu N, Yao B, Huang L, Rao C. A compact single-shot radial shearing interferometer with random phaseshift. Optics Letters. 2017;42:3622-3625. DOI: 10.1364/OL.42.003622
30. Lago E, Fuente R. Amplitude and phase reconstruction by radial shearing interferometry. Applied Optics. 2007; 47:372-376. DOI: 10.1364/AO.47.000372
31. Li D, Wen F, Wang Q, Zhao Y, Li F, Bao B. Improved formula of wavefront reconstruction from a radial shearing interferograms. Optics Letters. 2008;33: 210-212. DOI: 10.1364/OL.33.000210
32. Ragazzoni R, Marchetti E, Rigaut F. Modal tomography for adaptive optics. Astronomy and Astrophysics. 1999;342: L53-L56

CHAPTER 18

FIBER-OPTIC RING RESONATOR INTERFEROMETER

Ramón José Pérez Menéndez

UNED, Lugo, Spain

ABSTRACT

In this chapter, theoretical analysis and electro-optical characterization of a fiber optic passive ring resonator interferometer (FOPRRI) were realized. First, a theoretical description and analysis of all-pass and add-drop filter configurations were performed, hence obtaining their respective transfer function, taking into account, the physical properties of a standard single-mode optical fiber as resonator constitutive material. Second, computational numerical simulations of a typical FOPRRI were performed with different analysis methods: effective index method (EIM), coupled mode theory (CMT), finite element method (FEM) and finite difference time domain (FDTD). Third, a comparative analysis of the results obtained with these methods is realized, showing the most accurate and appropriate for filter electro-optical characterization as a consequence. Last, and as an ultimate step, taking as reference the previously obtained electro-optical characterization for both configurations, the main applications are derived.

Citation: (APA): Menéndez, R. J. P. (2019). Fiber-optic ring resonator interferometer. In Interferometry-Recent Developments and Contemporary Applications. (22 pages)

Keywords: passive optical ring resonator interferometer, single-mode optical fiber, all-pass filter, add-drop filter, EIM (effective index method), CMT (coupled mode theory), FEM (finite element method), FDTD (finite difference time domain)

INTRODUCTION

Optical interferometry constitutes an important technique used in high number of measurement processes for multiple physical magnitudes and quantitative phenomenon [1]. Particularly, fiber-optic waveguides can act as very useful and efficient transmission medium for light guidance in many interferometric sensor devices. The issue of fiber-optic interferometric devices has received an extensive treatment in the literature [2, 3]. Also, the general theory of optical resonators was extensively treated by authors, for example, Yariv and Yeh [4]. However, quantitative analytic treatment of a fiber optic passive ring resonator interferometer (FOPRRI) can be rather found in singular applications, particularly, in gyroscopes. A clear proof of it is that one of the pioneering and singular descriptions of fiber-optic ring interferometer devices was in 1976 when Vali and Shorthill designed the first fiber-optic gyroscope model [5, 6, 7]. Later, many other authors studied in detail this first design and improved some of its electro-optical parameters [8, 9]. Considered from a broader point of view, the FOPRRI device has now evolved toward what we could call the integrated optical passive ring resonator interferometer (IOPRRI). In this case, the fiber-optic ring resonator is replaced by a ring-integrated optical waveguide, made of high refractive index material, such as lithium niobate, indium phosphide, silicon nitride, or silicon on insulator (SOI) platform. This chapter focuses on the analysis and testing of fiber-optic ring resonator. Therefore, the integrated ring resonator study, its investigations and challenges are beyond the scope of it [10].

An all-fiber passive ring resonator interferometer is formed by a directional coupler and a fiber-optic loop, as shown in Figure 1. Two main geometrical configurations are possible to reach such an optical coupling: (a) direct-coupled fiber ring resonator or (b) cross-coupled fiber ring resonator, as it can be seen in figure. Superimposition of successive beams generated by re-coupling the first coupled beam into the fiber ring results in multi-beam optical interference in a similar way to a Fabry-Pérot interferometer. Resonance will occur when the increase of the phase of the beam around the ring equals 2πn−(π/2), π/2 being the phase shift due to cross-coupling and n a positive integer ($n = 1, 2, 3, \ldots$), so that the light reentering the ring after a single circulation around it is in phase with the light entering the ring for the first time. Following L. F. Stokes et al. [12, 13], the transfer function of the system for the optimum amplitude resonance condition is given by:

$$\frac{I_T}{I_0} = (1-\gamma)\left\{1-\left[\frac{(1-\kappa)^2}{1+\kappa^2+2\kappa\sin(\beta L)}\right]\right\} \tag{1}$$

where γ is the insertion loss of the coupler, $\kappa=(1-\gamma)\exp(-2\alpha L)$ is the optical resonance coupling coefficient, α is the fiber attenuation factor, L is the total length of the fiber ring and β is the propagation constant of the re-circulating mode. The transfer function is very similar to a Fabry-Pérot interferometer operating in back-reflection mode. A finesse F of more of 1000 has been demonstrated for very low excess loss coupler and low-loss fibers [11, 12]. At resonance, significant optical power is recirculating in the ring and nonlinear effects as stimulated Brillouin scattering (SBS) can be observed for input powers of less than 100 μW [13]. As it will be seen later, fiber-optic ring resonators have been used in a number of applications, including fiber ring lasers [8], sensors [14], fiber laser gyroscopes [15, 16], optical spectrum analyzers [17, 18] and optical delay lines [19], among others. One of the first and main sensing applications of fiber-optic ring resonator is as a gyroscope and the major attraction of this kind of resonator gyro is that the cavity length can be taken much shorter than the convectional gyro in the ratio of 2/F. In order to detect the phase shift induced by rotation, a highly coherent laser light source is required. A dual frequency modulation scheme is used to control the laser frequency and tracking the resonance condition [8, 9]. The difference between these two frequencies gives a measure of gyro output signal, which, in turn, is proportional to rotation rate to which the fiber-optic ring resonator system is subjected.

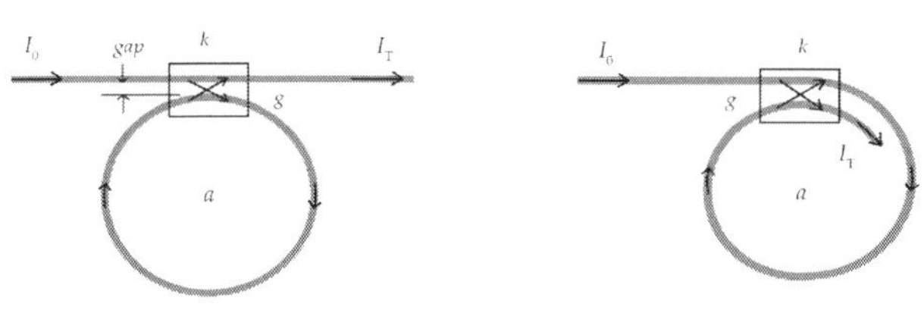

Figure 1. Fiber-optic-passive-ring-resonator-interferometer (FOPRRI) all-pass configuration: (a) direct-coupled fiber ring-resonator and (b) cross-coupled fiber ring-resonator.

ANALYSIS OF THE FIBER-OPTIC ALL-PASS FILTER RING RESONATOR CONFIGURATION

The all-pass ring resonator configuration is created by taking one output of a generic 2 × 2 directional coupler and feedback into one input, as it can be seen in Figure 2. Such a device exhibits a periodic cavity resonance when light traversing the ring acquires a phase shift corresponding to an integer multiple of 2π radians. The resonator can mathematically be formulated from two components: a coupling strength (from coupling coefficient κ) and a feedback path (via transmission coefficient t), in contrast to the infinite sum derivations made for Fabry-Perot and Gires-Tournois interferometers. Therefore, following the text of Rabus [20], we will derive the basic spectral properties by assuming the steady-state operation and the matching fields. Although both methods are equally valid, the field-matching method has the advantage of simplicity. The basic relations among the incident E_{i1}, transmitted E_{t1}, and circulating E_{i2}, E_{it2} fields of a single resonator are derived by combining the relations for the coupler with that of the feedback path. In the frequency domain, the fields exiting the coupling region are related to the input fields via the following equations:

$$E_{t1} = tE_{i1} + i\kappa E_{i2} \tag{2}$$

$$E_{t2} = tE_{i2} - i\kappa E_{i1} \tag{3}$$

where factor i is the imaginary unit ($i = \sqrt{-1}$) that takes into account, the phase shift caused by the cross-coupling of the waveguides. For simplicity, we assume the total symmetry of cross-coupling and bar-coupling coefficients for both waveguides, and therefore, the two approximate equations are ($\kappa_1 \approx \kappa_2 = \kappa$ and $t_1 \approx t_2 = t$). The two Eqs. (2) and (3) can be written in a compact way via the following unitary matrix:

$$\begin{pmatrix} E_{t1} \\ E_{t2} \end{pmatrix} = \begin{pmatrix} t & i\kappa \\ -i\kappa & t \end{pmatrix} \begin{pmatrix} E_{i1} \\ E_{i2} \end{pmatrix} \tag{4}$$

where sign and i-factor in iκ elements of first matrix on the right side of equation are due to cross-coupling phase shift caused by coupler. Also, since this matrix is an unitary matrix, we find that accomplishes the following relationship between bar- (t) and cross- (κ) coupling coefficients (assuming both are independent of frequency):

$$t^2 + \kappa^2 = 1 \tag{5}$$

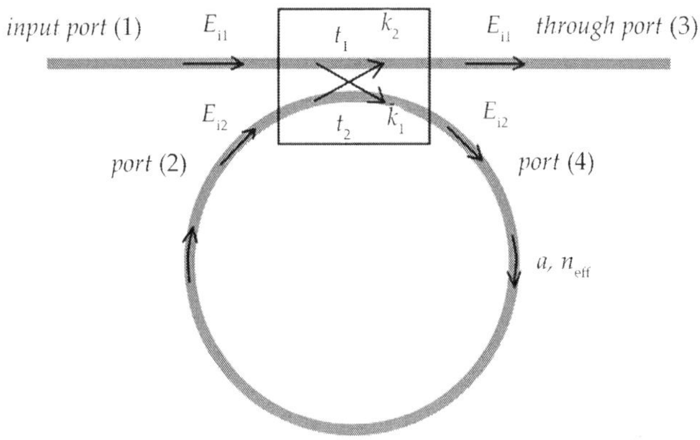

Figure 2. All-pass fiber-optic ring resonator configuration ($\kappa_1 \approx \kappa_2 = \kappa$ and $t_1 \approx t_2 = t$).

The feedback path of length $2\pi R$ (R being the ring radius) connects the output from port 4 back into input port 2, where the field can be expressed as:

$$E_{i2} = e^{-\alpha 2\pi R} e^{i\beta 2\pi R} E_{t2} \equiv a e^{i\phi} E_{t2} \tag{6}$$

where α is the attenuation factor of the fiber, a-factor represents the ring single-pass field amplitude transmission (taking into account the fiber attenuation), $\beta=(2\pi\lambda/)n_{eff}$ is the propagation constant of propagating mode in the fiber (where in time, n_{eff} is the effective index of the fiber) and ϕ represents the single-pass ring round-trip phase shift.

Since that adding or subtracting an integer number m of 2π radians from single-pass phase shift does not change the value of the function, the single-pass phase shift for all resonances is defined such that its value is zero for a local resonance of interest. Furthermore, because the single-pass phase shift is directly related to the radian frequency as $\phi=\omega T_R$, where T_R is the transit time of the round-trip of ring resonator and then ϕ is clearly representative of a *normalized frequency detuning*. Following the previous derivations, one can demonstrate that the transmission power P_{t1} in the output port (3) of the straight waveguide can be expressed as:

$$P_{t1} = |E_{t1}|^2 = \frac{a^2 + |t|^2 - 2a|t|\cos(\phi + \phi_c)}{1 + a^2|t|^2 - 2a|t|\cos(\phi + \phi_c)} \tag{7}$$

where $t = |t|\exp(j\phi_c)$, $|t|$ representing the coupling losses and ϕ_c represents the phase of the coupler. Making a similar deduction, we can reach the circulating power in the ring, which is:

$$P_{i2} = |E_{i2}|^2 = \frac{a^2\left(1-|t|^2\right)}{1+a^2|t|^2-2a|t|\cos\left(\phi+\phi_c\right)} \tag{8}$$

On resonance, condition $\phi+\phi_c=2\pi m$ is fulfilled, where *m* is an integer, and then Eqs. (7) and (8) becomes:

$$P_{t1} = |E_{t1}|^2 = \frac{(a-|t|)^2}{(1-a|t|)^2} \tag{9}$$

$$P_{i2} = |E_{i2}|^2 = \frac{a^2\left(1-|t|^2\right)}{(1-a|t|)^2} \tag{10}$$

where, as we have said above, *a* is the loss coefficient of the ring (for zero loss $\alpha = 0$, then $a = 1$). The special case that occurs when $a=|t|$ (or $\kappa_2=\alpha L$) is named *critical coupling* and is of high interest in applications. This happens when internal losses around the ring are equal to the coupling losses.

Then, the transmitted power to through port (3) becomes 0. The critical coupling is due to destructive interference between the direct beam and the cross-coupled beam of light, both after passing their respective coupling zone. A condition termed *undercoupling* does occur when the ring loss exceeds the coupling strength ($t>a$ or equally: $t^2>\alpha L$, also $\kappa^2<\alpha L$). An *overcoupling* condition takes place, when the round-trip loss does not exceed the coupling strength ($t<at<a$ or equally: $t^2<\alpha L$, also $\kappa^2>\alpha L$) and is the conventional mode of operation for an all-pass resonator. Figure 3(a) and (b) displays the effect of attenuation on the transmission and build-up factor for the overcoupled, critically coupled and undercoupled regimes. Experimental demonstration of the transmission characteristics in each of these regimes for a fiber all-pass filter ring resonator can be found in [22].

A good idea of the behavior of the system, when resonance condition accomplishes, can be obtained when one represents the T *transmission function* defined as the ratio P_{t1}/P_{i1} (i.e., the relationship between the through-port power and input-port power) versus normalized detuning ϕ , Figure 3(a). Starting from input field expressions and after some derivations, the following equation can be found:

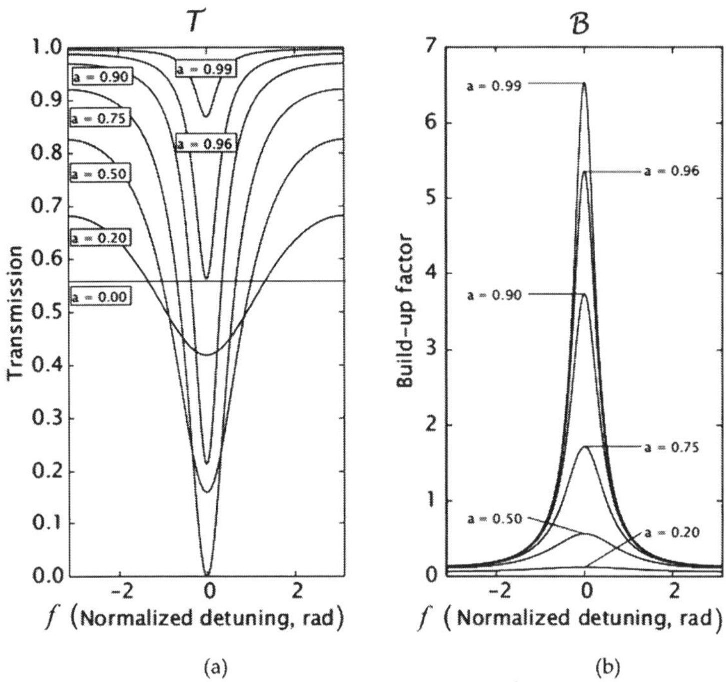

Figure 3. T transmission function (a) and B build-up factor (b) of ring-resonator, both as a function of normalized detuning ϕ , (curves extracted from [21], with permission).

$$\mathcal{T} = \frac{P_{t1}}{P_{i1}} = \left|\frac{E_{t1}}{E_{i1}}\right|^2 = \frac{a^2 - 2|t|a\cos\phi + |t|^2}{1 - 2|t|a\cos\phi + |t|^2 a^2} \qquad (11)$$

Figure 3(a) represents the T transmission function for an all-pass ring resonator with $t = 0.75$ and varying loss a as a parameter (from $a = 0.00$ up to $a = 0.99$). It can be seen that, in this case, critical coupling occurs when $a = t = 0.75$; therefore, the T transmission function drops to zero when ϕ=0 for this a-factor value. As a increases (up to 1) or decreases (to 0.00) from 0.75, the corresponding transmission curve arises its minimum point from 0. For $a = 0.00$, T transmission function reaches a constant value of $0.75^2 = 0.5625$ for all values of ϕ . Another function that allows to obtain a good idea of the behavior of the system is the so-called B *intensity build-up factor* as a function of ϕ (normalized detuning), defined as:

$$B = \frac{P_{i2}}{P_{i1}} = \left|\frac{E_{i2}}{E_{i1}}\right|^2 = \frac{\left(1 - |t|^2\right) a^2}{1 - 2|t|a\cos\phi + |t|^2 a^2} \Rightarrow \lim_{\phi = 2\pi m,\, a = 1} \frac{\left(1 - |t|^2\right) a^2}{1 - 2|t|a\cos\phi + |t|^2 a^2} = \frac{1 + |t|}{1 - |t|} \approx \frac{4}{\kappa^2} \quad (12)$$

B function can be thought as the ratio of the circulating light intensity inside the ring to incident light intensity. Figure 3(b) represents the B function (or intensity build-up factor) versus normalized detuning ϕ for the same all-pass ring than the one in Figure 3(a), when varying loss *a* is taken as a parameter (from $a = 0.20$ up to $a = 0.99$). The last result in Eq. (12) refers to case in which incident light is resonant with the ring ($\phi = 2m\pi$) and attenuation is negligible ($a = 1$). This happens when constructive interference is produced at the coupler port (4) between the coupled light entering the ring coming from waveguide and the recirculation light coming from input port (3). This way, the circulating intensity of light around the ring results reinforced. However, for this constructive interference to take place, a coherent source is essential at input port (1). A passive ring resonator under these conditions attains the maximum ratio of circulating power to incident power that can be achieved. Figure 4 shows two successive resonances versus normalized detuning with peak values given by right side of Eq. (12).

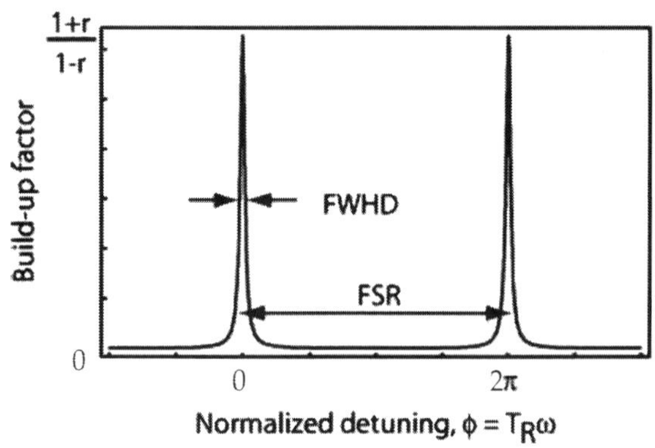

Figure 4. Build-up factor versus normalized detuning for all-pass fiber ring resonator (from [21], with permission).

The spectral characteristics of the all-pass fiber-ring resonator configuration (one waveguide, one resonator, as represented in Figure 2) as optical filter can better be viewed when spectral response of through port versus vacuum wavelength is represented. A notch type filter is obtained from observation of this resonance curve (Figure 5). For cross-coupling value of 10% ($\kappa^2=0.10$), the optical intensity in the ring can be 40 times higher than the intensity in the input port of the waveguide. This way, since the intensity in the fiber ring can be much higher than in the bus, fiber ring resonators can be used for nonlinear optics applications with low input intensities.

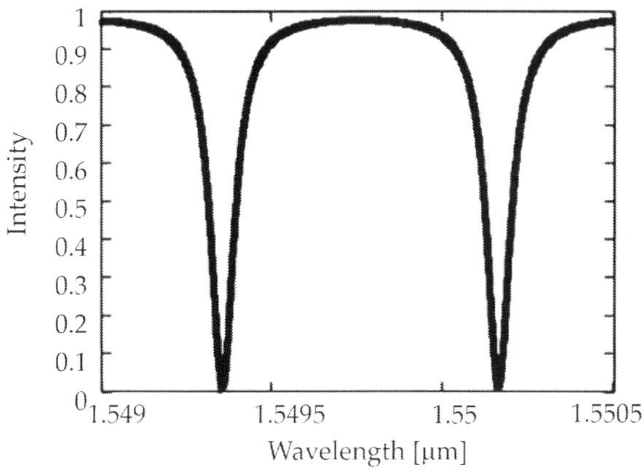

Figure 5. Spectral response of through port of an all-pass fiber ring resonator (notch-type filter).

Ring resonator filters can be characterized by four parameters, which are generally used to describe their performance, namely, *the free spectral range* (FSR), the *full width at half maximum* (FWHM), *the finesse* F and *the quality factor* Q. The FSR $\Delta\lambda$ is the distance between two successive resonance peaks (difference between the wavelengths corresponding to two successive resonant conditions). When neglecting the wavelength dependence of the fiber-optic effective refractive index n_{eff}, approximate calculation of FSR $\Delta\lambda$ is very simple [20], as follows:

$$FSR = \Delta\lambda = -\frac{2\pi}{L}\left(\frac{\partial\beta}{\partial\beta}\right)^{-1} \approx \frac{\lambda^2}{n_{eff}L} \tag{13}$$

This approximation is derived from the resonant condition next to a resonance found for the used propagation constant. A more accurate calculation is obtained taking into account the chromatic dispersion of the fiber. Then, the group refractive index of the fiber can be used instead of the effective index, so that the new expression for FRS to be:

$$FSR = \Delta\lambda = \frac{\lambda^2}{n_g L} \tag{14}$$

The second parameter in importance for describing the performance of the optical filter is the FWHM. It is defined as the full spectral width at half maximum (or 3 dB bandwidth $2\delta\lambda$). In the case of all-pass filter, FWHM must be redefined as full width at half depth (FWHD) and its value can be calculated as follows (in absence of ring internal losses, that is, $a = 1$):

$$FWHD = 2\delta\lambda = \frac{\lambda^2}{2\pi^2 n_g R} \frac{1-|t|}{\sqrt{|t|}} \xrightarrow{|t|\approx 1} \approx \frac{\lambda^2\kappa^2}{4\pi^2 n_g R} \tag{15}$$

Third parameter is the finesse F. The finesse is a figure of merit of the filter that can be defined as the ratio between FSR and FWHD. So that implementing this definition and in absence of ring internal losses, the resulting expression for the finesse of an all-pass fiber resonator is [23]:

$$\mathcal{F} = \frac{FSR}{FWHD} \approx \frac{\frac{\lambda^2}{2\pi n_g R}}{\frac{\lambda^2\kappa^2}{4\pi^2 n_g R}} = \frac{2\pi}{\kappa^2} \tag{16}$$

Finally, the fourth parameter is the quality factor Q. It represents a measure of the sharpness of the resonance. It can be defined as the ratio of the operation wavelength and the resonance width (FWHD). Its calculation is as follows:

$$\mathcal{Q} = \frac{\lambda}{FWHD} \approx \frac{\lambda}{\frac{\lambda^2\kappa^2}{4\pi^2 n_g R}} = \frac{\lambda}{\frac{\lambda^2\kappa^2}{2\pi n_g L}} = \frac{2\pi n_g L}{\lambda\kappa^2} = \frac{n_g L}{\lambda}\mathcal{F} \tag{17}$$

The quality factor Q can also be regarded as the total energy stored in the ring divided by the energy lost per optical cycle.

ANALYSIS OF THE FIBER-OPTIC ADD-DROP FILTER RING RESONATOR CONFIGURATION

The basic fiber-optic add-drop filter consists of one input-waveguide, one output-waveguide and one fiber-ring resonator such as the one represented in Figure 6. Both of the waveguides are side-coupled to ring resonator. The four ports of structure are defined as input port, throughput port, drop port and add port. Viewed from drop port acting as the optical output, this configuration behaves as a narrow-band amplitude filter that can add or drop a spectral band from an optical input. This configuration is mathematically equivalent to a well-known Fabry-Pérot interferometer. Therefore, the equations for transmission coefficients are easy to obtain. For simplification, E_{i1} is defined with unitary rms amplitude. Then, the throughput E_{t1} mode rms amplitude in the input-waveguide is given by:

$$E_{t1} = \left[t_1 + \frac{t_2 \kappa_1 \kappa_2 a_{1/2}^2 e^{j\theta(|t_1|^2+|\kappa_1|^2)}}{1 - t_1 t_2 a_{1/2}^2 e^{j\theta}} \right] E_{i1} = \frac{t_1 - t_2 a e^{j\theta}}{1 - t_1 t_2 a e^{j\theta}} \tag{18}$$

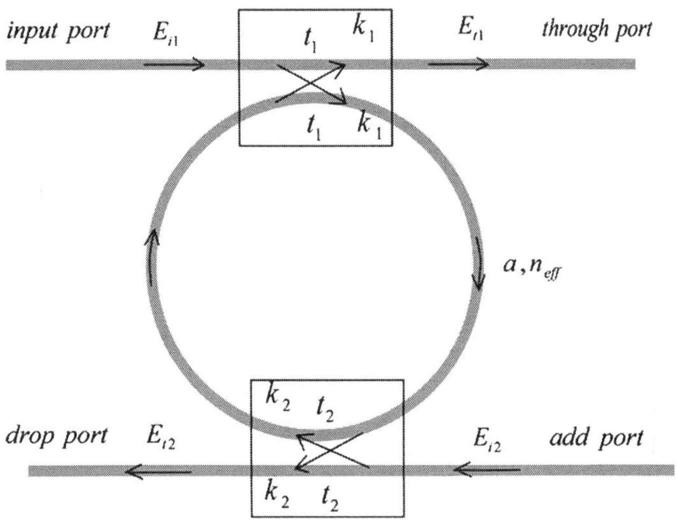

Figure 6. Add-drop fiber-optic ring resonator configuration (for symmetrical coupling: $\kappa_1 \approx \kappa_2 = \kappa$ and $t_1 \approx t_2 = t$).

In this equation, $a_{1/2}$ and $\theta_{1/2}$ are used, which are the half round-trip loss and phase, respectively. It is $a = a_{1/2}^2$ and $\theta = 2\theta_{1/2}$. The mode amplitude in the ring has to pass the second coupler to become the new dropped mode rms amplitude E_{t2}. Its rms amplitude value is given by:

$$E_{t2} = \frac{\kappa_1 \kappa_2 a_{1/2} e^{j\theta_{1/2}}}{1 - t_1 t_2 a e^{j\theta}} \tag{19}$$

At resonance condition ($\theta = m2\pi$, *m* integer), the output intensity from the drop port is given by:

$$P_{t2-resonance} = |E_{t2-resonance}|^2 = \frac{(1 - t_1^2)(1 - t_2^2) a}{(1 - t_1 t_2 a)^2} \tag{20}$$

The through port mode amplitude E_{t1} will be zero at resonance for identical symmetrical couplers ($t_1 = t_2$), when $a = 1$ (zero loss). This means that on resonance, all input optical power will be coupled to resonator and could be extracted by drop port. The value $a = 1$ (zero loss) can only be achieved by a given value of gain incorporated in the resonator ring to compensate the waveguide losses. In a pure passive ring resonator with defined waveguide constituents, the loss coefficient a is constant. By adjusting coupling parameters t_1 and t_2 to the loss coefficient a, minimum intensity can be achieved (ideally, $P_{t1} \approx 0$). From Eq. (18), one can obtain:

$$a = \frac{t_1}{t_2} \qquad (21)$$

According with Eq. (18), the transfer function of through port (relative to input port) is given by:

$$T_{through} = \left|\frac{E_{t1}}{E_{i1}}\right|^2 = \frac{t_2^2 a^2 - 2t_1 t_2 a \cos\phi + t_1^2}{1 - 2t_1 t_2 a \cos\phi + (t_1 t_2 a)^2} \xrightarrow[a=1,\, t_1=t_2=t]{} \frac{\frac{4t^2}{(1-t^2)^2}\sin^2\left(\frac{\phi}{2}\right)}{1 + \frac{4t^2}{(1-t^2)^2}\sin^2\left(\frac{\phi}{2}\right)} \qquad (22)$$

Taking into account, Eq. (19), the transfer function of drop port (relative to input port), will be:

$$T_{drop} = \left|\frac{E_{t2}}{E_{i1}}\right|^2 = \frac{\left(1 - t_1^2\right)\left(1 - t_2^2\right) a}{1 - 2t_1 t_2 a \cos\phi + (t_1 t_2 a)^2} \xrightarrow[a=1,\, t_1=t_2=t]{} \frac{1}{1 + \frac{4t^2}{(1-t^2)^2}\sin^2\left(\frac{\phi}{2}\right)} \qquad (23)$$

This way, the transmission spectral response of through port corresponds to an approximately Lorentzian curve centered on a resonance, where a narrow frequency band has been extracted (see Figure 7). These extracted band do appear (in inverted shape) in the spectrum response of drop port, as it can also be seen in same figure. In the case of add-drop filter, if attenuation is negligible ($a \approx 1$), *critical coupling* occurs at symmetric coupling ($\kappa_1 = \kappa_2 = \kappa$). For non-negligible single-pass resonator loss, critical coupling occurs when single-pass ring losses match the coupling, so it verified the following condition: $t_2 a = t_1$. Transfer functions, FWHM and extinction ratio (ER) relative to all-pass filter, through and dropt ports of add-drop filter are plotted.

It is worth to make an interpretation and comparison of the add-drop filter configuration with respect to the all-pass filter one. In the upper coupler (Figure 6), interference takes place between the coupled input field and ring circulating field. However, the lower coupler does not experience any interference (provided that excitation is from the upper waveguide only).

The lower coupler may thus be viewed simply as a tap waveguide that leaks power out of the ring into the drop port. For an add-drop filter, it is desirable to operate at critical coupling for achieving the complete extinguishment of a frequency band in the through port. In this case, the sum of all losses incurred in the resonator, even the loss at the out-coupled drop port, must be taken into account. Calculations of FSR, FWHM, B build-up factor, finesse F and quality factor Q are given below according to the results found in Refs. [21, 22].

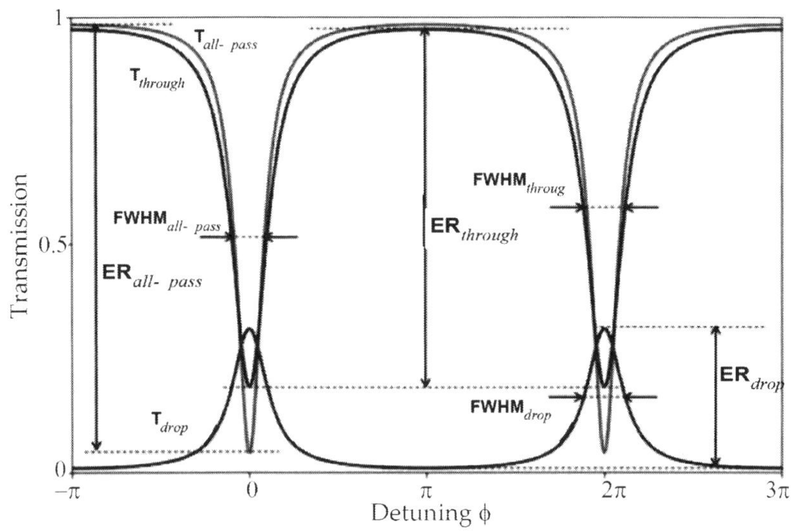

Figure 7. Spectral response of a fiber-optic add-drop filter (adapted from [24], with permission).

$$FSR = \Delta\lambda = \frac{\lambda^2}{n_{eff} L} \tag{24}$$

$$FWHD \approx \frac{\lambda^2 \kappa^2}{\pi L n_{eff}} \tag{25}$$

$$\mathcal{F} = \frac{FSR}{FWHD} \approx \frac{\pi}{\kappa^2} \tag{26}$$

$$\mathcal{Q} = \frac{\lambda}{FWHD} \approx \frac{\pi n_g L}{\lambda \kappa^2} = \frac{n_g L}{\lambda} \mathcal{F} \tag{27}$$

where the approximations again apply to symmetrically coupled, low-loss and weakly coupled microrings. Note that, comparing Eqs. (27) and (17), with the same optical parameters, add-drop filter has the half value of Q-factor (and also the finesse F) than that the all-pass one. This result could be expected since an all-pass filter configuration has half the loss due to external coupling than an add-drop one.

NUMERICAL SIMULATION OF FIBER-OPTIC ALL-PASS FILTER RING RESONATOR CONFIGURATION

As announced at the beginning of this chapter, this section will deal specifically with the numerical simulation of a typical FOPRRI in all-pass filter configuration. First, effective index method (EIM) [25, 26] will be applied to this structure in which main constituents are made of a standard single-mode optical fiber (coupler and ring resonator). Secondly, coupled mode theory (CMT) [27, 28, 29, 30, 31, 32, 33] will be performed to the same structure. In third time, finite element method (FEM) [34] is performed. Finally, 3D-FDTD (finite difference time domain) [35, 36] was applied to same structure. The results of all these numerical simulations are compared to each other to see which one best match each specific application.

Optical layout scheme for EIM and CMT numerical simulation is represented in Figure 8. OptSim® Circuit tool (from Synopsys™ Software) was applied to obtain the transmission power spectrum of this resonant structure, which takes the effective index method (EIM) as a numerical simulation algorithm. A side-polished fiber-optic 4 × 4 directional coupler is used to couple the waveguide fiber to fiber ring resonator (here implemented by a series of two waveguides, 1 and 2, respectively). A standard SMF-28 (Corning™ Glass) single-mode optical fiber is assumed for both, waveguide and ring resonator as constituent materials. Optical parameters of this fiber are: basis material = fused-silica (core = boron-silicate, cladding = silica), step refractive index, core diameter = 8.2 μm, cladding diameter = 125 μm, effective refractive indexes: $n_{core} = 1.467$, $n_{cladding} = 1.460$, numerical aperture (NA) = 0.143, maximum attenuation = 0.35 dB/km (at 1310 nm wavelength) and h-parameter = 2×10^{-6} m^{-1}. Gap between waveguide and ring resonator is 200 nm. Normalized Gaussian profile continuous wave (CW) optical field of 1310 nm vacuum wavelength was launched to the system from a wideband tunable laser source. To obtain the pass-through port power transmission spectrum, wavelength sweeping of laser source was made in order to fix each wavelength value for its respective measured optical power. The computation is iterated 30 times to obtain the steady-state solution. Obtained transmission power spectrum is represented in Figure 9.

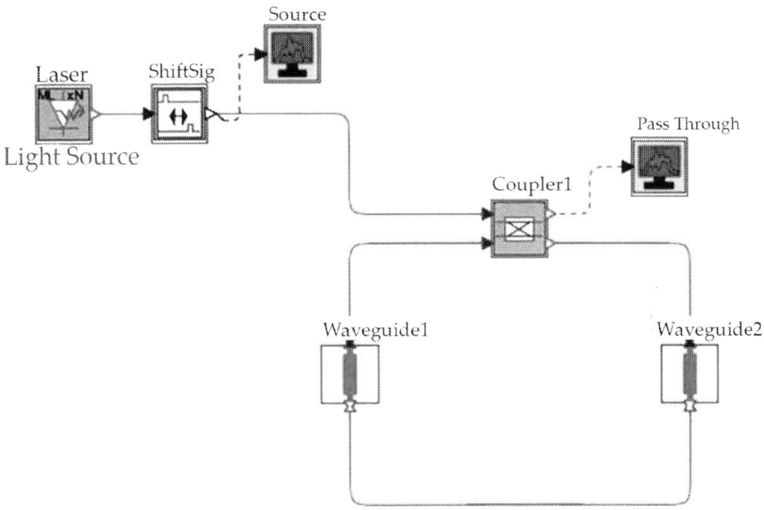

Figure 8. All-pass FOPRRI configuration: optical layout scheme for EIM and CMT numerical simulations.

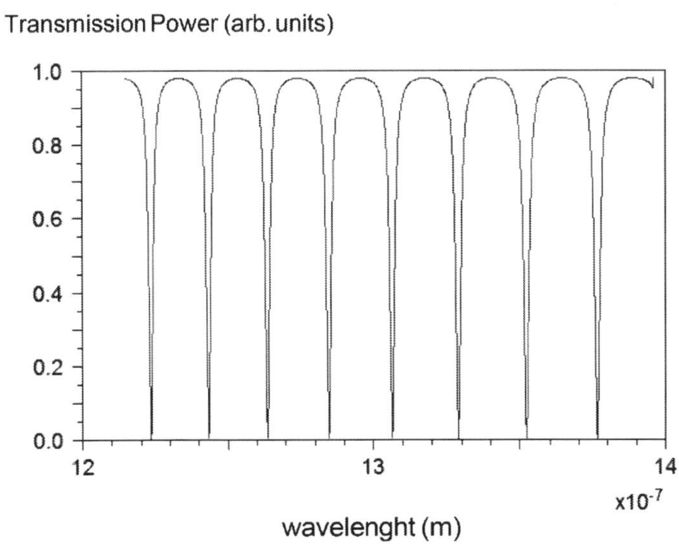

Figure 9. All-pass FOPRRI configuration: transmission power spectrum.

As it can be seen on expanded transmission spectrum showed in Figure 10, a resonance condition is achieved for 1285 nm wavelength. Since the effective refractive index of ring resonator is 1.467 and its optical path length is 51.420 μm, a FSR of 21.89 nm is obtained as it can also be observed in this same figure. This value agrees accurately with theoretical value as calculated from Eq. (13).

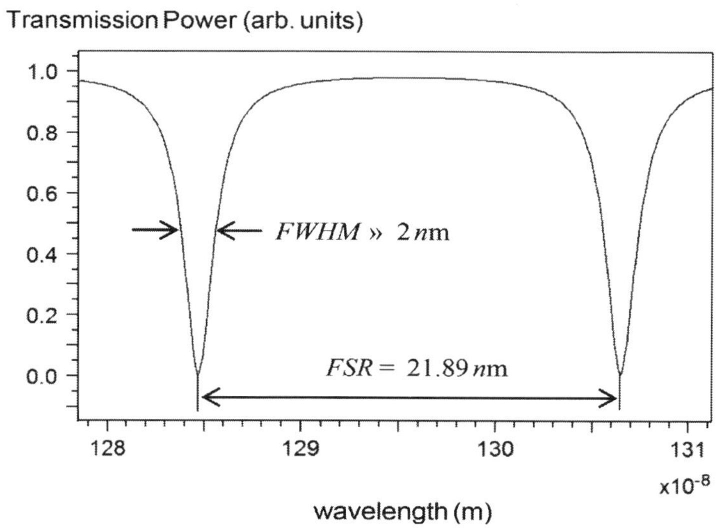

Figure 10. All-pass FOPRRI configuration: expanded transmission power spectrum around 1295 nm.

On the other hand, measured value of FWHM is approximately 2 nm. This value is also in frank agreement with value calculated analytically by means of Eq. (15), taking into account 8.184 μm as medium ring resonator radius. Then, a calculation from Eq. (16) gives the value of 10.95 for F finesse. As a consequence of last, calculation from Eq. (17) yields 642.80 for the Q quality-factor of the system.

Coupled mode theory (CMT), which uses BPM method, was applied to this structure by means of BeamPROP® tool (Synopsys™ Software) [32]. This method also takes into account the effective index of the ring resonator material, so the expected results from it will be improved compared with those obtained with the EIM method alone. Further in-depth readings on BPM method can be found in Refs. [37, 38, 39, 40, 41, 42]. Figure 11 shows the result of numerical simulation of all-pass ring resonator considered here.

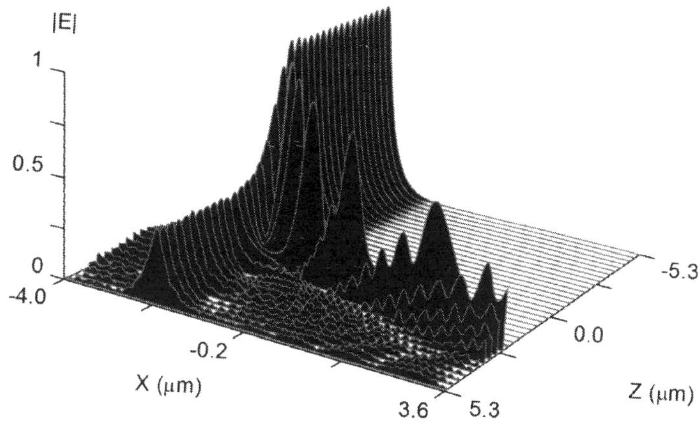

Figure 11. CMT method numerical simulation result (from BeamPROP® tool) for all-pass FOPRRI configuration (resonance at 1285 nm).

Finite element method (FEM) was applied to this structure by means of FemSIM® tool (Synopsys™ Software) [34]. This method also takes into account the effective index of the ring resonator material, so the expected results from it will be improved compared with those obtained with the EIM method alone. Additional treatment on the FEM method can be found in the Refs. [43, 44, 45, 46, 47]. The result of FemSIM® of all-pass ring resonator analyzed here is represented in Figure 12.

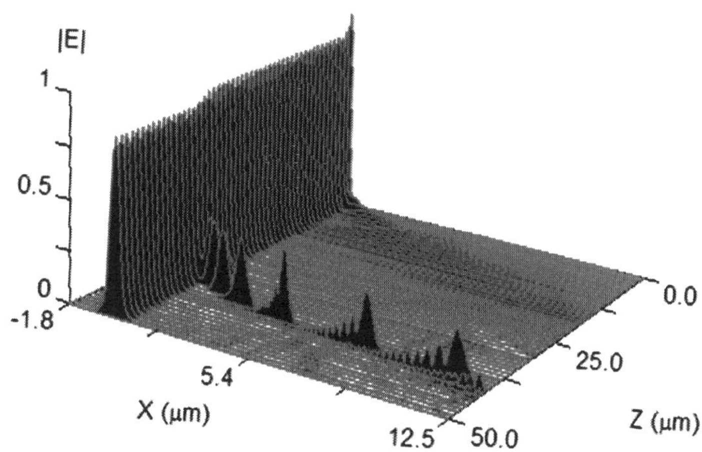

Figure 12. FEM numerical simulation (from FemSIM® tool) for all-pass FO-PRRI configuration (resonance at 1285 nm).

FullWAVE® (Synopsys™ Software) [36] is a full-vectorial numerical computation tool that allows the simulation of light wave propagation in an arbitrary geometry system. The simulation engine is based on the finite difference time domain (FDTD) method. The FDTD algorithm implements rigorous and complete time-space solution to Maxwell's equations. This method offers an efficient field wave propagation solution in a photonic-integrated structure when other methods such as beam propagation method (BPM) or finite element method (FEM) cannot cope with the structure geometry or do not provide a precise solution. Additional treatment on the FDTD method can be found in Refs. [48, 49, 50, 51, 52]. Figure 13 represents the result of FDTD numerical simulation of all-pass ring resonator studied here.

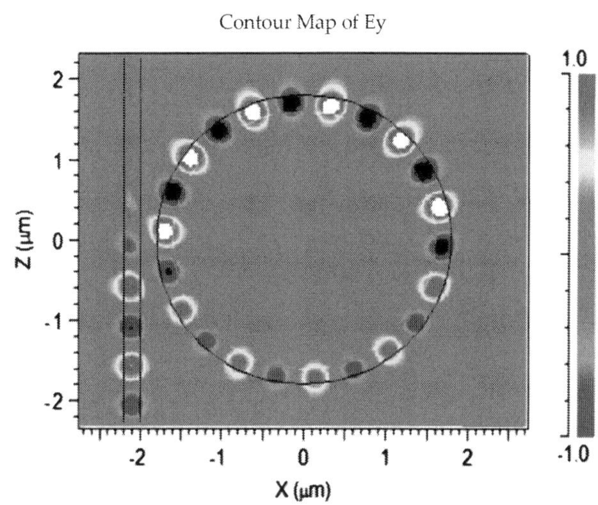

Figure 13. 2D-FTDT numerical simulation results for an all-pass FOPRRI (resonance at 1285 nm).

Table 1 shows a comparison of the results of the four methods used (EIM, CMT, FEM and FDTD) for all-pass filter and add-drop filter configurations, taking into account their suitability for accuracy and calculation time. In view of transmission spectrum of the all-pass filter configuration, this can find applications such as optical band-pass filter, optical delay line, optical switch and active ring modulator.

Table 1. Comparison of numerical simulation methods (EIM, CMT, FEM and FDTD) for all-pass/add-drop filters.

Comparison	EIM	CMT (BPM)	FEM	FDTD
All-pass filter	Good	Very good	Outstanding	Excellent
Add-drop filter	Good	Very good	Outstanding	Excellent

NUMERICAL SIMULATION OF FIBER-OPTIC ADD-DROP FILTER RING RESONATOR CONFIGURATION

As realized on all-pass configuration, this section will be dedicated to numerical simulation of a typical FOPRRI in add-drop filter configuration. Then, successive numerical simulations by applying effective index method (EIM), coupled mode theory (CMT), finite element method (FEM) and 3D finite difference time domain (FDTD) will be performed to this configuration. Finally, the comparison of all these numerical simulations will be carried out to verify, which is the most appropriate for each type of application.

Optical layout scheme for EIM and CMT numerical simulation is represented in Figure 14. OptSim® Circuit tool (from Synopsys™ Software) with the effective index method (EIM) as computational algorithm was applied here to obtain the transmission power spectrum of resonant structure. The configuration analyzed can be obtained from the one considered in Section 4 by adding a second fiber 4 × 4 directional coupler and a second side-coupled fiber waveguide. Also, standard SMF-28 (Corning™ Glass) single-mode optical fiber is assumed for both, upper and lower waveguides and ring resonator as constituent material. Gap between waveguide and ring resonator is 200 nm. Normalized Gaussian profile continuous wave (CW) optical field of 1310 nm vacuum wavelength was launched to the system from a wideband tunable laser source. To obtain the power transmission spectra of pass-through and drop ports, wavelength sweeping of laser source was made in order to fix each wavelength value for its respective measured optical power. The computation is iterated 30 times to obtain the steady-state solution. The resulting transmission power spectra are depicted in Figure 15(a)–(c), respectively. As it can be seen on expanded drop-port transmission spectrum showed in Figure 10(c), a resonance condition is achieved for 1285 nm wavelength. Since the effective refractive index of ring resonator is 1.467 and its optical path length is 51.420 μm, a FSR of 21.89 nm is obtained as it can also be observed in this same figure. This value agrees accurately with theoretical value as calculated from Eq. (24). On the other hand, measured value of FWHM is approximately 4 nm. This value is also in frank agreement with value calculated analytically by means of Eq. (25).

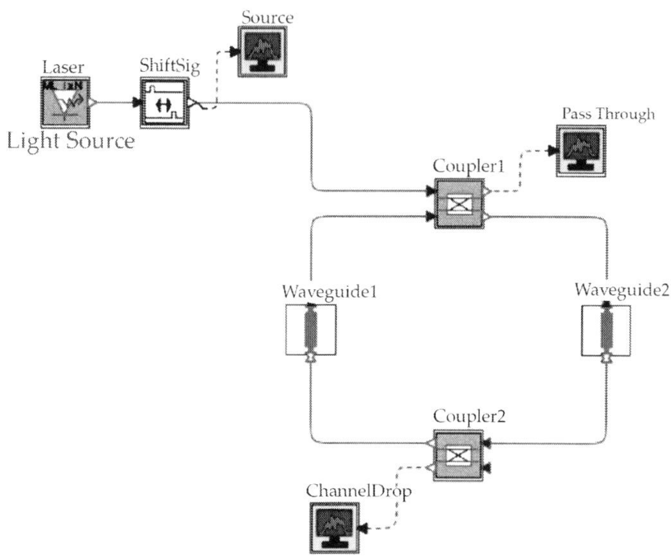

Figure 14. Add-drop FOPRRI configuration: optical layout scheme for EIM and CMT numerical simulations.

Then, a calculation from Eq. (26) gives the value of 5.475 for F finesse. As a consequence of last, calculation from Eq. (27) yields 321.40 for the Q quality-factor of the system. These results are the expected, taking into account that for the same coupling coefficients, waveguide and ring resonator geometry, the values F and Q for add-drop configuration must be half than those for the all-pass ones.

Comments about fundamentals and particularities of BPM, FEM and FDTD methods are given in the following. For an optical waveguide as cylindrical optical fiber, where light travels almost completely confined to its core medium, propagation mechanism can be simply described by means of the four Maxwell's equations. Then, BPM, FEM and FDTD methods assume that the solution to Maxwell's equations follows the slowly varying envelope approximation. This means that the field's amplitude variation is very small as compared to the variation in the phase and propagation distance. BPM is a fast method, which is applicable to a system whose physical structure and properties vary slowly with the geometry. The FDTD method uses the standard finite difference method to solve the Maxwell's equations. It is also known as the full wave method. It discretizes the equations and solves them using a linear algebraic algorithm.

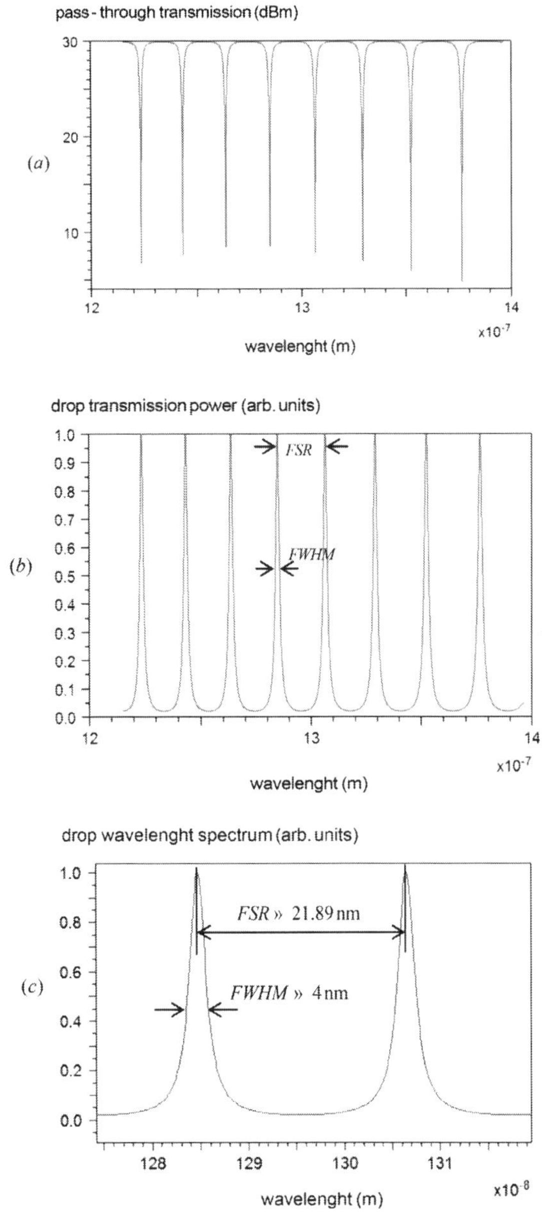

Figure 15. Transmission spectra of add-drop configuration: (a) pass-through port, (b) drop port and (c) drop-port (expanded).

To accelerate the finite difference method simulation, standard approaches such as Krylov subspace-based model order reduction or multi-grid analysis are used. It can be used for all types of structures. As compared with BPM, FDTD is at least an order of magnitude slower. The basic elements that are of application in the case of FDTD method allow as broad as the Maxwell's equations they embody in discretized form, the basic formulation using centered finite difference expressions, which can treat virtually any material type and geometry and the basic resources needed to implement the technique. The FDTD method treats transients in the time domain (e.g., pulses), and it is applicable over the computationally unpredictable resonance region in which a wavelength is comparable with to the object interaction size. The FDTD method offers many advantages as an electromagnetic modeling, simulation and analysis tool. Its capabilities include: 1) broadband response predictions centered about the system resonances, 2) arbitrary 2-D or 3-D model geometries, 3) interaction with an object of any conductivity from that of a perfect conductor, to that of a real metal, to that of low or zero conductivity, 4) frequency-dependent constitutive parameters for modeling most materials (lossy dielectrics, magnetic materials, anisotropic plasmas and magnetized ferrites) and 5) any type of response, including far fields derived from near fields, such as scattered fields, antenna patterns, radar-cross sections, surface response, current and power density and penetration/coupling. FTDT is an appropriate technique for time-domain analysis of passive microwave and millimeter wave structures. The strength of FDTD method lies in its capability to model any volumetric structure with what are typically rectangular cells. No requirements for symmetry or smooth surfaces are needed. Also, FDTD has been successfully used in modeling micro-structured micro-strip structures.

Coupled mode theory (CMT), which uses BPM method, was applied to this structure by means of BeamPROP® tool (Synopsys™ Software). This method also takes into account the effective index of the ring resonator material, so the expected results from it will be improved compared with those obtained with the EIM method alone. BPM numerical simulation results can be seen in Figure 16.

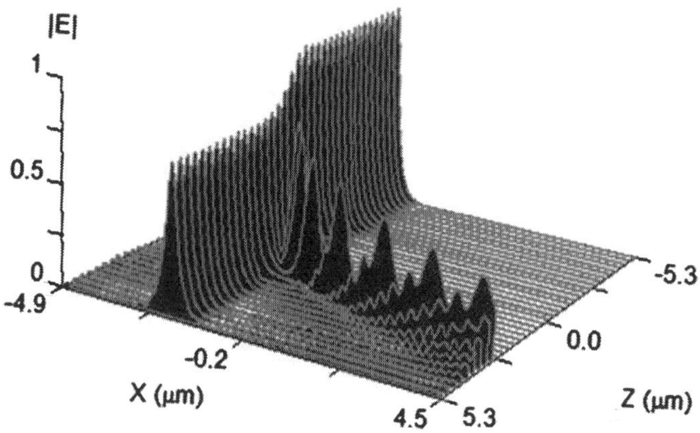

Figure 16. CMT method numerical simulation result (from BeamPROP® tool) for all-pass FOPRRI configuration (resonance at 1285 nm).

Finite element method (FEM) was applied to this structure by means of FemSIM® tool (Synopsys™ Software). This method also takes into account the effective index of the ring resonator material, so the expected results from it will be improved compared with those obtained with the EIM method alone. FEM numerical simulation of this case is represented in Figure 17.

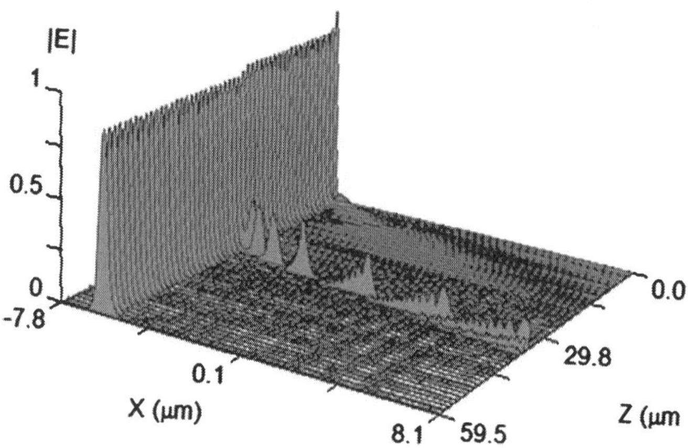

Figure 17. FEM numerical simulation (from FemSIM® tool) for all-pass FO-PRRI configuration (resonance at 1285 nm).

Result of a FDTD method numerical simulation applied to the add-drop filter configuration considered here is represented on Figure 18 (resonance takes place at 1285 nm wavelength). As it was said above, for this structure Eq. (25) gives the value of 5.475 for F finesse and Eq. (26) yields 321.40 for the Q quality-factor of the system.

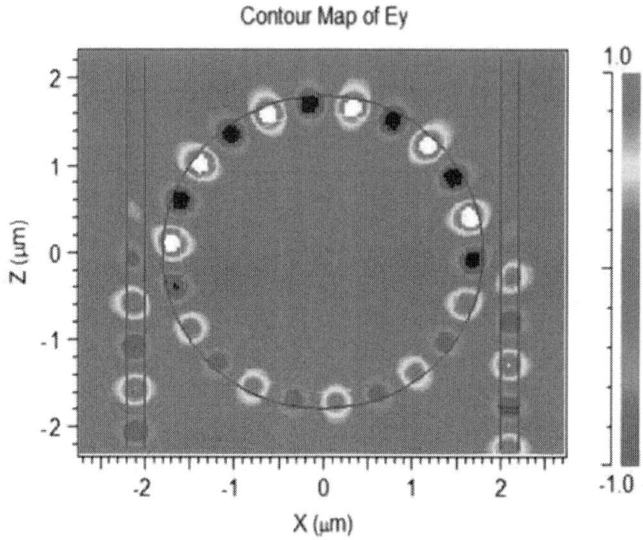

Figure 18. 2D-FTDT numerical simulation results for an add-drop FOPRRI (resonance at 1285 nm).

As a consequence of transmission spectrum of add-drop filter configuration, this can find applications such as optical band-pass filter, optical delay line, optical switch, inertial rotation sensor, photonic biosensor, tailored response filter and active ring modulator.

PERFORMANCE COMPARISON OF NUMERICAL SIMULATION METHODS USED

Table 2 shows performance comparison of all four methods used (EIM, CMT, FEM and FDTD) for both the configurations: all-pass filter and add-drop filter and taking into account three criteria: spent computational time (seconds), memory resources applied (MByte) and error found (%).

Table 2. Comparison of computational time, memory resources and error methods (EIM, CMT, FEM and FDTD).

Criterion/Method	EIM	CMT (BPM)	FEM	2D-FDTD/3D-FDTD
Computational time (sec) 2D	< 0.05	<0.10	1.7	4.3
Computational time (sec) 3D	68	100	547	1170
Memory (MByte)	16	17	17	17
Error (%)	<7	<1	<1	<1

CONCLUSIONS

A conventional model of a fiber optic passive ring resonator interferometer (FOPRRI) was theoretically analyzed in two main configurations: all-pass filter and add-drop filter. Then, four numerical simulation methods (EIM, CMT, FEM and 3D-FDTD) are successively applied on each of these two configurations. In both cases, results showed that EIM method is the most suitable when no very accuracy of results is needed (error < 7%) and low time of computation is desirable (<70 s). However, when high accuracy is considered the most important issue, 3D-FDTD is the best method (error < 1%) at the expense of a much larger computing time (≈1200 sec) than others. CMT (BPM) method provides substantial performance results with computational time similar than that EIM one, but an error < 1%. Finally, FEM is suitable for simulations that can be approximated with 2D techniques since it provides substantial performance results (error < 1%) with computational time only an order of magnitude above EIM or CMT.

ACKNOWLEDGMENTS

The author wishes to thank the Integrated Optics and Optoelectronics Laboratory of Faculty of Sciences of University of Oviedo-Spain for technical support and assistance provided in the realization of this work.

REFERENCES

1. Hariharan P. Optical Interferometry. 2nd ed. London: Elsevier, Academic Press; 2003
2. Sharma U, Wei X. Fiber optic interferometric devices. Chapter 2. In: Kang JU, editor. Fiber Optic Sensing and Imaging. New York: Springer; 2013. pp. 29-53
3. Yin S, Ruffin PB, Yu FTS, editors. Fiber Optic Sensors. 2nd ed. Boca Raton FL (USA): CRC Press; 2008. pp. 333-366
4. Yariv A, Yeh P. Optical Resonators from Photonics. Chapter 4. 6th ed. Oxford University Press, New York; 2007. pp. 156-210
5. Vali V, Shorthill RW. Fiber ring interferometer. Applied Optics. 1976; 15(5):1099-2000
6. Vali V, Shorthill RW. Fiber laser gyroscopes. Proceedings of SPIE. 1976; 76:110
7. Vali V, Shorthill RW. Ring interferometer 950 m long. Applied Optics. 1977;16:290
8. Ezequiel S, Balsamo SR. Passive ring resonator laser gyroscope. Applied Physics Letters. 1977;30:478
9. Meyer RE, Ezequiel S. Passive fiberoptic ring resonator for rotation sensing. Optics Letters. 1984;8:644-646
10. Rao YJ, Jackson DA. Principles of fiber-optic interferometry. In: Grattan LS, Meggitt BT, editors. Optical Fiber Sensor Technology. Boston, MA: Springer; 2000. pp. 167-191
11. Yve CY, Peng JD, Liao YB, Zhou BK. Fibre ring resonator with finesse of 1260. Electronics Letters. 1988;24:622-623
12. Stokes LF, Chodorow M, Shaw HJ. All-single-mode fiber resonator. Optics Letters. 1982;7:288-290
13. Stokes LF, Chodorow M, Shaw HJ. All-fiber stimulated Brillouin ring laser with sub-milliwatt pump threshold. Optics Letters. 1982;7:509-511
14. Jackson DA, Jones JDC. Fiber-optic sensors. Optica Acta. 1986;33:1469-1503
15. Zarinetchi F, Smith SP, Ezekiel S. Stimulated Brillouin fiber-optic laser gyroscope. Optics Letters. 1991;16: 229-231
16. Hill KO, Kawasaki BS, Johnson DC. CW Brillouin laser. Applied Physics Letters. 1976;28:608-609

17. Kalli K, Jackson DA. Ring resonator optical spectrum analyzer with 20 kHz resolution. Optics Letters. 1992;17: 1090-1092
18. Tai S, Kyuma K, Nakayama T. Novel measuring method for spectral linewidth of laser diodes using fiberoptic ring resonators. Electronics Letters. 1985;21:91-93
19. Newton SA, Howland RS, Jackson KP, Shaw HJ. High speed pulse-train generation using single-mode fiber recirculating delay lines. Electronics Letters. 1983;19:757-758
20. Rabus D. Integrated Ring Resonators. New York: Springer; 2007
21. Heebner JE et al. Optical MicroResonators: Theory, Fabrication and Applications. London: Springer; 2008
22. Heebner JE, Wong V, Schweingsberg A, Boyd RW, Jackson DJ. Optical transmission characteristics of fiber ring resonators. IEEE Journal of Quantum Electronics. 2004;40:736-730
23. Vien V. Optical Micro-Ring Resonators: Theory, Techniques and Applications. Boca Raton, FL: CRC Press; 2017
24. Bogaerts W et al. Silicon micro-ring resonators. Laser and Photonics Reviews. 2012;6(1):47-53
25. Knox RM, Thoulios PP. Integrated circuits for the millimeter through optical frequency range. In: Proceedings of Symposium on Submillimeter Waves. Vol. 20; Brooklyn-New York; 1970. pp. 497-515
26. Marcatili EA. Dielectric rectangular waveguide and directional coupler for integrated optics. The Bell System Technical Journal. 1969;48:2071-2102
27. Mode Solutions-Waveguide Mode Solver and Propagation Simulator. URL: http://www.lumerical.com/mode
28. WGMODES-Photonics Research Laboratory. URL: http://www.photonics.umd.edu/software/wgmodes/
29. COMSOL Multiphysics. URL: http:// comsol.com/mode
30. Phoenix Software. URL: http:// www.phoenixbv.com/FieldDesigner
31. Yariv A. Coupled-mode theory for guided-wave optics. IEEE Journal of Quantum Electronics. September 1973; QE-9(9):919-9333
32. RSoft Products-Synopsys Optical Solutions. URL: http://optics.synopsys. com/RSoft/BeamPROP
33. Phoenix Software. URL: http:// www.phoenixbv.com/OptoDesigner

34. RSoft Products-Synopsys Optical Solutions. URL: http://optics. synopsys. com/RSoft/FemSIM
35. FDTD Solutions-Lumerical's Nanophotonic FDTD Simulation Software. URL: http://lumerical.com/ fdtd
36. RSoft Products-Synopsys Optical Solutions. URL: http://optics. synopsys. com/RSoft/FullWAVE
37. Feit MD, Freck JA Jr. Light propagation in graded-index optical fibers. Applied Optics. 1978;17: 3990-3998
38. Thylén L. The beam propagation method: An analysis of its applicability. Optics Quantum Electron. 1983;15: 433-439
39. Yamauchi J, Shibayama J, Nakano H. Beam propagation method using Padé approximant operators. Transactions IEICE Japan. 1994;J-77-C-I: 490-494
40. Hadley GR. Wide-angle beam propagation using Padé approximant operators. Optics Letters. 1992;17: 1426-1428
41. Liu PL, Li BJ. Study on form birefringence in waveguide devices using the semi-vectorial beam propagation method. IEEE Photonics Technology Letters. 1991;3: 913-915
42. Koshiba M, Tsuji Y. A wide-angle finite element beam propagation method. IEEE Photonics Technology Letters. 1996;8:1208-1210
43. Zienckiewitz OC. The Finite Element Method. 3rd ed. New York: McGraw-Hill; 1973
44. Koshiba M, Saitoh H, Eguchi M, Hirayama K. Simple scalar finite element approach to optical waveguides. IEEE Proceedings Journal. 1992;139:166-171
45. Koshiba M. Optical Waveguide Theory by the Finite Element Method. Dordecht, Holland: KTK Scientific Publishers and Kluwer Academic Publishers; 1992
46. Meunier G. The Finite Element Method for Electromagnetic Modeling. River Street-Hoboken NJ: John Wiley & Sons; 2008
47. Kwon YW, Bang H. The Finite Element Method Using Matlab. CRC Press; 1977
48. Stern M. Semivectorial polarized finite difference method optical waveguides with arbitrary index profiles. IEEE Proceedings Journal. Piscataway NJ (USA); 1988;135:56-63

49. Stern M. Semivectorial polarized Hfield solutions for dielectric waveguides with arbitrary index profiles. IEEE Proceedings Journal. Piscataway NJ (USA); 1988;135:333-338

50. Kawano K, Kitoh T, Kohtuku M, Takeshita T, Hasumi Y. 3-D semivectorial analysis to calculate facet reflectivities of semiconductor optical waveguides based on the bi-directional method of line BPM (MoL-BPM). IEEE Photonics Technology Letters. 1998;10: 108-110

51. Gerdes B, Lunitz B, Benish D, Pregla R. Analysis of slab waveguide discontinuities including radiation and absorption effects. Electronics Letters. 1992;28:1013-1015

52. Taflove A. Computational Electrodynamics: Finite Difference Time Domain Method. Artech House: Norwood, MA; 1995

CHAPTER 19

NEURAL NETWORK-BASED LASER INTERFEROMETER COMPENSATION FOR SEISMIC SIGNAL DETECTION

Kyunghyun Lee, Hyungkwan Kwon, and Kwanho You

Department of Electrical Engineering, Sungkyunkwan University, Suwon, Republic of Korea

ABSTRACT

We propose a seismic wave detection method in the frequency domain using a heterodyne laser interferometer, which is used in ultraprecision fields as a displacement measurement device. In seismology, it is important to accurately measure seismic waves. To overcome the limited frequency range and low resolution of accelerometers and velocimeters and to enhance the precision of seismic data analysis, we use the heterodyne laser interferometer as a seismic detection apparatus. We apply the data fusion algorithm with the adaptive standard deviation ratio (ς) derived from the neural network to improve the laser interferometer's measurement precision. Moreover, by using the interferometric characteristics, we analyze the seismic data in the frequency domain. To determine the location of the epicenter from the body wave propagation analysis, we apply the STA/LTA algorithm to the

Citation: (APA): Lee, K., Kwon, H., & You, K. (2018). Neural network-based laser interferometer compensation for seismic signal detection. Journal of Sensors, 2018. (8 pages)

measurement data. The effectiveness of the proposed laser interferometric seismometer is shown through experiments to locate the precise epicenter.

INTRODUCTION

An earthquake is a devastating but common natural disaster around the world. To reduce and predict the damage caused by earthquakes, it is necessary to precisely measure and analyze the seismic signal. Hence, the study of earthquakes has been performed through various seismic analyses and predictions. Methods to analyze earthquakes include spectrum analysis, magnitude measurement, intensity interpretation, and epicenter decision. The short-time average/long-time average (STA/LTA) trigger algorithm is commonly applied in microseismic monitoring systems and has gained significant increased interest. In the STA/LTA algorithm, the most important factor for seismic wave analysis is to identify a seismic event that includes an external disturbance. Widely used seismic measurement devices include seismic velocimeters and accelerometers. These apparatuses detect seismic waves based on force and pressure sensors. They can be classified according to their measurable frequency range as short-period, long-period, and broadband seismographs. These apparatuses have defects such as limited frequency range and low resolution. Therefore, errors exist when calculating the exact P- and S-wave arrival time.

To improve seismic analyses, Colak et al. [1] proposed a new P- and S-wave arrival time decision method. The discrete wavelet transform was used to analyze the change of frequency and energy in high frequency bands. As a result, P- and S-wave arrival time can be measured using a series of data windows. Hloupis and Vallianatos [2] suggested an earthquake early warning system, based on a redundant wavelet transform as a processing tool to rapidly estimate earthquake magnitude. The suggested algorithm strengthens the ability to distinguish body waves from noise. Panagiotakis et al. [3] proposed an automatic identification method of P-phase arrival based on energy and frequency characteristics of the local maxima distribution. Moreover, the method performs a robust detection against the background noise of seismograms. Baziw [4] proposed a robust seismic event identification algorithm by obtaining the optimal estimate of a possible seismic event based on the Rao-Blackwellized particle filter and hidden Markov model. This enables obtaining the real-time phase estimate and the quantified frequency content of seismic data.

In this paper, we propose an interferometric seismometer that is based on a heterodyne laser interferometer and analyze seismic data in the frequency range to estimate epicenter location. Using the orthogonal characteristic of two beams from a laser head, the seismic signals are expressed as a form of intensity signal, which is represented as a trigonometrical function. Hence, the seismic signal can be detected using the laser interferometer. Each observatory has a different experimental environment since the observatory needs to be located in various regions for the seismic signal measurement. The nonuniformity of measurement environment such as temperature and humidity causes the measurement inaccuracy. To compensate for the environmental error that resides in seismic measurement data using a data fusion method, we measure the displacement with two independent laser heads set separately in opposite directions. Using the neural network, the standard deviation of errors is updated adaptively according to the experiment temperature change. With the measured data sets, the environmental error is reduced by projecting the measured data onto the points on the constraint set based on the ideal relationship between the two interferometers [5, 6]. The compensated data are converted into a signal in the frequency domain to extract the P- and S-wave arrival times, which are used to calculate the STA/LTA ratio. The difference between the two arrival times leads to the determination of the epicenter location.

A HETERODYNE LASER INTERFEROMETRIC SEISMOMETER

To measure displacement, a heterodyne laser interferometer uses the phase difference caused by the Doppler effect. The heterodyne laser interferometer system consists of a He-Ne laser head, a beam splitter, a polarization beam splitter, photodetectors, a moving mirror, and a fixed mirror. Figure 1 shows the heterodyne laser interferometer seismic system. The electric fields are measured by detector X. Likewise, the electric fields reflected by the fixed mirror and the moving mirror are measured by detector Y. The intensities obtained by detectors X and Y are expressed as follows.

$$I_{\mathrm{r}} \propto \frac{1}{2}\left(A^2 + B^2\right) + AB\cos\left[\Delta\omega t + (\phi_B - \phi_A)\right],$$

$$I_{\mathrm{m}} \propto \frac{1}{2}\left(A^2 + B^2\right) + AB\cos\left[\Delta\omega t + (\phi_B - \phi_A) + \Delta\phi\right]. \quad (1)$$

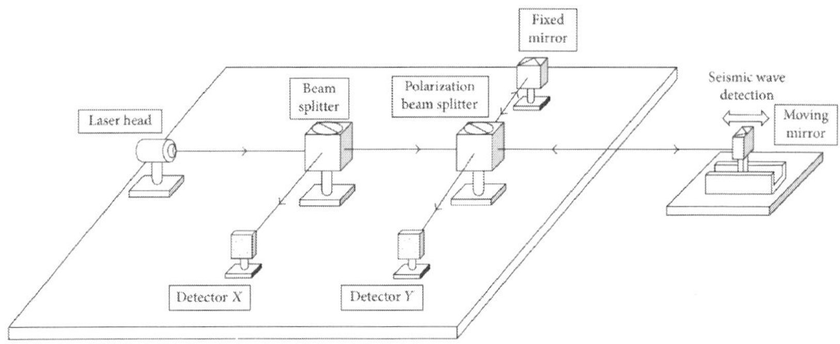

Figure 1: Seismic measurement system based on heterodyne laser interferometer.

In (1), A and B are the magnitudes of the electric fields and ϕ_A ϕ_B are the initial phases of laser sources. The signal I_r measured by detector X is the reference intensity, and the signal I_m obtained by detector Y is the measurement intensity. $\Delta\omega$ is the difference between the frequencies of the two orthogonal beams ω_1 and ω_2, and $\Delta\phi$ is the phase difference caused by the Doppler effect. We use a high-pass filter to eliminate DC components of the intensity signal I_m. The initial phase values ϕ_A and ϕ_B are ignored to obtain phase information. We obtain the AC signal of I_m as follows.

$$I_{m,AC} \propto AB\cos(\Delta\omega t + \Delta\phi). \tag{2}$$

To obtain the phase value, we use a lock-in amplifier. The phase value is derived by applying the I_r and $I_{m,AC}$ signals into a power divider, multiplier, and low pass filter, sequentially. As a result, two orthogonal intensity signals of I_x □ $I_{m,AC}$ □ I_r and I_y □ $I_{m,AC}$ □ $I_r e^{j\pi/2}$ can be represented as a trigonometric function.

$$I_x \propto \frac{AB}{2}\cos(\Delta\phi),$$

$$I_y \propto \frac{AB}{2}\sin(\Delta\phi). \tag{3}$$

By using an inverse trigonometric function, the phase information $\Delta\phi$ related to the displacement is obtained. We calculate the displacement change using $\Delta\phi = 4\pi\rho\Delta L/\lambda$. ρ is the refractive index, λ is the average wavelength, and ΔL is the difference in optical path length between the fixed mirror and the moving mirror.

DATA FUSION-BASED ERROR COMPENSATION

The environmental error causes a significant problem of a measurement accuracy in case of the nanoscale instrument [7]. In order to improve measurement precision, we use a data fusion method to reduce the error that exists in the measurement data. Data fusion uses two measurement data sets, which in this study are obtained by two laser systems installed separately in opposite directions. To utilize this measured data as a seismic signal, we compensate the error to improve the interferometer's accuracy [8].

We use two independent laser heads, which have the same performance characteristics, for the data fusion method. The two separate laser interferometers measure the same moving mirror with a displacement of^ d1 and^ d2, respectively. In the ideal case, the displacement d_1 and d_2 should satisfy the following constraint, because the two laser interferometers measure the identical moving mirror in the opposite direction. When we measure the displacement, the data includes environmental errors since environmental conditions are not uniform. Therefore, the measurement values $(\hat{d}_1, \hat{d}_2)$ do not satisfy the ideal constraint of $d_1(k) + d_2(k) = 0$.

We can represent the relationship between^ $\hat{d}_1$ and $\hat{d}_2$ in the Cartesian coordinate, where ^$\hat{d}_1$ and $\hat{d}_2$ are distributed around the constraint of $d_1(k) + d_2(k) = 0$. These patterns show that the measured displacement data contains environmental errors. We project the measured points in the Cartesian coordinate onto the points on the ideal constraint line to compensate for the environmental errors. An objective error function is utilized to minimize the environmental error.

$$J_e = \frac{1}{2}\sum_{k=1}^{N}\left\|(I_x(k) - I_x^*(k))^2 + \left(I_y(k) - I_y^*(k)\right)^2\right\|. \quad (4)$$

Here, I_x and I_y are measurement signals that contain environmental error, while I_x^* and I_y^* are reference signals without environmental error.

Even though the standard deviation is calculated for each measurement step, a compensation parameter (ς) represents one of the values derived under the fixed environment. To improve measurement accuracy, it is necessary to adaptively modify ς by considering temperature change, which is one of the primary environmental error factors. To estimate the adaptive standard deviations, a multilayer neural network is used. The network is composed of one input layer, two hidden layers, and one output layer. The neurons in the hidden layers and the output layer follow the hyperbolic tangent-sigmoid function f_1 (x) and the log-sigmoid function f_2 (x) as an activation function, respectively. The activation functions enable the neural network to obtain a complex nonlinear model.

$$f_1(x) = \frac{e^x - e^{-x}}{e^x + e^{-x}},$$

$$f_2(x) = \frac{1}{1 + e^{-x}}. \tag{5}$$

The backpropagation algorithm is applied to train and update weights in the network. The steepest gradient descent method is used during the training period. As a target vector, we obtain the error between the reference signals and the measurement signals.

$$\mathbf{V}_\mathrm{T} = \left[\begin{matrix} I_\mathrm{x} - I_\mathrm{x}^* \\ I_\mathrm{y} - I_\mathrm{y}^* \end{matrix} \right] \Bigg|_{I_{\mathrm{tp}}(k) = \tau}. \tag{6}$$

Here, $I_{\mathrm{tp}}(k) = \tau$ is the temperature value during the measurement process. The intensities from (3) are used as the inputs of the neural network during the training period. We define $\mathbf{V}_\mathrm{I} = [I_\mathrm{x}, I_\mathrm{y}, I_{\mathrm{tp}}]^T$ as an input vector and $\mathbf{V}_\mathrm{O} = [e_{I_\mathrm{x}}, e_{I_\mathrm{y}}]^T$ as an output vector. The objective function uses the errors between the output from the neural network and the target vector, V_T:

$$E = \frac{1}{2} \sum_{k=1}^{N} (\mathbf{V}_\mathrm{T}(k) - \mathbf{V}_\mathrm{O}(k))^2, \tag{7}$$

where m is the amount of sampled data. Weight values are calculated with a steepest gradient descent method and updated until minimizing the objective function.

$$w_j^{k+1} = w_j^k - \eta \frac{\partial E}{\partial w}. \tag{8}$$

Here, η is a learning rate and $-\partial E/\partial w$ is calculated using the chain rule. To guarantee a reliable performance, the error bound of the objective function is set as 10^{-9}. The maximum number of the training iterations is limited to 10,000 times for computational efficiency. Finally, we get the neural network modeled in Figure 2.

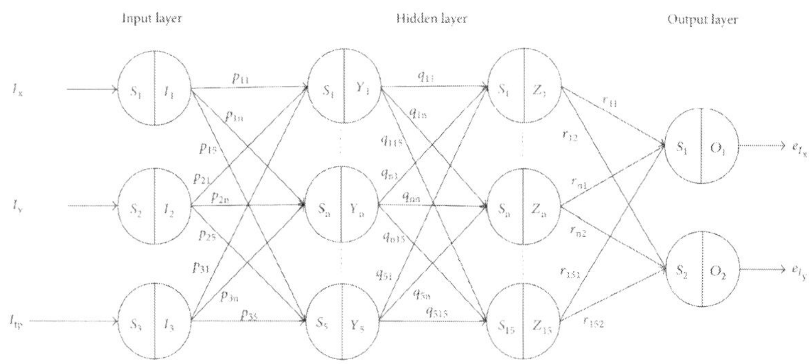

Figure 2: Neural network model to compensate for the environmental error.

The estimated error vector $\mathbf{e}_{\text{NN}} = [e_{I_x}, e_{I_y}]^T$ associated with temperature is obtained from the neural network by applying measurement signals (I_x, I_y) and temperature (I_{tp}) as an input vector.

$$\left[e_{I_x}, e_{I_y}\right]^T = \text{NN}\left(I_x, I_y, I_{tp}\right). \tag{9}$$

The adaptive standard deviations (σ_1^*, σ_2^*) are calculated for each measurement step with the neural network output, (e_{I_x}, e_{I_y}). The loci of the ellipsoidal equation with a centroid, $(\hat{d}_1(k), \hat{d}_2(k))$, and the contact points between $x_{m2} + x_{m1} = 0$ and each ellipse can be represented as follows:

$$\frac{\left(x_{m1} - \hat{d}_1(k)\right)^2}{\sigma_1^*(k)^2} + \frac{\left(x_{m2} - \hat{d}_2(k)\right)^2}{\sigma_2^*(k)^2} = 1,$$

$$d_1^*(k) = \pm \frac{\zeta_k^2 \hat{d}_1(k) - \hat{d}_2(k)}{1 + \zeta_k^2},$$

$$d_2^*(k) = \mp \frac{\zeta_k^2 \hat{d}_1(k) - \hat{d}_2(k)}{1 + \zeta_k^2}. \tag{10}$$

Here, ζ_k is the adaptive standard deviation ratio, $\zeta_k = \sigma_2^*(k)/\sigma_1^*(k)$. Finally, the environmental error is compensated by using the adaptive data fusion based on an artificial intelligence projection method.

SEISMIC DATA ANALYSIS AND STA/LTA ALGORITHM

To detect the seismic wave, we use the heterodyne laser interferometer system. Using the above equations, we eliminate the environmental error by the data fusion method, enabling more precise measurement of the seismic wave. With the seismic data, we obtain P-S wave arrival time (P-S time), which is an important factor for calculating the epicenter location. The methods that measure P-S time generally analyze seismic data in the time domain. However, if the amplitude of a seismic wave does not change enough during the measurement interval, it is difficult to distinguish between P-wave and S-wave arrival [9]. Hence, we suggest a P-S time measurement method in the frequency domain to more accurately identify the arrival times of P-wave and S-waves.

When we detect the seismic wave with a laser interferometer, the seismic signal can be expressed in terms of the intensity of the laser signal in a trigonometric function. The equation can be represented as follows:

$$\hat{I}_x \propto \frac{AB}{2}\cos(\Delta\hat{\phi}),$$
$$\hat{I}_y \propto \frac{AB}{2}\sin(\Delta\hat{\phi}). \qquad (11)$$

In (11),$\hat{I}_x$ and$\hat{I}_y$ are the intensity signals after applying the data fusion method to I_x and I_y in (3). As the refractive index n and wavelength λ are constants, the change in seismic wave movement, ΔL, is proportional to the phase change, $\Delta\hat{\phi}$. Therefore, the frequency data of the phase signal $\Delta\hat{\phi}$ is obtained by the polarity change of the seismic signal, and the Ix and I_y signals are both suitable for frequency analysis. However, the I_y signal is used to analyze the seismic wave as a sinusoidal wave to avoid a latent effect of nonlinearity [10]. The signals are measured in a discrete form. The frequency can be calculated from the signal's polarity change with the following equation.

$$\operatorname{sgn}\left[\hat{I}_y(k)\right] \cdot \operatorname{sgn}\left[\hat{I}_y(k+1)\right] < 0. \qquad (12)$$

Here, sgn (·) is the signum function. The final frequency value is represented as $f_c(k) = N_c(k)/2.$ $f_c(k)$is the frequency of the incoming intensity signal, and N_c (k) is the number of^$\hat{I}_y(k)$ sign changes in one second. Using this method, we can measure the seismic signal in the frequency domain.

An earthquake's epicenter can be determined using the frequency variation of^$\hat{I}_y(k)$ measured by the heterodyne laser interferometer. We analyze this data with the STA/ LTA ratio to calculate P-S time. With a threshold trigger algorithm, we can enhance the analyzing accuracy of seismic data. The STA/LTA algorithm parameters, such as shorttime window size, long-time window size, and threshold value, are determined by the seismic system environment. The STA/LTA algorithm analyzes the data using the difference between STA and LTA characteristics. Finally, we compare the measured STA/LTA ratio with the threshold value. The accuracy of the analysis data depends on the STA and LTA window sizes and threshold value. The STA and LTA are represented as follows [11–13].

$$S_{\text{v}}(n) = \frac{1}{\tau_{\text{S}}} \sum_{k=n-\tau_{\text{s}}}^{n} |f_{\text{c}}(k)|,$$

$$L_{\text{v}}(n) = \frac{1}{\tau_{1}} \sum_{k=n-\tau_{1}}^{n} |f_{\text{c}}(k)|. \tag{13}$$

Here, τ_s and τ_l are the size of the short- and long-time windows, respectively. We measure the time when the STA/LTA ratio becomes greater than the threshold value. The distance between a seismological observatory and an epicenter is calculated using measured P-S time by the following equation.

$$D = \frac{V_{\text{P}} V_{\text{S}}}{V_{\text{P}} - V_{\text{S}}} T_{\text{ps}}. \tag{14}$$

Here, D is the epicentral distance, V_P is the P-wave velocity, V_S is the S-wave velocity, and T_{ps} is the P-S time.

PERFORMANCE EVALUATION

In this section, we show the performance of the proposed seismic detection system based on a heterodyne laser interferometer. To prove the performance and effectiveness of the interferometric seismometer, we use measured seismic data detected by the laser interferometric seismometer and the accelerometer, respectively. We use the He-Ne laser head with an intermode beat of 632.9912 nm (Wavetronics: WT307B). The laser head emits two-frequency beams of λ_1 = 632.9912576 nm for vertical polarization and λ_2 = 632.9912604 nm for horizontal polarization, respectively. The linear stage (PI: M531.5IM) is used as a vibrational source to generate a seismic signal. An accelerometer (Mitutoyo: JEP-8A3) is used as a reference. Figure 3 shows the seismic wave measurement system.

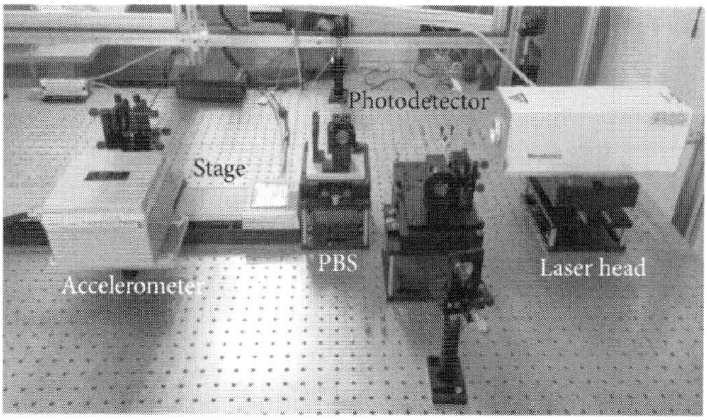

Figure 3: Laser interferometric seismometer.

First, we prove the environmental error compensation by data fusion method for the measurement precision enhancement in a laser interferometer. In our simulation, we set the wavelength λ as 632.9912 nm and the refractive index as 1.00000026653516. Figure 4 shows the compensated data by the data fusion algorithm. The thick line and thin line represent the compensated seismic data and the measurement seismic data, respectively. To verify the accuracy of compensated seismic data using the data fusion in Figure 4, we compare the root-mean-square (RMS) of the compensated measurement error with the uncompensated one. The RMS error comparison results appear in Figure 5. The environmental errors that occurred in the seismic signal detection are compensated by data fusion. Figure 6 shows the seismic data measured by the laser interferometric seismometer and the accelerometer, respectively. Figure 7 denotes the difference between the true amplitude and the measurement value at each sampling time. The thick solid line represents the RMS error of the compensated interferometric seismometer amplitude data. The thin dotted line means the RMS error measured from an accelerometer. As shown in this figure, the proposed neural network-based compensated seismic wave is more precise than the measurement of accelerometer.

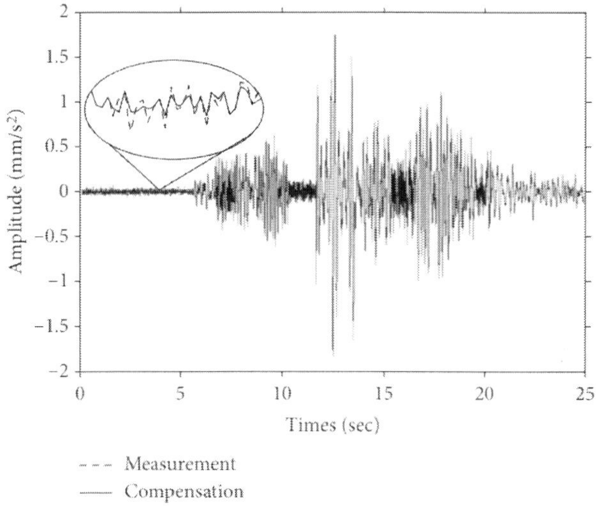

Figure 4: Seismic data compensation with data fusion.

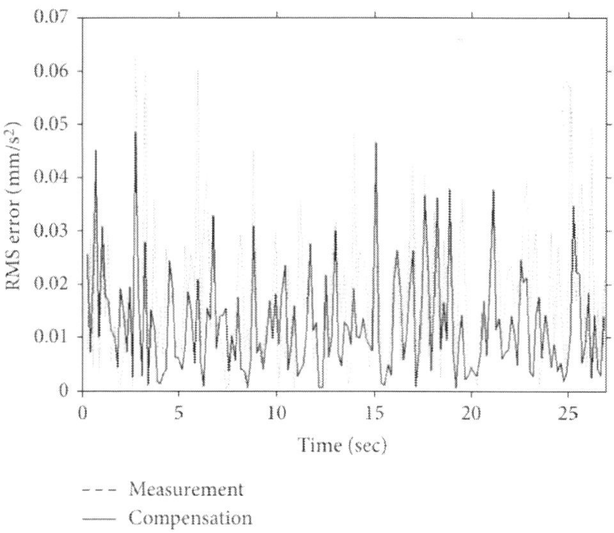

Figure 5: RMS of displacement error.

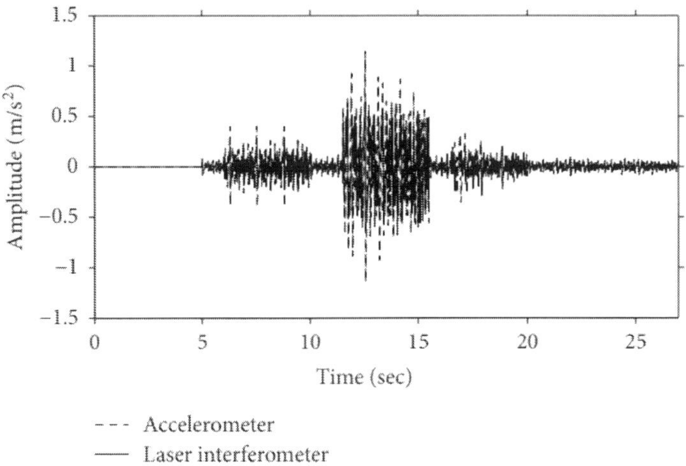

Figure 6: Seismic data detection by interferometric seismometer and accelerometer.

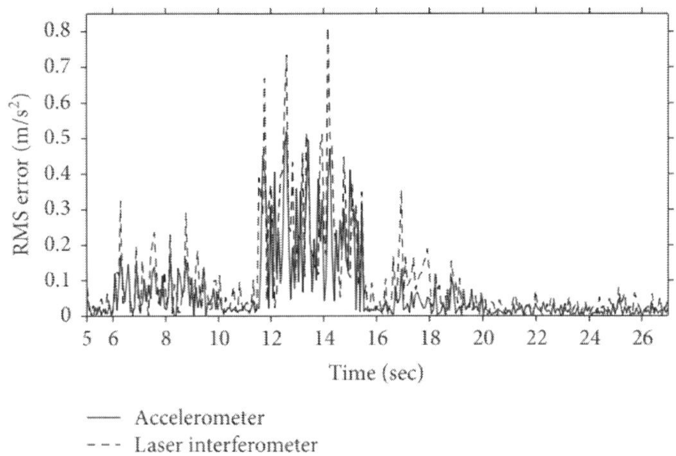

Figure 7: RMS error comparison between a laser interferometer and accelerometer.

Figure 8 shows the intensity signal I_y (k) of the seismic wave and the signal in a time-frequency plane measured by the interferometric seismometer. We obtain the intensity signal (I_y) to minimize the nonlinearity error. The displacement change of the seismic wave is measured from the phase difference between I_r and I_m of the heterodyne interferometer. To analyze the seismic data, we represent the seismic signal in a time-frequency plane and observe sudden frequency changes around 6 and 12 seconds, which are generated by P-wave and S-wave, respectively.

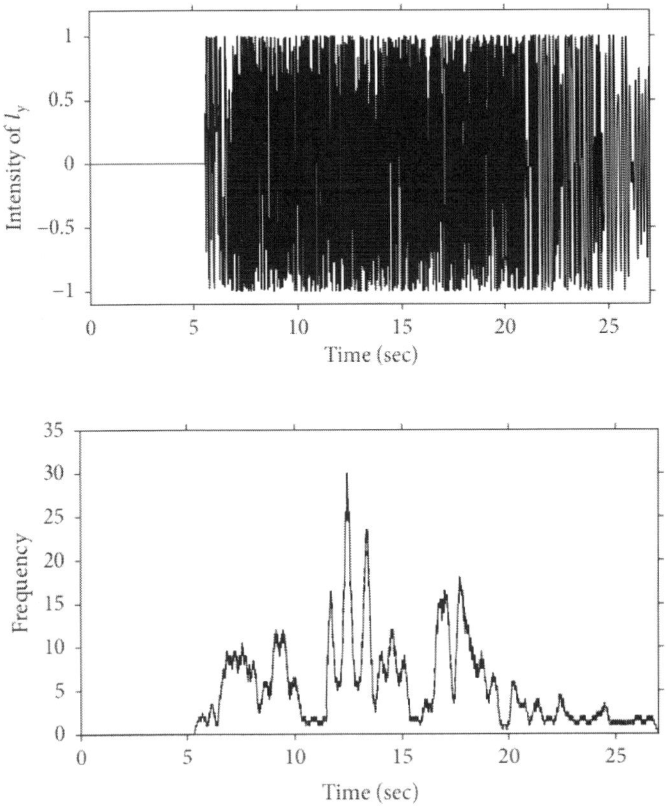

Figure 8: Intensity and frequency variation of the seismic signal.

We use the STA/LTA algorithm to measure P-S time for calculating the epicenter location. First, we determine the STA and LTA window size. The LTA window size is set as 14 times the STA. Next, the threshold value used for the P-wave and S-wave arrival time is set as 1.3 times the average of the STA/LTA ratio. The STA/LTA ratio and the threshold value are shown in Figure 9.

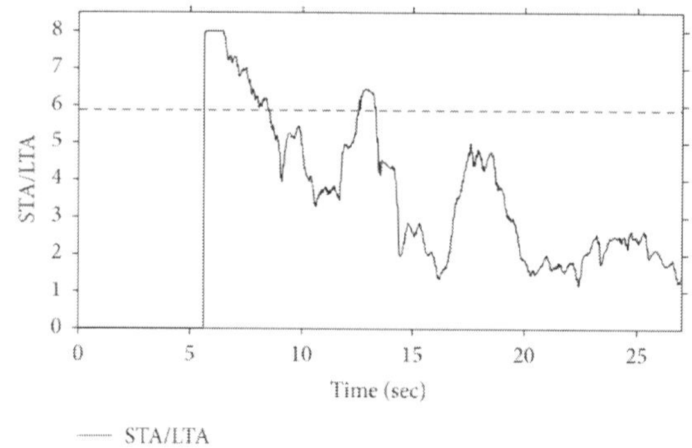

Figure 9: STA/LTA ratio.

From the STA/LTA ratio in Figure 8, we measure P-wave and S-wave arrival time as 5.66 and 12.71 seconds, respectively. With the P-wave velocity and the S-wave velocity as 8 km/s and 4 km/s, respectively, we calculate the epicentral distance as 24.20 km using (14). Figure 10 shows the epicenter location. n_e means the maximum absolute value of the environmental noises. We suppose the parameter ne as 0.05 m/s^2 in Figure 10. The distance from the observatory to the epicenter is measured by the STA/LTA algorithm.

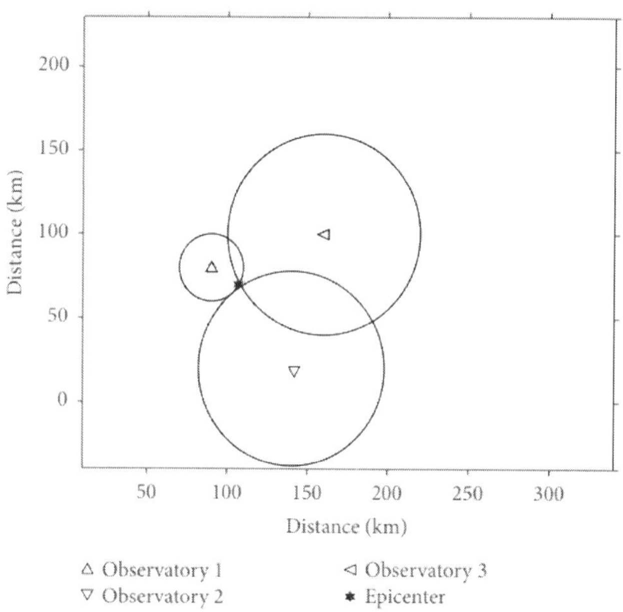

Figure 10: Epicenter location.

With three distinct sets of observatory data, we can estimate the epicenter location in the overlapping region. Table 1 shows the RMS error comparison of epicenter localization method based on the laser interferometric seismometer and accelerometer. Although the localization accuracy is decreased as the parameter n_e increases, we can confirm that the localization result based on the laser interferometric seismometer shows better performance than accelerometer.

Table 1: RMS error comparison for various environmental errors (n_e).

Measurement	RMS error (km)				
	n_e = 0 05	n_e = 0 07	n_e = 0 1	n_e = 0 15	n_e = 0 2
Laser interferometric seismometer	2.54	2.81	4.27	5.83	7.19
Accelerometer	4.18	5.22	7.19	9.45	13.21

CONCLUSION

In this paper, we proposed a seismic signal detection method using a heterodyne laser interferometer in the frequency domain. We applied the data fusion algorithm to minimize the objective error function. To prove the performance of the interferometric seismometer, we compared seismic data measured by a laser interferometric seismometer with data detected by an accelerometer. Moreover, we measured the seismic signal data in the frequency range with the laser interferometer. Using the STA/LTA algorithm, the P-S time was derived. The epicentral distance was calculated using the measured P-S time. We confirmed the effectiveness of the laser interferometer-based seismometer through simulation and experiment.

Conflicts of Interest

The authors declare that they have no conflicts of interest regarding the publication of this paper.

ACKNOWLEDGMENTS

This work was supported by the National Research Foundation of Korea (NRF) grant funded by the Korea government (MSIP) (NRF-2016R1A2B4011369).

REFERENCES

1. O. Colak, T. Destici, S. Ozen, H. Arman, and O. Cerezci, "Detection of P- and S- wave arrival times using the discrete wavelet transform in real seismograms," *The Arabian Journal for Science and Engineering*, vol. 34, no. 1A, pp. 79–89, 2009.
2. G. Hloupis and F. Vallianatos, "Wavelet-based rapid estimation of earthquake magnitude oriented to early warning," *IEEE Geoscience and Remote Sensing Letters*, vol. 10, no. 1, pp. 43–47, 2013.
3. C. Panagiotakis, E. Kokinou, F. Vallianatos, and P. Automatic, "Automatic *P*-phase picking based on local-maxima distribution," *IEEE Transactions on Geoscience and Remote Sensing*, vol. 46, no. 8, pp. 2280–2287, 2008.
4. E. Baziw, "Real-time seismic signal enhancement utilizing a hybrid Rao-Blackwellized particle filter and hidden Markov model filter," *IEEE Geoscience and Remote Sensing Letters*, vol. 2, no. 4, pp. 418–422, 2005.
5. A. Effler, R. M. S. Schofield, V. V. Frolov et al., "Environmental influences on the LIGO gravitational wave detectors during the 6th science run," *Classical and Quantum Gravity*, vol. 32, no. 3, article 035017, 2015.
6. L. Yan, B. Chen, E. Zhang, S. Zhang, and Y. Yang, "Precision measurement of refractive index of air based on laser synthetic wavelength interferometry with Edlén equation estimation," *Review of Scientific Instruments*, vol. 86, no. 8, article 085111, 2015.
7. S. G. Gevorgyan, V. S. Gevorgyan, H. G. Shirinyan, G. H. Karapetyan, and A. G. Sarkisyan, "A radically new principle of operation seismic detector of nano-scale vibrations," *IEEE Transactions on Applied Superconductivity*, vol. 17, no. 2, pp. 629–632, 2007.
8. W. Lee, J. Lee, and K. You, "Improved measurement accuracy in heterodyne laser interferometer using WDF," *Electronics Letters*, vol. 45, no. 21, pp. 1085–1087, 2009.
9. S. Gabarda and G. Cristobal, "Detection of events in seismic time series by time-frequency methods," *IET Signal Processing*, vol. 4, no. 4, pp. 413–420, 2010.
10. W. Hou, "Optical parts and the nonlinearity in heterodyne interferometers," *Precision Engineering*, vol. 30, no. 3, pp. 337–346, 2006.

11. B. Sharma, A. Kumar, and V. Murthy, "Evaluation of seismic events detection algorithms," *Journal of the Geological Society of India*, vol. 75, no. 3, pp. 533–538, 2010.
12. A. Velasco, S. Hernandez, T. Parsons, and K. Pankow, "Global ubiquity of dynamic earthquake triggering," *Nature Geoscience*, vol. 1, no. 6, pp. 375–379, 2008.
13. A. Maggi, C. Tape, M. Chen, D. Chao, and J. Tromp, "An automated time-window selection algorithm for seismic tomography," *Geophysical Journal International*, vol. 178, no. 1, pp. 257–281, 2009.

Index

D

E

F

G

V

W